Linear Algebra and Its Applications

BICENTENNIAL
1807
WILEY
2007
BICENTENNIAL

THE WILEY BICENTENNIAL–KNOWLEDGE FOR GENERATIONS

*E*ach generation has its unique needs and aspirations. When Charles Wiley first opened his small printing shop in lower Manhattan in 1807, it was a generation of boundless potential searching for an identity. And we were there, helping to define a new American literary tradition. Over half a century later, in the midst of the Second Industrial Revolution, it was a generation focused on building the future. Once again, we were there, supplying the critical scientific, technical, and engineering knowledge that helped frame the world. Throughout the 20th Century, and into the new millennium, nations began to reach out beyond their own borders and a new international community was born. Wiley was there, expanding its operations around the world to enable a global exchange of ideas, opinions, and know-how.

For 200 years, Wiley has been an integral part of each generation's journey, enabling the flow of information and understanding necessary to meet their needs and fulfill their aspirations. Today, bold new technologies are changing the way we live and learn. Wiley will be there, providing you the must-have knowledge you need to imagine new worlds, new possibilities, and new opportunities.

Generations come and go, but you can always count on Wiley to provide you the knowledge you need, when and where you need it!

WILLIAM J. PESCE
PRESIDENT AND CHIEF EXECUTIVE OFFICER

PETER BOOTH WILEY
CHAIRMAN OF THE BOARD

Linear Algebra and Its Applications

Second Edition

PETER D. LAX

New York University
Courant Institute of Mathematical Sciences
New York, NY

WILEY-INTERSCIENCE
A JOHN WILEY & SONS, INC., PUBLICATION

Library of Congress Cataloging-in-Publication Data:

Lax, Peter D.
 Linear algebra and its applications / Peter D. Lax. – 2nd ed.
     p. cm. – (Pure and applied mathematics. A Wiley-Interscience of texts, monographs and tracts)
 Previous ed.: Linear algebra. New York : Wiley, c1997.
 Includes bibliographical references and index.
 ISBN 978-0-471-75156-4 (cloth)
 1. Algebras, Linear. I. Lax, Peter D. Linear algebra. II. Title.
 QA184.2.L38 2008
 512'.5–dc22

                                                                          2007023226

10 9 8 7 6 5 4 3 2 1

Contents

Preface

The outlook of this second edition is the same as that of the original: to present linear algebra as the theory and practice of linear spaces and linear mappings. Where it aids understanding and calculations, I don't hesitate to describe vectors as arrays of numbers and to describe mappings as matrices. Render onto Caesar the things which are Caesar's.

If you can reduce a mathematical problem to a problem in linear algebra, you can most likely solve it, provided that you know enough linear algebra. Therefore, a thorough grounding in linear algebra is highly desirable. A sound undergraduate education should offer a second course on the subject, at the senior level. I wrote this book as a suitable text for such a course. The changes made in this second edition are partly to make it more suitable as a text. Terse descriptions, especially in the early chapters, were expanded, more problems were added, and a list of solutions to selected problems has been provided.

In addition, quite a bit of new material has been added, such as the compactness of the unit ball as a criterion of finite dimensionality of a normed linear space. A new chapter discusses the QR algorithm for finding the eigenvalues of a self-adjoint matrix. The Householder algorithm for turning such matrices into tridiagonal form is presented. I describe in some detail the beautiful observation of Deift, Nanda, and Tomei of the analogy between the convergence of the QR algorithm and Moser's theorem on the asymptotic behavior of the Toda flow as time tends to infinity.

Eight new appendices have been added to the first edition's original eight, including the Fast Fourier Transform, the spectral radius theorem, proved with the help of the Schur factorization of matrices, and an excursion into the theory of matrix-valued analytic functions. Appendix 11 describes the Lorentz group, 12 is an interesting application of the compactness criterion for finite dimensionality, 13 is a characterization of commutators, 14 presents a proof of Liapunov's stability criterion, 15 presents the construction of the Jordan Canonical form of matrices, and 16 describes Carl Pearcy's elegant proof of Halmos' conjecture about the numerical range of matrices.

I conclude with a plea to include the simplest aspects of linear algebra in high-school teaching: vectors with two and three components, the scalar product, the cross product, the description of rotations by matrices, and applications to geometry. Such modernization of the high-school curriculum is long overdue.

I acknowledge with pleasure much help I have received from Ray Michalek, as well as useful conversations with Albert Novikoff and Charlie Peskin. I also would like to thank Roger Horn, Beresford Parlett, and Jerry Kazdan for very useful comments, and Jeffrey Ryan for help in proofreading.

PETER D. LAX

New York, New York

Preface to the First Edition

This book is based on a lecture course designed for entering graduate students and given over a number of years at the Courant Institute of New York University. The course is open also to qualified undergraduates and on occasion was attended by talented high school students, among them Alan Edelman; I am proud to have been the first to teach him linear algebra. But, apart from special cases, the book, like the course, is for an audience that has some—not much—familiarity with linear algebra.

Fifty years ago, linear algebra was on its way out as a subject for research. Yet during the past five decades there has been an unprecedented outburst of new ideas about how to solve linear equations, carry out least square procedures, tackle systems of linear inequalities, and find eigenvalues of matrices. This outburst came in response to the opportunity created by the availability of ever faster computers with ever larger memories. Thus, linear algebra was thrust center stage in numerical mathematics. This had a profound effect, partly good, partly bad, on how the subject is taught today.

The presentation of new numerical methods brought fresh and exciting material, as well as realistic new applications, to the classroom. Many students, after all, are in a linear algebra class only for the applications. On the other hand, bringing applications and algorithms to the foreground has obscured the structure of linear algebra—a trend I deplore; it does students a great disservice to exclude them from the paradise created by Emmy Noether and Emil Artin. One of the aims of this book is to redress this imbalance.

My second aim in writing this book is to present a rich selection of analytical results and some of their applications: matrix inequalities, estimates for eigenvalues and determinants, and so on. This beautiful aspect of linera algebra, so useful for working analysts and physicists, is often neglected in texts.

I strove to choose proofs that are revealing, elegant, and short. When there are two different ways of viewing a problem, I like to present both.

The Contents describes what is in the book. Here I would like to explain my choice of materials and their treatment. The first four chapters describe the abstract theory of linear spaces and linear transformations. In the proofs I avoid elimination of the unknowns one by one, but use the linear structure; I particularly exploit quotient spaces as a counting device. This dry material is enlivened by some nontrivial applications to quadrature, to interpolation by polynomials, and to solving the Dirichlet problem for the discretized Laplace equation.

In Chapter 5, determinants are motivated geometrically as signed volumes of ordered simplices. The basic algebraic properties of determinants follow immediately.

Chapter 6 is devoted to the spectral theory of arbitrary square matrices with complex entries. The completeness of eigenvectors and generalized eigenvectors is proved without the characteristic equation, relying only on the divisibility theory of the algebra of polynomials. In the same spirit we show that two matrices A and B are similar if and only if $(A - kI)^m$ and $(B - kI)^m$ have nullspaces of the same dimension for all complex k and all positive integer m. The proof of this proposition leads to the Jordan canonical form.

Euclidean structure appears for the first time in Chapter 7. It is used in Chapter 8 to derive the spectral theory of selfadjoint matrices. We present two proofs, one based on the spectral theory of general matrices, the other using the variational characterization of eigenvectors and eigenvalues. Fischer's minmax theorem is explained.

Chapter 9 deals with the calculus of vector- and matrix-valued functions of a single variable, an important topic not usually discussed in the undergraduate curriculum. The most important result is the continuous and differentiable character of eigenvalues and normalized eigenvectors of differentiable matrix functions, provided that appropriate nondegeneracy conditions are satisfied. The fascinating phenomenon of "avoided crossings" is briefly described and explained.

The first nine chapters, or certainly the first eight, constitute the core of linear algebra. The next eight chapters deal with special topics, to be taken up depending on the interest of the instructor and of the students. We shall comment on them very briefly.

Chapter 10 is a symphony of inequalities about matrices, their eigenvalues, and their determinants. Many of the proofs make use of calculus.

I included Chapter 11 to make up for the unfortunate disappearance of mechanics from the curriculum and to show how matrices give an elegant description of motion in space. Angular velocity of a rigid body and divergence and curl of a vector field all appear naturally. The monotonic dependence of eigenvalues of symmetric matrices is used to show that the natural frequencies of a vibrating system increase if the system is stiffened and the masses are decreased.

Chapters 12, 13, and 14 are linked together by the notion of convexity. In Chapter 12 we present the descriptions of convex sets in terms of gauge functions and support functions. The workhorse of the subject, the hyperplane separation theorem, is proved by means of the Hahn–Banach procedure. Carathéodory's theorem on extreme points is proved and used to derive the König–Birkhoff theorem on doubly stochastic matrices; Helly's theorem on the intersection of convex sets is stated and proved.

Chapter 13 is on linear inequalities; the Farkas–Minkowski theorem is derived and used to prove the duality theorem, which then is applied in the usual fashion to a maximum–minimum problem in economics, and to the minmax theorem of von Neumann about two-person zero-sum games.

Chapter 14 is on normed linear spaces; it is mostly standard fare except for a dual characterization of the distance of a point from a linear subspace. Linear mappings of normed linear spaces are discussed in Chapter 15.

Chapter 16 presents Perron's beautiful theorem on matrices all of whose entries are positive. The standard application to the asymptotics of Markov chains is described. In conclusion, the theorem of Frobenius about the eigenvalues of matrices with nonnegative entries is stated and proved.

The last chapter discusses various strategies for solving iteratively systems of linear equations of the form $Ax = b$, A a self-adjoint, positive matrix. A variational formula is derived and a steepest descent method is analyzed. We go on to present several versions of iterations employing Chebyshev polynomials. Finally we describe the conjugate gradient method in terms of orthogonal polynomials.

It is with genuine regret that I omit a chapter on the numerical calculation of eigenvalues of self-adjoint matrices. Astonishing connections have been discovered recently between this important subject and other seemingly unrelated topics.

Eight appendices describe material that does not quite fit into the flow of the text, but that is so striking or so important that it is worth bringing to the attention of students. The topics I have chosen are special determinants that can be evaluated explicity, Pfaff's theorem, symplectic matrices, tensor product, lattices, Strassen's algorithm for fast matrix multiplication, Gershgorin's theorem, and the multiplicity of eigenvalues. There are other equally attractive topics that could have been chosen: the Baker–Campbell–Hausdorff formula, the Kreiss matrix theorem, numerical range, and the inversion of tridiagonal matrices.

Exercises are sprinkled throughout the text; a few of them are routine; most require some thinking and a few of them require some computing.

My notation is neoclassical. I prefer to use four-letter Anglo-Saxon words like "into," "onto" and "1-to-1," rather than polysyllabic ones of Norman origin. The end of a proof is marked by an open square.

The bibliography consists of the usual suspects and some recent texts; in addition, I have included Courant–Hilbert, Volume I, unchanged from the original German version in 1924. Several generations of mathematicians and physicists, including the author, first learned linear algebra from Chapter 1 of this source.

I am grateful to my colleagues at the Courant Institute and to Myron Allen at the University of Wyoming for reading and commenting on the manuscript and for trying out parts of it on their classes. I am grateful to Connie Engle and Janice Want for their expert typing.

I have learned a great deal from Richard Bellman's outstanding book, *Introduction to Matrix Analysis*; its influence on the present volume is considerable. For this reason and to mark a friendship that began in 1945 and lasted until his death in 1984, I dedicate this book to his memory.

PETER D. LAX

New York, New York

CHAPTER 1

Fundamentals

This first chapter aims to introduce the notion of an abstract linear space to those who think of vectors as arrays of components. I want to point out that the class of abstract linear spaces is no larger than the class of spaces whose elements are arrays. So what is gained by this abstraction?

First of all, the freedom to use a single symbol for an array; this way we can think of vectors as basic building blocks, unencumbered by components. The abstract view leads to simple, transparent proofs of results.

More to the point, the elements of many interesting vector spaces are not presented in terms of components. For instance, take a linear ordinary differential equation of degree n; the set of its solutions form a vector space of dimension n, yet they are not presented as arrays.

Even if the elements of a vector space are presented as arrays of numbers, the elements of a subspace of it may not have a natural description as arrays. Take, for instance, the subspace of all vectors whose components add up to zero.

Last but not least, the abstract view of vector spaces is indispensable for infinite-dimensional spaces; even though this text is strictly about finite-dimensional spaces, it is a good preparation for functional analysis.

Linear algebra abstracts the two basic operations with vectors: the addition of vectors, and their multiplication by numbers (scalars). It is astonishing that on such slender foundations an elaborate structure can be built, with romanesque, gothic, and baroque aspects. It is even more astounding that linear algebra has not only the right theorems but also the right language for many mathematical topics, including applications of mathematics.

A *linear space X* over a *field K* is a mathematical object in which two operations are defined:

Addition, denoted by +, as in

$$x + y \tag{1}$$

Linear Algebra and Its Applications, Second Edition, by Peter D. Lax
Copyright © 2007 John Wiley & Sons, Inc.

and assumed to be *commutative*:

$$x + y = y + x, \tag{2}$$

and *associative*:

$$x + (y + z) = (x + y) + z, \tag{3}$$

and to form a *group*, with the neutral element denoted as 0:

$$x + 0 = x. \tag{4}$$

The inverse of addition is denoted by $-$:

$$x + (-x) \equiv x - x = 0. \tag{5}$$

EXERCISE I.   Show that the zero of vector addition is unique.

The second operation is *multiplication* of elements of X *by elements k of the field K*:

$$kx.$$

The result of this multiplication is a vector, that is, an element of X.
   Multiplication by elements of K is assumed to be *associative*:

$$k(ax) = (ka)x \tag{6}$$

and *distributive*:

$$k(x + y) = kx + ky, \tag{7}$$

as well as

$$(a + b)x = ax + bx. \tag{8}$$

We assume that multiplication by the *unit* of K, denoted as 1, acts as the identity:

$$1x = x. \tag{9}$$

These are the *axioms* of linear algebra. We proceed to draw some deductions:
   Set $b = 0$ in (8); it follows from Exercise 1 that for all x

$$0x = 0. \tag{10}$$

Set $a = 1, b = -1$ in (8); using (9) and (10) we deduce that for all x

$$(-1)x = -x.$$

EXERCISE 2.   Show that the vector with all components zero serves as the zero element of classical vector addition.

In this analytically oriented text the field K will be either the field $\mathbb{R}$ of real numbers or the field $\mathbb{C}$ of complex numbers.

An interesting example of a linear space is the set of all functions $x(t)$ that satisfy the differential equation

$$\frac{d^2}{dt^2}x + x = 0.$$

The sum of two solutions is again a solution, and so is the constant multiple of one. This shows that the set of solutions of this differential equation form a linear space.

Solutions of this equation describe the motion of a mass connected to a fixed point by a spring. Once the initial position $x(0) = p$ and initial velocity $\frac{d}{dt}x(0) = v$ are given, the motion is completely determined for all t. So solutions can be described by a pair of numbers (p, v).

The relation between the two descriptions is *linear*; that is, if (p, v) are the initial data of a solution $x(t)$, and (q, w) the initial data of another solution $y(t)$, then the initial data of the solution $x(t) + y(t)$ are $(p + q, v + w) = (p, v) + (q, w)$. Similarly, the initial data of the solution $kx(t)$ are $(kp, kv) = k(p, v)$.

This kind of relation has been abstracted into the notion of *isomorphism*.

Definition.   A one-to-one correspondence between two linear spaces over the same field that maps sums into sums and scalar multiples into scalar multiples is called an *isomorphism*.

Isomorphism is a basic notion in linear algebra. Isomorphic linear spaces are indistinguishable by means of operations available in linear spaces. Two linear spaces that are presented in very different ways can be, as we have seen, isomorphic.

Examples of Linear Spaces.   (i) Set of all row vectors: $(a_1, \ldots, a_n), a_j$ in K; addition, multiplication defined componentwise. This space is denoted as K^n.
 (ii) Set of all real-valued functions $f(x)$ defined on the real line, $K = \mathbb{R}$.
 (iii) Set of all functions with values in K, defined on an arbitrary set S.
 (iv) Set of all polynomials of degree less than n with coefficients in K.

EXERCISE 3.   Show that (i) and (iv) are isomorphic.

EXERCISE 4.   Show that if S has n elements, (i) and (iii) are isomorphic.

EXERCISE 5. Show that when $K = \mathbb{R}$, (iv) is isomorphic with (iii) when S consists of n distinct points of $\mathbb{R}$.

Definition. A subset Y of a linear space X is called a *subspace* if sums and scalar multiples of elements of Y belong to Y.

Examples of Subspaces. (**a**) X as in Example (i), Y the set of vectors $(0, a_2, \ldots, a_{n-1}, 0)$ whose first and last component is zero.
  (**b**) X as in Example (ii), Y the set of all periodic functions with period π.
  (**c**) X as in Example (iii), Y the set of constant functions on S.
  (**d**) X as in Example (iv), Y the set of all even polynomials.

Definition. The *sum* of two subsets Y and Z of a linear space X, denoted as $Y + Z$, is the set of all vectors of form $y + z$, y in Y, z in Z.

EXERCISE 6. Prove that $Y + Z$ is a linear subspace of X if Y and Z are.

Definition. The *intersection* of two subsets Y and Z of a linear space X, denoted as $Y \cap Z$, consists of all vectors x that belong to both Y and Z.

EXERCISE 7. Prove that if Y and Z are linear subspaces of X, so is $Y \cap Z$.

EXERCISE 8. Show that the set $\{0\}$ consisting of the zero element of a linear space X is a subspace of X. It is called the *trivial subspace*.

Definition. A *linear combination of* j vectors $x_1, \ldots, x_j$ of a linear space is a vector of the form

$$k_1 x_1 + \cdots + k_j x_j, \quad k_1, \ldots, k_j \in K.$$

EXERCISE 9. Show that the set of *all* linear combinations of $x_1, \ldots, x_j$ is a subspace of X, and that it is the smallest subspace of X containing $x_1, \ldots, x_j$. This is called the *subspace spanned* by $x_1, \ldots, x_j$.

Definition. A set of vectors $x_1, \ldots, x_m$ in X *span* the whole space X if every x in X can be expressed as a linear combination of $x_1, \ldots, x_m$.

Definition. The vectors $x_1, \ldots, x_j$ are called *linearly dependent* if there is a nontrivial linear relation between them, that is, a relation of the form

$$k_1 x_1 + \cdots + k_j x_j = 0,$$

where not all $k_1, \ldots, k_j$ are zero.

Definition. A set of vectors $x_1, \ldots, x_j$ that are not linearly dependent is called *linearly independent*.

EXERCISE 10. Show that if the vectors $x_1, \ldots, x_j$ are linearly independent, then none of the x_i is the zero vector.

Lemma 1. Suppose that the vectors $x_1, \ldots, x_n$ span a linear space X and that the vectors $y_1, \ldots, y_j$ in X are linearly independent. Then

$$j \leq n.$$

Proof. Since $x_1, \ldots, x_n$ span X, every vector in X can be written as a linear combination of $x_1, \ldots, x_n$. In particular, y_1:

$$y_1 = k_1 x_1 + \cdots + k_n x_n.$$

Since $y_1 \neq 0$ (see Exercise 10), not all k are equal to 0, say $k_i \neq 0$. Then x_i can be expressed as a linear combination of y_1 and the remaining x_s. So the set consisting of the x's, with x_i replaced by y_1 span X. If $j \geq n$, repeat this step $n - 1$ more times and conclude that $y_1, \ldots, y_n$ span X: if $j > n$, this contradicts the linear independence of the y's for then y_{n+1} is a linear combination of $y_1, \ldots, y_n$. ☐

Definition. A finite set of vectors which span X and are linearly independent is called a *basis* for X.

Lemma 2. A linear space X which is spanned by a finite set of vectors $x_1, \ldots, x_n$ has a basis.

Proof. If $x_1, \ldots, x_n$ are linearly dependent, there is a nontrivial relation between them; from this one of the x_i can be expressed as a linear combination of the rest. So we can drop that x_i. Repeat this step until the remaining x_j are linear independent: they still span X, and so they form a basis. ☐

Definition. A linear space X is called *finite dimensional* if it has a basis.

A finite-dimensional space has many, many bases. When the elements of the space are represented as arrays with n components, we give preference to the special basis consisting of the vectors that have one component equal to 1, while all the others equal 0.

Theorem 3. All bases for a finite-dimensional linear space X contain the same number of vectors. This number is called the dimension of X and is denoted as

$$\dim X.$$

Proof. Let $x_1, \ldots, x_n$ be one basis, and let $y_1, \ldots, y_m$ be another. By Lemma 1 and the definition of basis we conclude that $m \leq n$, and also $n \leq m$. So we conclude that n and m are equal. $\square$

We define the dimension of the trivial space consisting of the single element 0 to be zero.

Theorem 4. Every linearly independent set of vectors $y_1, \ldots, y_j$ in a finite-dimensional linear space X can be completed to a basis of X.

Proof. If $y_1, \ldots, y_j$ do not span X, there is some x_1 that cannot be expressed as a linear combination of $y_1, \ldots, y_j$. Adjoin this x_1 to the y's. Repeat this step until the y's span X. This will happen in less than n steps, $n = \dim X$, because otherwise X would contain more than n linearly independent vectors, impossible for a space of dimension n. $\square$

Theorem 4 illustrates the many different ways of forming a basis for a linear space.

Theorem 5. **(a)** Every subspace Y of a finite-dimensional linear space X is finite dimensional.
(b) Every subspace Y has a complement in X, that is, another subspace Z such that every vector x in X can be decomposed uniquely as

$$x = y + z, \qquad y \text{ in } Y, z \text{ in } Z. \tag{11}$$

Furthermore

$$\dim X = \dim Y + \dim Z. \tag{11$'$}$$

Proof. We can construct a basis in Y by starting with any nonzero vector y_1, and then adding another vector y_2 and another, as long as they are linearly independent. According to Lemma 1, there can be no more of these y_i than the dimension of X. A maximal set of linearly independent vectors $y_1, \ldots, y_j$ in Y spans Y, and so forms a basis of Y. According to Theorem 4, this set can be completed to form a basis of X by adjoining $Z_{j+1}, \ldots, Z_n$. Define Z as the space spanned by $Z_{j+1}, \ldots, Z_n$; clearly Y and Z are complements, and

$$\dim X = n = j + (n - j) = \dim Y + \dim Z. \qquad \square$$

Definition. X is said to be the *direct sum of* two subspaces Y and Z that are complements of each other. More generally X is said to be the direct sum of its subspaces $Y_1, \ldots, Y_m$ if every x in X can be expressed *uniquely* as

$$x = y_1 + \cdots + y_m, \qquad y_j \text{ in } Y_j, \tag{12}$$

This relation is denoted as

$$X = Y_1 \oplus \cdots \oplus Y_m.$$

EXERCISE 11. Prove that if X is finite dimensional and the direct sum of $Y_1, \ldots, Y_m$, then

$$\dim X = \sum \dim Y_j. \tag{12}'$$

Definition. An $(n-1)$-dimensional subspace of an n-dimensional space is called a hyperplane.

EXERCISE 12. Show that every finite-dimensional space X over K is isomorphic to $K^n, n = \dim X$. Show that this isomorphism is not unique when n is > 1.

Since every n-dimensional linear space over K is isomorphic to K^n, it follows that *two linear spaces over the same field and of the same dimension are isomorphic.*

Note: There are many ways of forming such an isomorphism; it is not unique.

The concept of congruence modulo a subspace, defined below, is a very useful tool.

Definition. For X a linear space, Y a subspace, we say that two vectors x_1, x_2 in X are *congruent modulo Y*, denoted

$$x_1 \equiv x_2 \bmod Y,$$

if $x_1 - x_2 \in Y$. Congruence mod Y is an equivalence relation, that is, it is

(i) symmetric: if $x_1 \equiv x_2$, then $x_2 \equiv x_1$.
(ii) reflexive: $x \equiv x$ for all x in X.
(iii) transitive: if $x_1 \equiv x_2, x_2 \equiv x_3$, then $x_1 \equiv x_3$.

EXERCISE 13. Prove (i)–(iii) above. Show furthermore that if $x_1 \equiv x_2$, then $kx_1 \equiv kx_2$ for every scalar k.

We can divide elements of X into *congruence classes* mod Y. The congruence class containing the vector x is the set of all vectors congruent with X; we denote it by $\{x\}$.

EXERCISE 14. Show that two congruence classes are either identical or disjoint.

The set of congruence classes can be made into a linear space by defining addition and multiplication by scalars, as follows:

$$\{x\} + \{z\} = \{x + z\}$$

and

$$k\{x\} = \{kx\}.$$

That is, the sum of the congruence class containing x and the congruence class containing z is the class containing $x + z$. Similarly for multiplication by scalars.

EXERCISE 15.   Show that the above definition of addition and multiplication by scalars is independent of the choice of representatives in the congruence class.

The linear space of congruence classes defined above is called the *quotient space* of X mod Y and is denoted as

$$X(\text{mod } Y) \quad \text{or} \quad X/Y.$$

The following example is illuminating: Take X to be the linear space of all row vectors $(a_1, \ldots, a_n)$ with n components, and take Y to be all vectors $y = (0, 0, a_3, \ldots, a_n)$ whose first two components are zero. Then two vectors are congruent mod Y iff their first two components are equal. Each equivalence class can be represented by a vector with two components, the common components of all vectors in the equivalence class.

This shows that forming a quotient space amounts to throwing away information contained in those components that pertain to Y. This is a very useful simplification when we do not need the information contained in the neglected components.

The next result shows the usefulness of quotient spaces for counting the dimension of a subspace.

Theorem 6. Y is a subspace of a finite-dimensional linear space X; then

$$\dim Y + \dim(X/Y) = \dim X. \tag{13}$$

Proof. Let $y_1, \ldots, y_j$ be a basis for Y, $j = \dim Y$. According to Theorem 4, this set can be completed to form a basis for X by adjoining $x_{j+1}, \ldots, x_n, n = \dim X$. We claim that

$$\{x_{j+1}\}, \ldots, \{x_n\} \tag{13\prime}$$

form a basis for X/Y. To show this we have to verify two properties of the cosets $(13)'$:

  (i) They span X/Y.
  (ii) They are linearly independent.

(i) Since $y_1, \ldots, x_n$ form a basis for X, every x in X can be expressed as

$$x = \sum a_i y_i + \sum b_k x_k.$$

It follows that

$$\{x\} = \sum b_k \{x_k\}.$$

(ii) Suppose that

$$\sum c_k \{x_k\} = 0.$$

This means that

$$\sum c_k x_k = y, \qquad y \text{ in } Y.$$

Express y as $\sum d_i y_i$; we get

$$\sum c_k x_k - \sum d_i y_i = 0.$$

Since $y_1, \ldots, x_n$ form a basis, they are linearly independent, and so all the c_k and d_i are zero.

It follows that

$$\dim X/Y = \# \text{ of } x_k = n - j.$$

So

$$\dim Y + \dim X/Y = j + n - j = n = \dim X. \qquad \square$$

EXERCISE 16. Denote by X the linear space of all polynomials $p(t)$ of degree $< n$, and denote by Y the set of polynomials that are zero at $t_1, \ldots, t_j$, $j < n$.

(i) Show that Y is a subspace of X.
(ii) Determine $\dim Y$.
(iii) Determine $\dim X/Y$.

The following corollary is a consequence of Theorem 6.

Corollary 6'. A subspace Y of a finite-dimensional linear space X whose dimension is the same as the dimension of X is all of X.

EXERCISE 17. Prove Corollary 6'.

Theorem 7. Suppose X is a finite-dimensional linear space, U and V two subspaces of X such that X is the sum of U and V:

$$X = U + V.$$

Denote by W the intersection of U and V:

$$W = U \cap V.$$

Then

$$\dim X = \dim U + \dim V - \dim W. \qquad (14)$$

Proof. When the intersection W of U and V is the trivial space $\{0\}$, $\dim W = 0$, and (14) is relation (11)' of Theorem 5. We show now how to use the notion of quotient space to reduce the general case to the simple case $\dim W = 0$.

Define $U_0 = U/W$, $V_0 = V/W$; then $U_0 \cap V_0 = \{0\}$, and so $X_0 = X/W$ satisfies

$$X_0 = U_0 + V_0.$$

So according to (11)',

$$\dim X_0 = \dim U_0 + \dim V_0. \qquad (14)'$$

Applying (13) of Theorem 6 three times, we get

$$\dim X_0 = \dim X - \dim W, \qquad \dim U_0 = \dim U - \dim W,$$
$$\dim V_0 = \dim V - \dim W.$$

Setting this into relation (14)' gives (14). $\qquad \square$

Definition. The *Cartesian sum* of two linear spaces over the same field is the set of pairs

$$(x_1, x_2); \quad x_1 \text{ in } X_1, x_2 \text{ in } X_2,$$

where addition and multiplication by scalars is defined componentwise. The direct sum is denoted as

$$X_1 \oplus X_2.$$

It is easy to verify that $X_1 \oplus X_2$ is indeed a linear space.

EXERCISE 18. Show that

$$\dim X_1 \oplus X_2 = \dim X_1 + \dim X_2.$$

EXERCISE 19. X a linear space, Y a subspace. Show that $Y \oplus X/Y$ is isomorphic to X.

Note: The most frequently occurring linear spaces in this text are our old friends $\mathbb{R}^n$ and $\mathbb{C}^n$, the spaces of vectors $(a_1, \ldots, a_n)$ with n real, respectively complex, components.

So far the only means we have for showing that a linear space X is finite dimensional is to find a finite set of vectors that span it. In Chapter 7 we present another, powerful criterion for a Euclidean space to be finite dimensional. In Chapter 14 we extend this criterion to all normed linear spaces.

We have been talking about sets of vectors being linearly dependent or independent, but have given no indication how to decide which is the case. Here is an example:

Decide if the four vectors

$$\begin{pmatrix} 1 \\ 1 \\ 0 \\ 1 \end{pmatrix}, \begin{pmatrix} 1 \\ -1 \\ 1 \\ 1 \end{pmatrix}, \begin{pmatrix} 2 \\ 1 \\ 1 \\ 3 \end{pmatrix}, \begin{pmatrix} 2 \\ -1 \\ 2 \\ 3 \end{pmatrix}$$

are linearly dependent or not. That is, are there four numbers k_1, k_2, k_3, k_4, not all zero, such that

$$k_1 \begin{pmatrix} 1 \\ 1 \\ 0 \\ 1 \end{pmatrix} + k_2 \begin{pmatrix} 1 \\ -1 \\ 1 \\ 1 \end{pmatrix} + k_3 \begin{pmatrix} 2 \\ 1 \\ 1 \\ 3 \end{pmatrix} + k_4 \begin{pmatrix} 2 \\ -1 \\ 0 \\ 3 \end{pmatrix} = \begin{pmatrix} 0 \\ 0 \\ 0 \\ 0 \end{pmatrix}?$$

This vector equation is equivalent to four scalar equations:

$$\begin{aligned} k_1 + k_2 + 2k_3 + 2k_4 &= 0, \\ k_1 - k_2 + k_3 - k_4 &= 0, \\ k_2 + k_3 &= 0, \\ k_1 + k_2 + 3k_3 + 3k_4 &= 0. \end{aligned} \tag{15}$$

The study of such systems of linear equations is the subject of Chapters 3 and 4. There we describe an algorithm for finding all solutions of such systems of equations.

EXERCISE 20.   Which of the following sets of vectors $x = (x_1, \ldots, x_n)$ in $\mathbb{R}^n$ are a subspace of $\mathbb{R}^n$? Explain your answer.

(a) All x such that $x_1 \geq 0$.

(b) All x such that $x_1 + x_2 = 0$.

(c) All x such that $x_1 + x_2 + 1 = 0$.

(d) All x such that $x_1 = 0$.

(e) All x such that x_1 is an integer.

EXERCISE 21.   Let U, V, and W be subspaces of some finite-dimensional vector space X. Is the statement

$$\dim(U + V + W) = \dim U + \dim V + \dim W - \dim(U \cap V) - \dim(U \cap W)$$
$$- \dim(V \cap W) + \dim(U \cap V \cap W),$$

true or false? If true, prove it. If false, provide a counterexample.

CHAPTER 2

Duality

Readers who are meeting the concept of an abstract linear space for the first time may balk at the notion of the dual space as piling an abstraction on top of an abstraction. I hope that the results presented at the end of this chapter will convince such skeptics that the notion is not only natural but useful for expeditiously deriving interesting concrete results. The dual of a *normed* linear space, presented in Chapter 14, is a particularly fruitful idea.

The dual of an *infinite-dimensional* normed linear space is indispensable for their study.

Let X be a linear space over a field K. A scalar valued function l,

$$l : X \to K,$$

defined on X, is called *linear* if

$$l(x + y) = l(x) + l(y) \tag{1}$$

for all x, y in X, and

$$l(kx) = kl(x) \tag{1}'$$

for all x in X and all k in K. Note that these two properties, applied repeatedly, show that

$$l(k_1 x_1 + \cdots + k_n x_n) = k_1 l(x_1) + \cdots + k_n l(x_n). \tag{1}''$$

We define the sum of two functions by pointwise addition; that is,

$$(l + m)(x) = l(x) + m(x).$$

Linear Algebra and Its Applications, Second Edition, by Peter D. Lax
Copyright © 2007 John Wiley & Sons, Inc.

Multiplication of a function by a scalar is defined similarly. It is easy to verify that the sum of two linear functions is linear, as is the scalar multiple of one. Thus the set of linear functions on a linear space X itself forms a linear space, called the *dual* of X and denoted by X'.

EXAMPLE 1. $X = \{\text{continuous functions } f(s), 0 \leq s \leq 1\}$. Then for any point s_1 in $[0, 1]$,

$$l(f) = f(s_1)$$

is a linear function. So is

$$l(f) = \sum_1^n k_j f(s_j),$$

where s_j is an arbitrary collection of points in $[0, 1]$, k_j arbitrary scalars. So is

$$l(f) = \int_0^1 f(s)ds.$$

EXAMPLE 2. $X = \{\text{Differentiable functions f on } [0, 1]\}$. For s in $[0, 1]$,

$$l(f) = \sum_1^n a_j \partial^j f(s)$$

is a linear function, where ∂^j denotes the jth derivative.

Theorem 1. Let X be a linear space of dimension n. The elements x of X can be represented as arrays of n scalars:

$$x = (c_1, \ldots, c_n), \tag{3}$$

Addition and multiplication by a scalar is defined componentwise. Let $a_1, \ldots, a_n$ be any array of n scalars; the function l be defined by

$$l(x) = a_1 c_1 + \cdots + a_n c_n \tag{4}$$

is a linear function of x. Conversely, every linear function l of x can be so represented.

Proof. That $l(x)$ defined by (4) is a linear function of x is obvious. The converse is not much harder. Let l be any linear function defined on X. Define x_j to be the vector whose jth component is 1, with all other components zero. Then x defined by (3) can be expressed as

$$x = c_1 x_1 + \cdots + c_n x_n.$$

Denote $l(x_j)$ by a_j; it follows from formula (1)″ that l is of form (4). $\qquad \square$

Theorem 1 shows that if the vectors in X are regarded as arrays of n scalars, then the elements l of X' can also be regarded as arrays of n scalars. It follows from (4) that the sum of two linear functions is represented by the sum of the two arrays representing the summands.

Similarly, multiplication of l by a scalar is accomplished by multiplying each component. We deduce from all this the following theorem.

Theorem 2. The dual X' of a finite-dimensional linear space X is a finite-dimensional linear space, and

$$\dim X' = \dim X.$$

The right-hand side of (4) depends symmetrically on the two arrays representing x and l. Therefore we ought to write the left-hand side also symmetrically, we accomplish that by the *scalar product* notation

$$(l, x)^{\text{def}} = l(x). \tag{5}$$

We call it a product because it is a *bilinear* function of l and x: for fixed l it is a linear function of x, and for fixed x it is a linear function of l.

Since X' is a linear space, it has its own dual X'' consisting of all linear functions on X'. For fixed x, (l, x) is such a linear function. By Theorem 1, all linear functions are of this form. This proves the following theorem.

Theorem 3. The bilinear function (l, x) defined in (5) gives a natural identification of X with X''.

EXERCISE 1. Given a nonzero vector x_1 in X, show that there is a linear function l such that

$$l(x_1) \neq 0.$$

Definition. Let Y be a subspace of X. The set of linear functions l that vanish on Y, that is, satisfy

$$l(y) = 0 \quad \text{for all } y \text{ in } Y, \tag{6}$$

is called the *annihilator* of the subspace Y; it is denoted by $Y^\perp$.

EXERCISE 2. Verify that $Y^\perp$ is a subspace of X'.

Theorem 4. Let Y be a subspace of a finite-dimensional space X, $Y^\perp$ its annihilator. Then

$$\dim Y^\perp + \dim Y = \dim X. \tag{7}$$

Proof. We shall establish a natural isomorphism between $Y^\perp$ and the dual $(X/Y)'$ of X/Y. Given l in $Y^\perp$ we define L in $(X/Y)'$ as follows: for any congruence class $\{x\}$ in X/Y, we define

$$L\{x\} = l(x). \tag{8}$$

It follows from (6) that this definition of L is unequivocal, that is, does not depend on the element x picked to represent the class.

Conversely, given any L in $(X/Y)'$, (8) defines a linear function l on X that satisfies (6). Clearly, the correspondence between l and L is one-to-one and an isomorphism. Thus since isomorphic linear spaces have the same dimension,

$$\dim Y^\perp = \dim(X/Y)'.$$

By Theorem 2, $\dim(X/Y)' = \dim X/Y$, and by Theorem 6 of Chapter 1, $\dim X/Y = \dim X - \dim Y$, so Theorem 4 follows.   □

The dimension of $Y^\perp$ is called the *codimension* of Y as a subspace of X. By Theorem 4,

$$\text{codim } Y + \dim Y = \dim X.$$

Since $Y^\perp$ is a subspace of X', its annihilator, denoted by $Y^{\perp\perp}$, is a subspace of X''.

Theorem 5. Under the identification (5) of X'' and X, for every subspace Y of a finite-dimensional space X,

$$Y^{\perp\perp} = Y.$$

Proof. It follows from definition (6) of the annihilator of Y that all y in Y belong to $Y^{\perp\perp}$, the annihilator of $Y^\perp$. To show that Y is all of $Y^{\perp\perp}$, we make use of (7) applied to X' and its subspace $Y^\perp$:

$$\dim Y^{\perp\perp} + \dim Y^\perp = \dim X'. \tag{7'}$$

Since $\dim X' = \dim X$, it follows by comparing (7) and (7)' that

$$\dim Y^{\perp\perp} = \dim Y.$$

So Y is a subspace of $Y^{\perp\perp}$ that has the same dimension as $Y^{\perp\perp}$; but then according to Corollary 6' in Chapter 1, $Y = Y^{\perp\perp}$.   □

The following notion is useful:

Definition. Let X be a finite-dimensional linear space, and let S be a *subset* of X. The *annihilator* $S^\perp$ of S is the set of linear functions l that are zero at all vectors s of S:

$$l(s) = 0 \qquad \text{for } s \text{ in } S.$$

Theorem 6. Denote by Y the smallest subspace containing S:

$$S^\perp = Y^\perp.$$

EXERCISE 3. Prove Theorem 6.

According to formalist philosophy, all of mathematics is tautology. Chapter 2 might strike the reader—as it does the author—as quintessential tautology. Yet even this trivial-looking material has some interesting consequences:

Theorem 7. Let l be an interval on the real axis, $t_1, \ldots, t_n$ n distinct points. Then there exist n numbers $m_1, \ldots, m_n$ such that the *quadrature formula*,

$$\int_l p(t)dt = m_1 p(t_1) + \cdots + m_n p(t_n) \tag{9}$$

holds for all polynomials p of degree less than n.

Proof. Denote by X the space of all polynomials $p(t) = a_0 + a_1 t + \cdots + a_{n-1}t^{n-1}$ of degree less than n. Since X is isomorphic to the space $(a_0, a_1, \ldots, a_{n-1}) = \mathbb{R}^n$, $\dim X = n$. We define l_j as the linear function

$$l_j(p) = p(t_j) \tag{10}$$

The l_j are elements of the dual space of X; we claim that they are linearly independent. For suppose there is a linear relation between them:

$$c_1 l_1 + \cdots + c_n l_n = 0. \tag{11}$$

According to the definition of the l_j, (11) means that

$$c_1 p(t_1) + \cdots + c_n p(t_n) = 0 \tag{12}$$

for all polynomials p of degree less than n. Define the polynomial q_k as the product

$$q_k(t) = \prod_{j \neq k} (t - t_j).$$

Clearly, q_k is of degree $n - 1$, and is zero at all points $t_j, j \neq k$. Since the points t_j are distinct, q_k is nonzero at t_k. Set $p = q_k$ in (12); since $q_k(t_j) = 0$ for $j \neq k$, we obtain that $c_k q_k(t_k) = 0$; since $q_k(t_k)$ is not zero, c_k must be. This shows that all coefficients c_k are zero, that is, that the linear relation (11) is trivial. Thus the $l_j, j = 1, \ldots, n$ are n linearly independent elements of X'. According to Theorem 2, $\dim X' = \dim X = n$;

therefore the l_j form a basis of X'. This means that any other linear function l on X can be represented as a linear combination of the l_j:

$$l = m_1 l_1 + \cdots + m_n l_n.$$

The integral of p over I is a linear function of p; therefore it can be represented as above. This proves that given any n distinct points $t_1, \ldots, t_n$, there is a formula of form (9) that is valid for all polynomials of degree less than n. □

EXERCISE 4.　In Theorem 6 take the interval I to be $[-1, 1]$, and take n to be 3. Choose the three points to be $t_1 = -a, t_2 = 0$, and $t_3 = a$.

(i) Determine the weights m_1, m_2, m_3 so that (9) holds for all polynomials of degree <3.

(ii) Show that for $a > \sqrt{1/3}$, all three weights are positive.

(iii) Show that for $a = \sqrt{3/5}$, (9) holds for all polynomials of degree <6.

EXERCISE 5.　In Theorem 6 take the interval I to be $[-1, 1]$, and take $n = 4$. Choose the four points to be $-a, -b, b, a$.

(i) Determine the weights m_1, m_2, m_3, and m_4 so that (9) holds for all polynomials of degree <4.

(ii) For what values of a and b are the weights positive?

EXERCISE 6.　Let $\mathcal{P}_2$ be the linear space of all polynomials

$$p(x) = a_0 + a_1 x + a_2 x^2$$

with real coefficients and degree ≤ 2. Let ξ_1, ξ_2, ξ_3 be three distinct real numbers, and then define

$$\ell_j = p(\xi_j) \quad \text{for} \quad j = 1, 2, 3.$$

(a) Show that ℓ_1, ℓ_2, ℓ_3 are linearly independent linear functions on $\mathcal{P}_2$.

(b) Show that ℓ_1, ℓ_2, ℓ_3 is a basis for the dual space $\mathcal{P}'_2$.

(c) (1) Suppose $\{e_1, \ldots, e_n\}$ is a basis for the vector space V. Show there exist linear functions $\{\ell_1, \ldots \ell_n\}$ in the dual space V' defined by

$$\ell_i(e_j) = \begin{cases} 1 & \text{if } i = j, \\ 0 & \text{if } i \neq j, \end{cases}$$

Show that $\{\ell_1, \ldots, \ell_n\}$ is a basis of V', called the *dual basis*.

(2) Find the polynomials $p_1(x), p_2(x), p_3(x)$ in $\mathcal{P}_2$ for which ℓ_1, ℓ_2, ℓ_3 is the dual basis in $\mathcal{P}'_2$.

EXERCISE 7.　Let W be the subspace of $\mathbb{R}^4$ spanned by $(1, 0, -1, 2)$ and $(2, 3, 1, 1)$. Which linear functions $\ell(x) = c_1 x_1 + c_2 x_2 + c_3 x_3 + c_4 x_4$ are in the annihilator of W?

CHAPTER 3

Linear Mappings

Chapter 3 abstracts the concept of a matrix as a linear mapping of one linear space into another. Again I point out that no greater generality is achieved, so what has been gained?

First of all, simplicity of notation; we can refer to mappings by single symbols, instead of rectangular arrays of numbers. The abstract view leads to simple, transparent proofs. This is strikingly illustrated by the proof of the associative law of matrix multiplication and by the proof of the basic result that the column rank of a matrix equals its row rank.

Many important mappings are not presented in matrix form; see, for example, the first two applications presented in this chapter.

Last but not least, the abstract view is indispensable for infinite-dimensional spaces. There the view of mappings as infinite matrices has held up progress until it was replaced by an abstract concept.

A mapping from one set X into another set U is a function whose arguments are points of X and whose values are points of U:

$$f(x) = u.$$

In this chapter we discuss a class of very special mappings:

(i) Both X, called the *domain space*, and U, called the *target space*, are linear spaces over the same field.

(ii) A mapping T: $X \rightarrow U$ is called *linear* if it is *additive*, that is, satisfies

$$T(x + y) = T(x) + T(y)$$

for all x, y in X, and if it is *homogeneous*, that is, satisfies

$$T(kx) = kT(x)$$

Linear Algebra and Its Applications, Second Edition, by Peter D. Lax
Copyright © 2007 John Wiley & Sons, Inc.

for all x in X and all k in K. The value of T at x is written multiplicatively as Tx; the additive property becomes the distributive law: $T(x + y) = Tx + Ty$.

Other names for linear mapping are linear transformation and linear operator.

Example 1.   Any isomorphism.

Example 2. $X = U$ polynomials of degree less than n in s; $T = d/ds$.

Example 3. $X = U = \mathbb{R}^2$, T rotation around the origin by angle θ.

Example 4. X any linear space, U the one dimensional space K, T any linear function on X.

Example 5. $X = U = $ Differentiable functions, T linear differential operator.

Example 6. $X = U = C_0(\mathbb{R})$, $(Tf)(x) = \int\limits_{-1}^{1} f(y)(x - y)^2 dy$.

Example 7. $X = \mathbb{R}^n$, $U = \mathbb{R}^m$, $u = Tx$ defined by

$$u_i = \sum_{1}^{n} t_{ij}x_j, \qquad i = 1, \ldots, m.$$

Here $u = (u_1, \ldots, u_m)$, $x = (x_1, \ldots, x_n)$.

Theorem 1.   (a) The image of a subspace of X under a linear map T is a subspace of U.
   (b) The inverse image of a subspace of U, that is the set of all vectors in X mapped by T into the subspace, is a subspace of X.

EXERCISE I.   Prove Theorem 1.

Definition.   The *range* of T is the image of X under T; it is denoted as R_T. By part (a) of Theorem 1, it is a subspace of U.

Definition.   The *nullspace* of T is the set X mapped into 0 by T: $Tx = 0$; it is denoted as N_T. By part (b) of Theorem 1, it is a subspace of X.

The following result is a workhorse of the subject, a fundamental result about linear maps.

Theorem 2.   Let T: $X \to U$ be a linear map; then

$$\dim N_T + \dim R_T = \dim X.$$

Proof. Since T maps N_T into 0, $Tx_1 = Tx_2$ when x_1 and x_2 are equivalent mod N_T. So we can define T acting on the quotient space X/N_T by setting

$$T\{x\}. = Tx.$$

T is an isomorphism between X/N_T and R_T; since isomorphic spaces have the same dimension,

$$\dim X/N_T = \dim R_T.$$

According to Theorem 6 of Chapter 1, $\dim X/N = \dim X - \dim N$; combined with the relation above we get Theorem 2. □

Corollaries. **A** Suppose $\dim U < \dim X$; then

$$Tx = 0 \qquad \text{for some } x \neq 0.$$

B Suppose $\dim U = \dim X$ and the only vector satisfying $Tx = 0$ is $x = 0$. Then

$$R_T = U.$$

Proof. **A** $\dim R_T \leq \dim U < \dim X$; it follows therefore from Theorem 2 that $\dim N_T > 0$, that is, that N_T contains some vector not equal to 0.
B By hypothesis, $N_T = \{0\}$, so $\dim N_T = 0$. It follows then from Theorem 2 and from the assumption in **B** that

$$\dim R_T = \dim X = \dim U.$$

By Corollary 6' of Chapter 1, a subspace whose dimension is the dimension of the whole space is the whole space; therefore $R_T = U$. □

Theorem 2 and its corollaries have many applications, possibly more than any other theorem of mathematics. It is useful to have concrete versions of them.

Corollary A'. $X = \mathbb{R}^n$, $U = \mathbb{R}^m$, $m < n$. Let T be any mapping of $\mathbb{R}^n \to \mathbb{R}^m$ as in Example 7; since $m = \dim U < \dim X = n$, by Corollary A, the system of linear equations

$$\sum_1^n t_{ij}x_j = 0, \qquad i = 1, \ldots, m \tag{1}$$

has a nontrivial solution, that is one where at least one $x_j \neq 0$.

Corollary B'. $X = \mathbb{R}^n$, $U = \mathbb{R}^n$, T given by

$$\sum_1^n t_{ij}x_j = u_i, \qquad i = 1, \ldots, n. \tag{2}$$

If the homogeneous system of equations

$$\sum_1^n t_{ij}x_j = 0, \qquad i = 1, \ldots, n \tag{3}$$

has only the trivial solution $x_1 = \cdots = x_n = 0$, then the inhomogeneous system (2) has a solution for all $u_1, \ldots, u_n$. Since the homogeneous system (3) has only the trivial solution, the solution of (2) is uniquely determined.

Application I. Take X equal to the space of all polynomials $p(s)$ with complex coefficients of degree less than n, and take $U = \mathbb{C}^n$. We choose $s_1, \ldots, s_n$ as n distinct complex numbers, and define the linear mapping T: $X \to U$ by

$$Tp = (p(s_1), \ldots, p(s_n)).$$

We claim that N_T is trivial; for $Tp = 0$ means that $p(s_1) = 0, \ldots, p(s_n) = 0$, that is, that p has zeros at $s_1, \ldots, s_n$. But a polynomial p of degree less than n cannot have n distinct zeros, unless $p \equiv 0$. Then by Corollary B, the range of T is all of U; that is, the values of p at $s_1, \ldots, s_n$ can be prescribed arbitrarily.

Application 2. X is the space of polynomials with real coefficients of degree $< n$, $U = \mathbb{R}^n$. We choose n pairwise disjoint intervals $S_1, \ldots, S_n$ on the real axis. We define $\bar{p}_j$ to be the average value of p over S_j:

$$\bar{p}_j = \frac{1}{|S_j|} \int_{S_j} p(s)ds, \qquad |S_j| = \text{length of } S_j. \tag{4}$$

We define the linear mapping T: $X \to U$ by

$$Tp = (\bar{p}_1, \ldots, \bar{p}_n).$$

We claim that the nullspace of T is trivial; for, if $\bar{p}_j = 0$, p changes sign in S_j and so vanishes somewhere in S_j. Since the S_j are pairwise disjoint, p would have n distinct zeros, too many for a polynomial of degree less than n. Then by Corollary B the range of T is all of U; that means that the average values of p over the intervals $S_1, \ldots, S_n$ can be prescribed arbitrarily.

Application 3. In constructing numerical approximations to solutions of the Laplace equation in a bounded domain G of the plane,

$$\Delta u = u_{xx} + u_{yy} = 0 \quad \text{in } G, \tag{5}$$

with u prescribed on the boundary of G, one fills G, approximately, with a lattice and replaces the second partial derivatives with centered differences:

$$u_{xx} \simeq \frac{u_W - 2u_o + u_E}{h^2},$$

$$u_{yy} \simeq \frac{u_N - 2u_o + u_S}{h^2}, \tag{6}$$

where

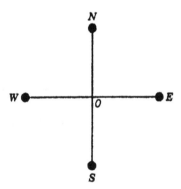

and h is the mesh spacing. Setting (6) into (5) gives the following relations:

$$u_o = \frac{u_W + u_N + u_E + u_S}{4}. \tag{7}$$

This equation relates the value u_o of u at each lattice point O in the domain G to the values of u at the four lattice neighbors of u. In case any lattice neighbor of O lies outside G, we set the value of u there equal to the boundary value of u at the nearest boundary point. The resulting set of equations (7) is a system of n equations for n unknowns of the form (2); n is equal to the number of lattice points in G.

We claim that the corresponding homogeneous equations have only the trivial solution $u_o = 0$ for all lattice points. The homogeneous equations correspond to taking the boundary values to be zero. Now take any solution of the homogeneous equations and denote by u_{max} the maximal value of u_o over all lattice points in G. That maximum is assumed at some point O of G; it follows from (7) that then $u = u_{max}$ at all four lattice neighbors of O. Repeating this argument we eventually reach a lattice neighbor which falls outside G. Since u was set to zero at all such points, we conclude that $u_{max} = 0$. Similarly we show that $u_{min} = 0$; together these imply that $u_0 = 0$ for all lattice points for a solution of the homogeneous equation.

By Corollary B', the system of equations (7), with arbitrary boundary data, has a unique solution.

EXERCISE 2. Let

$$\sum_{1}^{n} t_{ij}x_j = u_i, \qquad i = 1, \ldots, m$$

be an overdetermined system of linear equations—that is, the number m of equations is greater than the number n of unknowns $x_1, \ldots, x_n$. Take the case that in spite of the overdeterminacy, this system of equations has a solution, and assume that this solution is unique. Show that it is possible to select a subset of n of these equations which uniquely determine the solution.

We turn now to the rudiments of the *algebra* of linear mappings, that is, their addition and multiplication. Suppose that T and S are both linear maps of $X \to U$; then we define their sum $T + S$ by setting for each vector x in X,

$$(T + S)(x) = Tx + Sx.$$

Clearly, under this definition $T + S$ is again a linear map of $X \to U$. We define kT similarly, and we get another linear map.

It is not hard to show that under the above definition the set of linear mappings of $X \to U$ themselves forms a linear space. This space is denoted by $\mathbb{L}(X, U)$.

Let T, S be maps, not necessarily linear, of X into U, and U into V, respectively, X, U, V arbitrary sets. Then we can define the *composition* of T with S, a mapping of X into V obtained by letting T act first, followed by S, schematically

$$V \xleftarrow{\;S\;} U \xleftarrow{\;T\;} X.$$

The composite is denoted by $S \circ T$:

$$S \circ T(x) = S(T(x)).$$

Note that composition is *associative*: if R maps V into Z, then

$$R \circ (S \circ T) = (R \circ S) \circ T.$$

Theorem 3. (i) The composite of linear mappings is also a linear mapping.
(ii) Composition is distributive with respect to the addition of linear maps, that is,

$$(R + S) \circ T = R \circ T + S \circ T$$

and

$$S \circ (T + P) = S \circ T + S \circ P,$$

where R and S map $U \to V$ and P and T map $X \to U$.

EXERCISE 3.   Prove Theorem 3.

On account of this distributive property, coupled with the associative law that holds generally, composition of linear maps is denoted as *multiplication*:

$$S \circ T \equiv ST.$$

We warn the reader that this kind of multiplication is generally *not* commutative; for example, TS may not even be defined when ST is, much less equal to it.

Example 8. $X = U = V =$ polynomials in s, $T = d/ds$, $S =$ multiplication by s.

Example 9. $X = U = V = R^3$.

S: rotation around x_1 axis	T: rotation around x_2 axis
by 90 degrees	by 90 degrees

EXERCISE 4.   Show that S and T in Examples 8 and 9 are linear and that $ST \neq TS$.

Definition.   A linear map is called *invertible* if it is 1-to-1 and onto, that is, if it is an isomorphism. The inverse is denoted as T^{-1}.

EXERCISE 5.   Show that if T is invertible, TT^{-1} is the identity.

Theorem 4.   (i) The inverse of an invertible linear map is linear.
(ii) If S and T are both invertible, and if ST is defined, then ST also is invertible, and

$$(ST)^{-1} = T^{-1}S^{-1}.$$

EXERCISE 6.   Prove Theorem 4.

Let T be a linear map $X \to U$, and l a linear function, that is, l is an element of U'. Then the product (i.e., composite) lT is a linear mapping of X into K, that is, an element of X'; denote this element by m:

$$m(x) = l(Tx). \tag{8}$$

This defines an assignment of an element m of X' to every element l of U'. It is easy to deduce from (8) that this assignment is a linear mapping $U' \to X'$; it is called the *transpose* of T and is denoted by T'.

Using the notation (6) in Chapter 2 to denote the value of a linear function, we can rewrite (8) as

$$(m, x) = (l, \text{T}x). \tag{8'}$$

Using the notation $m = \text{T}'l$, this can be written as

$$(\text{T}'l, x) = (l, \text{T}x). \tag{9}$$

EXERCISE 7.   Show that whenever meaningful,

$$(\text{ST})' = \text{T}'\text{S}', \qquad (\text{T} + \text{R})' = \text{T}' + \text{R}' \quad \text{and} \quad (\text{T}^{-1})' = (\text{T}')^{-1}.$$

Example 10. $X = \mathbb{R}^n$, $U = \mathbb{R}^m$, T as in Example 7.

$$u_i = \sum t_{ij} x_j. \tag{10}$$

U' is then also $\mathbb{R}^m$, $X' = \mathbb{R}^n$, with $(l, u) = \sum_1^m l_i u_i$, $(m, x) = \sum_1^n m_j x_j$. Then with $u = \text{T}x$, using (10) we have

$$(l, u) = \sum_i l_i u_i = \sum_i \sum_j l_i t_{ij} x_j$$

$$= \sum_j \left(\sum_i l_i t_{ij} \right) x_j = \sum m_j x_j = (m, x),$$

where $m = \text{T}'l$, with

$$m_j = \sum_l l_i t_{ij}. \tag{11}$$

EXERCISE 8.   Show that if X'' is identified with X and U'' with U via (5) in Chapter 2, then

$$\text{T}'' = \text{T}.$$

We shall show in Chapter 4 that if a mapping T is interpreted as a matrix, its transpose T' is obtained by making the columns of T the rows of T'.

We recall from Chapter 2 the notion of the annihilator of a subspace.

Theorem 5. The annihilator of the range of T is the nullspace of its transpose:

$$R_T^{\perp} = N_{T'}. \tag{12}$$

Proof. By the definition in Chapter 2 of annihilator, the annihilator of the range R_T consists of those linear functions l defined on the target space U for which

$$(l, u) = 0 \qquad \text{for all } u \text{ in } R_T.$$

Since u in R_T consists of $u = Tx$, x in X, we can rewrite the above as

$$(l, Tx) = 0 \qquad \text{for all } x.$$

Using (9), we can rewrite this as

$$(T'l, x) = 0 \qquad \text{for all } x.$$

It follows that l is in $R_T^{\perp}$ iff $T'l = 0$; this proves Theorem 5. □

Now take the annihilator of both sides of (12). According to Theorem 5 of Chapter 2, the annihilator of $R^{\perp}$ is R itself. In this way we obtain the following theorem.

Theorem 5'. The range of T is the annihilator of the nullspace of T'.

$$R_T = N_{T'}^{\perp}. \tag{12'}$$

(12)' is a very useful characterization of the range of a mapping. Next we give another consequence of Theorem 5.

Theorem 6.

$$\dim R_T = \dim R_{T'}. \tag{13}$$

Proof. We apply Theorem 4 of Chapter 2 to U and its subspace R_T:

$$\dim R_T^{\perp} + \dim R_T = \dim U.$$

Next we use Theorem 2 of this chapter applied to $T': U' \to X'$:

$$\dim N_{T'} + \dim R_{T'} = \dim U'.$$

According to Theorem 2, Chapter 2, $\dim U = \dim U'$; according to Theorem 5 of this chapter, $R_T^{\perp} = N_{T'}$, and so $\dim R_T^{\perp} = \dim N_{T'}$. So we deduce (13) from the last two equations. □

The following is an easy consequence of Theorem 6.

Theorem 6′.   Let T be a linear mapping of X into U, and assume that X and U have the same dimension. Then

$$\dim N_T = \dim N_{T'}. \tag{13}'$$

Proof. According to Theorem 2, applied to both T and T′,

$$\dim N_T = \dim X - \dim R_T,$$
$$\dim N_{T'} = \dim U' - \dim R_{T'}.$$

Since $\dim U = \dim U'$ is assumed to be the same as $\dim X$, (13)′ follows from the above relations and (13).   □

Theorem 6 is an abstract version of the classical result that the column rank and row rank of a matrix are equal. The usual proofs of this result are abstruse and unclear.

We turn now to linear mappings of a linear space X into itself. The aggregate of such mappings is denoted as $\mathscr{L}(X,X)$; they are a particularly important and interesting class of maps. Any two such maps can be added and multiplied, that is, composed, and can be multiplied by a scalar. Thus $\mathscr{L}(X,X)$ is an *algebra*. We investigate now briefly some of the algebraic aspects of $\mathscr{L}(X,X)$.

First we remark that $\mathscr{L}(X,X)$ is an associative, but not commutative algebra, with a unit; the role of the unit is played by the identity map I, defined by $Ix = x$. The zero map 0 is defined by $0x = 0$. $\mathscr{L}(X,X)$ contains *divisors* of *zero*, that is, pairs of mappings S and T whose product ST is 0, but neither of which is 0. To see this, choose T to be any nonzero mapping with a nontrivial nullspace N_T, and S to be any nonzero mapping whose range R_S is contained in N_T. Clearly, $TS = 0$.

There are mappings $D \neq 0$ whose square D^2 is zero. As an example, take X to be the linear space of polynomials of degree less than 2. Differentiation D maps this space into itself. Since the second derivative of every polynomial of degree less than 2 is zero, $D^2 = 0$, but clearly $D \neq 0$.

EXERCISE 9.   Show that if A in $\mathscr{L}(X,X)$ is a left inverse of B in $\mathscr{L}(X,X)$, that is $AB = I$, then it is also a right universe: $BA = I$.

We have seen in Theorem 4 that the product of invertible elements is invertible. Therefore the set of *invertible* elements of $\mathscr{L}(X,X)$ forms a *group* under multiplication. This group depends only on the dimension of X, and the field K of scalars. It is denoted as $GL(n, K)$, $n = \dim X$.

Given an invertible element S of $\mathscr{L}(X,X)$, we assign to each M in $\mathscr{L}(X,X)$ the element M_S constructed as follows:

$$M_S = SMS^{-1}. \tag{14}$$

This assignment $M \to M_S$ is called a *similarity transformation*; M is said to be *similar* to M_S.

Theorem 7. **(a)** Every similarity transformation is an automorphism of $L(X,X)$, maps sums into sums, products into products, scalar multiples into scalar multiples:

$$(kM)_S = kM_S. \tag{15}$$

$$(M + K)_S = M_S + K_S. \tag{15}'$$

$$(MK)_S = M_S K_S. \tag{15}''$$

(b) The similarity transformations form a group.

$$(M_S)_T = M_{TS}. \tag{16}$$

Proof. (15) and (15)′ are obvious; to verify (15)″ we use the definition (14):

$$M_S K_S = SMS^{-1}\ SKS^{-1} = SMKS^{-1} = (MK)_S,$$

where we made use of the associative law.

The verification of (16) is analogous; by (14),

$$(M_S)_T = T(SMS^{-1})T^{-1} = TSM(TS)^{-1} = M_{TS};$$

here we made use of the associative law, and that $(TS)^{-1} = S^{-1}T^{-1}$. $\qquad\square$

Theorem 8. Similarity is an equivalence relation; that is, it is:

(i) Reflexive. M is similar to itself.
(ii) Symmetric. If M is similar to K, then K is similar to M.
(iii) Transitive. If M is similar to K, and K is similar to L, then M is similar to L.

Proof. **(i)** is true because we can in the definition (14) choose $S = I$.
(ii) M similar to K means that

$$K = SMS^{-1}. \tag{14}'$$

Multiply both sides by S on the right and S^{-1} on the left, and we see that K is similar to M.
(iii) If K is similar to L, then

$$L = TKT^{-1}, \tag{14}''$$

where T is some invertible mapping. Multiply both sides of (14)' by T^{-1} on the right and by on the left; we get

$$TKT^{-1} = TSMS^{-1}T^{-1}.$$

According to (14)'', the left-hand side is L. The right-hand side can be written as

$$(TS)\, M(TS)^{-1},$$

which is similar to M. $\square$

EXERCISE 10.   Show that if M is invertible, and similar to K, then K also is invertible, and K^{-1} is similar to M^{-1}.

Multiplication in $\mathscr{L}(X,X)$ is not commutative, that is, AB is in general not equal to BA. Yet they are not totally unrelated.

Theorem 9.   If either A or B in $\mathscr{L}(X,X)$ is invertible, then AB and BA are similar.

EXERCISE 11.   Prove Theorem 9.

Given any element A of $\mathscr{L}(X,X)$ we can, by addition and multiplication, form all polynomials in A:

$$a_N A^N + a_{N-1} A^{N-1} + \cdots + a_0 I; \tag{17}$$

we can write (17) as $p(A)$, where

$$p(s) = a_N s^N + \cdots + a_0. \tag{17'}$$

The set of all polynomials in A forms a *subalgebra* of $\mathscr{L}(X,X)$; this subalgebra is *commutative*. Such commutative subalgebras play a big role in spectral theory, discussed in Chapters 6 and 8.

An important class of mappings of a linear space X into itself are *projections*.

Definition.   A linear mapping P: $X \rightarrow X$ is called a projection if it satisfies

$$P^2 = P.$$

Example 11.   X is the space of vectors $x = (a_1, a_2, \ldots, a_n)$, P defined as

$$Px = (0, 0, a_3 \ldots, a_n).$$

That is, the action of P is to set the first two components of x equal to zero.

EXERCISE 12.   Show that P defined above is a linear map, and that it is a projection.

Example 12.   Let X be the space of continuous functions f in the interval $[-1, 1]$; define Pf to be the *even part* of f, that is,

$$(Pf)(x) = \frac{f(x) + f(-x)}{2}.$$

EXERCISE 13.   Prove that P defined above is linear, and that it is a projection.

Definition.   The commutator of two mappings A and B of X into X is AB-BA. Two mappings of X into X commute if their commutator is zero.

Remark.   We can prove Corollary A' directly by induction on the number of equations m, using one of the equations to express one of the unknowns x_j in terms of the others. By substituting this expression for x_j into the remaining equations, we have reduced the number of equations and the number of unknowns by one.

The practical execution of such a scheme has pitfalls when the number of equations and unknowns is large. One has to pick intelligently the unknown to be eliminated and the equation that is used to eliminate it. We shall take up these matters in the next chapter.

Definition.   The *rank* of a linear mapping is the dimension of its range.

EXERCISE 14.   Suppose **T** is a linear map of rank 1 of a finite dimensional vector space into itself.
(a) Show there exists a unique number c such that $\mathbf{T}^2 = c\mathbf{T}$.
(b) Show that if $c \neq 1$ then $\mathbf{I} - \mathbf{T}$ has an inverse. (As usual **I** denotes the identity map $\mathbf{Ix} = \mathbf{x}$.)

EXERCISE 15.   Suppose **T** and **S** are linear maps of a finite dimensional vector space into itself. Show that the rank of **ST** is less than or equal the rank of **S**. Show that the dimension of the nullspace of **ST** is less than or equal the sum of the dimensions of the nullspaces of **S** and of **T**.

CHAPTER 4

Matrices

In Example 7 of Chapter 3 we defined a class of mappings T: $\mathbb{R}^n \to \mathbb{R}^m$ where the ith component of $u = Tx$ is expressed in terms of the components x_j of x by the formula

$$u_i = \sum_{i}^{n} t_{ij} x_j, \qquad i = 1, \ldots, m \tag{1}$$

and the t_{ij} are arbitrary scalars. These mappings are linear; conversely, we have the following theorem.

Theorem 1. Every linear map $Tx = u$ from $\mathbb{R}^n$ to $\mathbb{R}^m$ can be written in form (1).

Proof. The vector x can be expressed as a linear combination of the unit vectors $e_1, \ldots, e_n$ where e_j has jth component 1, all others 0:

$$x = \sum x_j e_j. \tag{2}$$

Since T is linear

$$u = Tx = \sum x_j Te_j. \tag{3}$$

Denote the ith component of Te_j by t_{ij}:

$$t_{ij} = (Te_j)_i. \tag{4}$$

Linear Algebra and Its Applications, Second Edition, by Peter D. Lax
Copyright © 2007 John Wiley & Sons, Inc.

It follows from (3) and (4) that the ith component u_i of u is

$$u_i = \sum x_j t_{ij},$$

exactly as in formula (1).   □

It is convenient and traditional to arrange the coefficients t_{ij} appearing in (1) in a rectangular array,

$$\begin{pmatrix} t_{11} & t_{12} & \cdots & t_{1n} \\ t_{21} & & & \vdots \\ t_{m1} & & \cdots & t_{mn} \end{pmatrix} \tag{5}$$

Such an array is called an m by n $(m \times n)$ *matrix*, m being the number of rows, n the number of columns. A matrix that has the same number of rows and columns is called a square matrix. The numbers t_{ij} are called the entries of the matrix T.

According to Theorem 1, there is a 1-to-1 correspondence between $m \times n$ matrices and linear mappings T: $\mathbb{R}^n \rightarrow \mathbb{R}^m$. We shall denote the (ij)th entry t_{ij} of the matrix identified with T by

$$T_{ij} = (T)_{ij}. \tag{5$'$}$$

A matrix T can be thought of as a *row of column vectors*, or a *column of row vectors*:

$$T = (c_1, \ldots, c_n) = \begin{pmatrix} r_1 \\ \vdots \\ r_m \end{pmatrix}, \qquad c_j = \begin{pmatrix} t_{1j} \\ \vdots \\ t_{mj} \end{pmatrix}, \qquad r_i = (t_{i1}, \ldots, t_{in}). \tag{6}$$

According to (4), the ith component of Te_j is t_{ij}; according to (6), the ith component of c_j is t_{ij}. Thus

$$Te_j = c_j. \tag{7}$$

This formula shows that, as consequence of the decision to put t_{ij} in the ith row and jth column, the image of e_j under T appears as a column vector. To be consistent, we shall write all vectors in $U = \mathbb{R}^m$ as column vectors:

$$u = \begin{pmatrix} u_1 \\ \vdots \\ u_m \end{pmatrix}.$$

We shall also write elements of $X = \mathbb{R}^n$ as column vectors:

$$x = \begin{pmatrix} x_1 \\ \vdots \\ x_n \end{pmatrix}.$$

The matrix representation (6) of a linear map l from $\mathbb{R}^n$ to $\mathbb{R}$ is a *single row vector* of n components:

$$l = (l_1, \ldots, l_n), \qquad l(x) = l_1 x_1 + \cdots + l_n x_n. \tag{8}$$

We define by (8) the product of a row vector r with a column vector x, in this order. It can be used to give a compact description of formula (1) giving the action of a matrix on a column vector:

$$Tx = \begin{pmatrix} r_1 x \\ \vdots \\ r_m x \end{pmatrix}, \tag{9}$$

where $r_1, \ldots, r_m$ are the rows of the matrix T.

In Chapter 3 we have described the algebra of linear mappings. Since matrices represent linear mappings of $\mathbb{R}^m$ into $\mathbb{R}^n$, there is a corresponding algebra of matrices.

Let S and T be $m \times n$ matrices, representing mappings of $\mathbb{R}^m$ to $\mathbb{R}^n$. Their sum $T + S$ represents the sum of these mappings. It follows from formula (4) that the entries of $T + S$ are the sums of the corresponding entries of T and S:

$$(T + S)_{ij} = T_{ij} + S_{ij}.$$

Next we show how to use (8) and (9) to calculate the elements of the product of two matrices. Let T, S be matrices

$$T: \mathbb{R}^n \to \mathbb{R}^m, \qquad S: \mathbb{R}^m \to \mathbb{R}^l.$$

Since the target space of T is the domain space of S, the product ST is well-defined. According to formula (7) applied to ST, the jth column of ST is

$$STe_j.$$

According to (7), $Te_j = c_j$; applying (9) to $x = Te_j$, and S in place of T gives

$$STe_j = Sc_j = \begin{pmatrix} s_1 c_j \\ \vdots \\ s_j c_j \end{pmatrix},$$

where s_k denotes the kth row of S. Thus we deduce this rule:

Rule of matrix multiplication. Let T be an $m \times n$ matrix and S an $l \times m$ matrix. Then the product of ST is an $l \times n$ matrix whose (kj)th entry is the product of the kth row of S and the jth column of T:

$$(ST)_{kj} = s_k c_j, \tag{10}$$

where s_k is the kth row of S and c_j is the jth column of T.
In terms of entries,

$$(ST)_{kj} = \sum_i S_{ki} T_{ij}. \tag{10}'$$

Example 1. $\begin{pmatrix} 1 & 2 \\ 3 & 4 \end{pmatrix} \begin{pmatrix} 5 & 6 \\ 7 & 8 \end{pmatrix} = \begin{pmatrix} 19 & 22 \\ 43 & 50 \end{pmatrix}.$

Example 2. $\begin{pmatrix} 1 \\ 2 \end{pmatrix} (3 \quad 4) = \begin{pmatrix} 3 & 4 \\ 6 & 8 \end{pmatrix}.$

Example 3. $(3 \quad 4) \begin{pmatrix} 1 \\ 2 \end{pmatrix} = (11).$

Example 4. $(1 \quad 2) \begin{pmatrix} 3 & 4 \\ 5 & 6 \end{pmatrix} = (13 \quad 16).$

Example 5. $\begin{pmatrix} 3 & 4 \\ 5 & 6 \end{pmatrix} \begin{pmatrix} 1 \\ 2 \end{pmatrix} = \begin{pmatrix} 11 \\ 17 \end{pmatrix}.$

Example 6. $(1 \quad 2) \begin{pmatrix} 3 & 4 \\ 5 & 6 \end{pmatrix} \begin{pmatrix} 1 \\ 2 \end{pmatrix} = (1 \quad 2) \begin{pmatrix} 11 \\ 17 \end{pmatrix} = (45);$

$(1 \quad 2) \begin{pmatrix} 3 & 4 \\ 5 & 6 \end{pmatrix} \begin{pmatrix} 1 \\ 2 \end{pmatrix} = (13 \quad 16) \begin{pmatrix} 1 \\ 2 \end{pmatrix} = (45);$

Example 7. $\begin{pmatrix} 5 & 6 \\ 7 & 8 \end{pmatrix} \begin{pmatrix} 1 & 2 \\ 3 & 4 \end{pmatrix} = \begin{pmatrix} 23 & 34 \\ 31 & 46 \end{pmatrix}.$

Examples 1 and 7 show that matrix multiplication of square matrices need not be commutative. Example 6 is an illustration of the associative property of matrix multiplication.

EXERCISE I. Let A be an arbitrary $m \times n$ matrix, and let D be an $m \times n$ diagonal matrix,

$$D_{ij} = \begin{cases} d_i & \text{if } i = j, \\ 0 & \text{if } i \neq j. \end{cases}$$

Show that the ith row of DA equals d_i times the ith row of A, and show that the jth column of AD equals d_j times the jth column of A.

An $n \times n$ matrix A represents a mapping of $\mathbb{R}^n$ into $\mathbb{R}^n$. If this mapping is invertible, the matrix A is called invertible.

Remark. Since the composition of linear mappings is associative, matrix multiplication, which is the composition of mappings from $\mathbb{R}^n$ to $\mathbb{R}^m$ with mappings from $\mathbb{R}^m$ to $\mathbb{R}^l$, also is associative.

We shall identify the dual of the space $\mathbb{R}^n$ of all column vectors with n components as the space $(\mathbb{R}^n)'$ of all row vectors with n components.

The action of a vector l in the dual space $(\mathbb{R}^n)'$ on a vector x of $\mathbb{R}^n$, denoted by brackets in formula (6) of Chapter 2, shall be taken to be the matrix product (8):

$$(l, x) = l_1 x_1 + \cdots + l_n x_n. \tag{11}$$

Let x, T and l be linear mappings as follows:

$$l: \mathbb{R}^m \to \mathbb{R}, \qquad T: \mathbb{R}^n \to \mathbb{R}^m, \qquad x: \mathbb{R} \to \mathbb{R}^n.$$

According to the associative law,

$$(lT)x = l(Tx). \tag{12}$$

We identify l with an element of $(\mathbb{R}^m)'$, and lT with an element of $(\mathbb{R}^n)'$. Using the notation (11) we can rewrite (12) as

$$(lT, x) = (l, Tx). \tag{13}$$

We recall now the definition of the transpose T' of T, defined by formula (9) of Chapter 3,

$$(T'l, x) = (l, Tx). \tag{13$'$}$$

Comparing (13) and (13)$'$ we see that *the matrix T acting from the right on row vectors is the transpose of the matrix T acting from the left on column vectors.*

To represent the transpose T' as a matrix acting on column vectors, we change its rows into columns, its columns into rows, and denote the resulting matrix as T^T:

$$(T^T)_{ij} = T_{ji}. \tag{13$''$}$$

Given a row vector $r = (r_1, \ldots, r_n)$, we denote by r^T the column vector with the same components. Similarly, given a column vector c, c^T denotes the row vector with the same components.

Next we turn to expressing the range of T in matrix language. Setting (7), $c_j = Tl_j$, into (3), $Tx = \sum x_j Te_j$, gives

$$u = Tx = x_1 c_1 + \cdots + x_n c_n.$$

This gives the following theorem.

Theorem 2. The range of T consists of all linear combinations of the columns of the matrix T.

The dimension of this space is called in old-fashioned texts the *column rank* of T. The row rank is defined similarly; $(13)''$ shows that the row rank of T is the dimension of the range of T^T. Since according to Theorem 6 of Chapter 3.

$$\dim R_T = \dim R_{T^T},$$

we conclude that *the column rank and row rank of a matrix are equal.*

EXERCISE 2. Look up in any text the proof that the row rank of a matrix equals its column rank, and compare it to the proof given in the present text.

We show now how to represent a linear mapping $T: X \to U$ by a matrix. We have seen in Chapter 1 that X is isomorphic to $\mathbb{R}^n$, $n = \dim X$, and U isomorphic to $\mathbb{R}^m$, $m = \dim U$. The isomorphisms are accomplished by choosing a basis in X, $y_1, \ldots, y_n$, and then mapping $y_j \leftrightarrow e_j, j = 1, \ldots, n$:

$$B: X \to \mathbb{R}^n; \tag{14}$$

similarly,

$$C: U \to \mathbb{R}^m. \tag{14}'$$

Clearly, there are as many isomorphisms as there are bases. We can use any of these isomorphisms to represent T as $\mathbb{R}^n \to \mathbb{R}^m$, obtaining a matrix representation M:

$$CTB^{-1} = M. \tag{15}$$

When T is a mapping of a space X into itself, we use the same isomorphism in (14) and $(14)'$, that is, we take $B = C$. So in this case the matrix representing T has the form

$$BTB^{-1} = M. \tag{15}'$$

Suppose we change the isomorphism B. How does the matrix representing T change? If C is another isomorphism $X \to \mathbb{R}^n$, the new matrix N representing T is $N = CTC^{-1}$. We can write, using the associative rule and $(15)'$,

$$N = CTC^{-1} = CB^{-1}BTB^{-1}BC^{-1} = SMS^{-1}, \qquad (16)$$

where $S = CB^{-1}$. Since B and C both map X into $\mathbb{R}^n$, $CB^{-1} = S$ maps $\mathbb{R}^n$ onto $\mathbb{R}^n$, that is, S is an invertible $n \times n$ matrix.

Two square matrices N and M related to each other as in (16) are called *similar*. Our analysis shows that similar matrices describe the *same* mapping of a space into itself, in different bases. Therefore we expect similar matrices to have the same intrinsic properties; we shall make the meaning of this more precise in Chapter 6.

We can write any $n \times n$ matrix A in 2×2 block form:

$$A = \begin{pmatrix} A_{11} & A_{12} \\ A_{21} & A_{22} \end{pmatrix},$$

where A_{11} is the submatrix of A contained in the first k rows and columns, A_{12} the submatrix contained in the first k rows and the last $n - k$ columns, and so on.

EXERCISE 3.   Show that the product of two matrices in 2×2 block form can be evaluated as

$$\begin{pmatrix} A_{11} & A_{12} \\ A_{21} & A_{22} \end{pmatrix} \begin{pmatrix} B_{11} & B_{12} \\ B_{21} & B_{22} \end{pmatrix} = \begin{pmatrix} A_{11}B_{11} + A_{12}B_{21} & A_{11}B_{12} + A_{12}B_{22} \\ A_{21}B_{11} + A_{22}B_{21} & A_{21}B_{12} + A_{22}B_{22} \end{pmatrix}.$$

The inversion of matrices will be discussed from a theoretical point of view in Chapter 5, and from a numerical point of view in Chapter 17.

A matrix that is not invertible is called *singular*.

Definition.   The square matrix 1 whose elements are $I_{ij} = 0$ when i is $\neq j$, $I_{jj} = 1$ is called the *unit* matrix.

Definition.   A square matrix (t_{ij}) for which $t_{ij} = 0$ for $i > j$ is called *upper triangular*. Lower triangular is defined similarly.

Definition.   A square matrix (t_{ij}) for which $t_{ij} = 0$ when $|i - j| > 1$ is called *tridiagonal*.

EXERCISE 4.   Construct two 2×2 matrices A and B such that $AB = 0$ but $BA \neq 0$.

We turn now to the most important, certainly the oldest, way to solve sets of linear equations, *Gaussian elimination*. We illustrate it on a simple example of four linear equations for four unknowns x_1, x_2, x_3, and x_4:

$$
\begin{aligned}
x_1 + 2x_2 + 3x_3 - x_4 &= -2, \\
2x_1 + 5x_2 + 4x_3 - 3x_4 &= 1, \\
2x_1 + 3x_2 + 4x_3 + x_4 &= 1, \\
x_1 + 4x_2 + 2x_3 - 2x_4 &= 3.
\end{aligned}
\tag{17}
$$

We solve this system of equations by eliminating the unknowns one by one; here is how it is done. We use the first equation in (17) to eliminate x_1 from the rest of the equations. To accomplish this, subtract two times the first equation from the second and the third equations, obtaining

$$
x_2 - 2x_3 - x_4 = 5,
\tag{18}$_1$
$$

and

$$
-x_2 - 2x_3 + 3x_4 = 5.
\tag{18}$_2$
$$

Subtract the first equation from the fourth one, obtaining

$$
2x_2 - x_3 - x_4 = 5.
\tag{18}$_3$
$$

We use the same technique to eliminate x_2 from the set of three equations (18). We obtain

$$
\begin{aligned}
-4x_3 + 2x_4 &= 10, \tag{19}$_1$ \\
3x_3 + x_4 &= -5. \tag{19}$_2$
\end{aligned}
$$

Finally we eliminate x_3 from equations (19) by adding 3/4 times $(19)_1$ to $(19)_2$; we get

$$
\frac{5}{2}x_4 = 5/2,
$$

which yields

$$
x_4 = 1.
\tag{20}$_4$
$$

We proceed in the reverse order, by backward substitution, to determine the other unknowns. Setting the value of x_4 from $(20)_4$ into equation $(19)_1$ gives

$$
-4x_3 + 2 = 10,
$$

which yields

$$x_3 = -2. \tag{20_3}$$

We could have used equation $(19)_2$ and would have gotten the same answer.

We determine x_2 from any of the equations (18), say $(18)_1$, by using $(20)_3$ and $(20)_4$ for x_3 and x_4. We get

$$x_2 + 4 - 1 = 5,$$

so

$$x_2 = 2. \tag{20_2}$$

Finally we determine x_1 from, say, the first equation (17), using the previously determined values of x_4, x_3, and x_2:

$$x_1 + 4 - 6 - 1 = -2,$$
$$x_1 = 1. \tag{20_1}$$

EXERCISE 5.   Show that x_1, x_2, x_3, and x_4 given by $(20)_j$ satisfy all four equations (20).

Notice that the order in which we eliminate the unknowns, along with the equations which we use to eliminate them, is arbitrary. We shall return to these points.

A system of n equations

$$\sum_1^n t_{ij} x_i = u_j, \qquad j = 1, \ldots, n \tag{21}$$

for n unknowns $x_1, \ldots, x_n$ may have a unique solution, may have no solution, or may have many solutions. We show now how to use Gaussian elimination to determine all solutions, or conclude that no solution exists. Here is an example that illustrates the last two possibilities.

$$\begin{aligned}
x_1 + x_2 + 2x_3 + 3x_4 &= u_1, \\
x_1 + 2x_2 + 3x_3 + x_4 &= u_2, \\
2x_1 + x_2 + 2x_3 + 3x_4 &= u_3, \\
3x_1 + 4x_2 + 6x_3 + 2x_4 &= u_4.
\end{aligned} \tag{22}$$

We eliminate x_1 from the last three equations by subtracting from them an appropriate multiple of the first equation:

$$x_2 + x_3 - 2x_4 = u_2 - u_1$$
$$-x_2 - 2x_3 - 3x_4 = u_3 - 2u_1$$
$$x_2 - 7x_4 = u_4 - 3u_1$$

We use the first equation above to eliminate x_2 from the last two:

$$-x_3 - 5x_4 = u_3 + u_2 - 3u_1$$
$$-x_3 - 5x_4 = u_4 - u_2 - 2u_1$$

We eliminate x_3 by subtracting the last two equations from each other. We find that thereby we have eliminated x_4 as well, and we get

$$0 = u_4 - u_3 - 2u_2 + u_1. \tag{23}$$

This is the necessary and sufficient condition for the system of equations (22) to have a solution.

EXERCISE 6.   Choose values of u_1, u_2, u_3, u_4 so that condition (23) is satisfied, and determine all solutions of equations (22).

Equation (22) can be written in matrix notation as

$$Mx = u, \tag{22$'$}$$

where x and u are column vectors with components x_1, x_2, x_3, x_4 and u_1, u_2, u_3, u_4, and

$$M = \begin{pmatrix} 1 & 1 & 2 & 3 \\ 1 & 2 & 3 & 1 \\ 2 & 1 & 2 & 3 \\ 3 & 4 & 6 & 2 \end{pmatrix}.$$

EXERCISE 7.   Verify that $l = (1, -2, -1, 1)$ is a left nullvector of M:

$$lM = 0.$$

Multiply equation (22)$'$ on the left by l; using the result of Exercise 7, we get that

$$lMx = lu = 0,$$

a rederivation of (23) as a necessary condition for (22)$'$ to have a solution.

EXERCISE 8. Show by Gaussian elimination that the only left nullvectors of M are multiples of l in Exercise 7, and then use Theorem 5 of Chapter 3 to show that condition (23) is sufficient for the solvability of the system (22).

Next we show how to use Gaussian elimination to prove Corollary A' in Chapter 3:

A system of homogeneous linear equations

$$\sum_{j=1}^{n} t_{ij}x_j = 0, \qquad i = 1, \ldots, m, \tag{24}$$

with fewer equations than unknowns, $m < n$, has a nontrivial solution—that is, one where at least one of the x_j is nonzero.

Proof. We use one of the equations (24) to express x_1 as a linear function of the rest of the x's:

$$x_1 = l_1(x_2, \ldots, x_n). \tag{25}_1$$

We replace x_1 by l_1 in the remaining equations, and we use one of them to express x_2 as a linear function of the remaining x's:

$$x_2 = l_2(x_3, \ldots, x_n). \tag{25}_2$$

We proceed in this fashion until we reach x_m:

$$x_m = l_m(x_{m+1}, \ldots, x_n). \tag{25}_m$$

Since there were only m equations and $m < n$, there are no more equations left to be satisfied. So we choose the values of $x_{m+1}, \ldots, x_n$ arbitrarily, and we use equations $(25)_m, (25)_{m-1}, \ldots, (25)_1$, in this order, to determine the values of $x_m, x_{m-1}, \ldots, x_1$.

This procedure may break down at the ith step if none of the remaining equations contain x_i. In this case we set $x_{i+1}, \ldots, x_n$ equal to zero, assign an arbitrary value to x_i, and determine $x_{i-1}, \ldots, x_1$ from equations $(25)_{i-1}, \ldots, (25)_1$, in this order. $\square$

We conclude this chapter with some observations on how Gaussian elimination works for determined systems of n inhomogeneous equations

$$\sum_{j=1}^{n} t_{ij}x_j = u_i, \qquad i = 1, \ldots, n \tag{26}$$

for n unknowns $x_1, \ldots, x_n$. In its basic form the first equation is used to eliminate x_1, that is, express it as

$$x_1 = v + l_1(x_2, \ldots, x_n). \tag{27}_1$$

Then x_1 is replaced in the remaining equations by $v_1 + l_1$. The first of these equations is used to express x_2 as

$$x_2 = v_2 + l_2(x_3, \ldots, x_n). \tag{27}_2$$

We proceed in this fashion until after $(n - 1)$ steps we find the value of x_n. Then we determine the values of $x_{n-1}, \ldots, x_1$, in this order, from the relations $(27)_{n-1}, \ldots, (27)_1$.

This procedure may break down right at the start if the coefficient t_{11} of x_1 in the first equation is zero. Even if t_{11} is not zero but very small, using the first equation to express x_1 in terms of the rest of the x's involves division by t_{11} and produces very large coefficients in formula $(27)_1$. This wouldn't matter if all arithmetic operations were carried out exactly, but they never are; they are carried out in finite digit floating point arithmetic, and when $(27)_1$ is substituted in the remaining equations, the coefficients t_{ij}, $i > 1$, are swamped.

A natural remedy is to choose another unknown, x_j, for elimination and another equation to accomplish it, so chosen that t_{ij} is not small compared with the other coefficients. This strategy is called *complete pivoting* and is computationally expensive. A compromise is to keep the original order of the unknowns for elimination, but use another equation for elimination, for which t_{i1} is not small compared to the other coefficients. This strategy, called *partial pivoting*, works very well in practice (see, e.g., the text entitled *Numerical Linear Algebra*, by Trefethen and Bau.)

CHAPTER 5

Determinant and Trace

In this chapter we shall use the intuitive properties of volume to define the determinant of a square matrix. According to the precepts of elementary geometry, the concept of volume depends on the notions of length and angle and, in particular, perpendicularity, concepts that will be defined only in Chapter 8. Nevertheless, it turns out that volume is independent of all these things, except for an arbitrary multiplicative constant that can be fixed by specifying that the unit cube have volume one.

We start with the geometric motivation and meaning of determinants. A *simplex* in $\mathbb{R}^n$ is a polyhedron with $n + 1$ vertices. We shall take one of the vertices to be the origin and denote the rest as $a_1, \ldots, a_n$. The order in which the vertices are taken matters, so we call $0, a_1, \ldots, a_n$ the vertices of an *ordered simplex*.

We shall be dealing with two geometrical attributes of ordered simplices, their *orientation* and *volume*. An ordered simplex S is called *degenerate* if it lies on an $(n - 1)$-dimensional subspace.

An ordered simplex $(0, a_1, \ldots, a_n) = S$ that is nondegenerate can have one of two orientations: positive or negative. We call S *positively oriented* if it can be deformed continuously and nondegenerately into the *standard ordered simplex* $(0, e_1, \ldots, e_n)$, where e_j is the jth unit vector in the standard basis of $\mathbb{R}^n$. By such deformation we mean n vector-valued continuous functions $a_j(t)$ of $t, 0 \leq t \leq 1$, such that (i) $S(t) = (0, a_1(t), \ldots, a_n(t))$ is nondegenerate for all t and (ii) $a_j(0) = a_j$, $a_j(1) = e_j$. Otherwise S is called negatively oriented.

For a nondegenerate oriented simplex S we define $O(S)$ as $+1$ or -1, depending on the orientation of S, and zero when S is degenerate.

The *volume* of a simplex is given by the elementary formula

$$\mathrm{Vol}(S) = \frac{1}{n} \mathrm{Vol}_{n-1}(\mathrm{Base})\mathrm{Altitude}. \tag{1}$$

Linear Algebra and Its Applications, Second Edition, by Peter D. Lax
Copyright © 2007 John Wiley & Sons, Inc.

By base we mean any of the $(n-1)$-dimensional faces of S, and by altitude we mean the distance of the opposite vertex from the hyperplane that contains the base.

A more useful concept is *signed volume*, denoted as $\sum(S)$, and defined by

$$\sum(S) = O(S)\mathrm{Vol}(S). \tag{2}$$

Since S is described by its vertices, $\sum(S)$ is a function of $a_1, \ldots, a_n$. Clearly, when two vertices are equal, S is degenerate, and therefore we have the following:

(i) $\sum(S) = 0$ if $a_j = a_k$, $j \neq k$.

A second property of $\sum(S)$ is its dependence on a_j when the other vertices are kept fixed:

(ii) $\sum(S)$ is a *linear function* of a_j when the other a_k, $k \neq j$, are kept fixed.

Let us see why we combine formulas (1) and (2) as

$$\sum(S) = \frac{1}{n}\mathrm{Vol}_{n-1}(\text{base})k, \tag{1}'$$

where

$$k = O(S)\text{Altitude}.$$

The altitude is the *distance* of the vertex a_j; we call k the *signed distance* of the vertex from the hyperplane containing the base, because $O(S)$ has one sign when a_j lies on one side of the base and the opposite sign when a_j lies on the opposite side.

We claim that when the base is fixed, k is a linear function of a_j. To see why this is so we introduce Cartesian coordinate axes so that first axis is perpendicular to the base and the rest lie in the base plane. By definition of Cartesian coordinates, the first coordinate $k_1(a)$ of a vector a is its signed distance from the hyperplane spanned by the other axes. According to Theorem 1 (i) in Chapter 2, $k_1(a)$ is a linear function of a. Assertion (ii) now follows from formula (1)'.

Determinants are related to the signed volume of ordered simplices by the classical formula,

$$\sum(S) = \frac{1}{n!}D(a_1, \ldots, a_n), \tag{3}$$

where D is the abbreviation of the determinant whose columns are $a_1, \ldots, a_n$. Rather than start with a formula for the determinant, we shall deduce it from the properties forced on it by the geometric properties of signed volume. This approach to determinants is due to E. Artin.

Property (i). $D(a_1, \ldots, a_n) = 0$ if $a_i = a_j$, $i \neq j$.

Property (ii). $D(a_1, \ldots, a_n)$ is a *multilinear* function of its arguments, in the sense that if all a_i, $i \neq j$ are fixed, D is a linear function of the remaining argument a_j.

Property (iii). Normalization:

$$D(e_1, \ldots, e_n) = 1. \tag{4}$$

We show now that all remaining properties of D can be deduced from those so far postulated.

Property (iv). D is an *alternating* function of its arguments, in the sense that if a_i and a_j are interchanged, $i \neq j$, the value of D changes by the factor (-1).

Proof. Since only the ith and jth argument change, we shall indicate only these. Setting $a_i = a$, $a_j = b$ we can write, using Properties (i) and (ii):

$$\begin{aligned} D(a, b) &= D(a, b) + D(a, a) = D(a, a + b) \\ &= D(a, a + b) - D(a + b, a + b) \\ &= -D(b, a + b) = -D(b, a) - D(b, b) = -D(b, a). \end{aligned} \qquad \square$$

Property (v). If $a_1, \ldots, a_n$ are linearly dependent, then $D(a_1, \ldots, a_n) = 0$.

Proof. If $a_1, \ldots, a_n$ are linearly dependent, then one of them, say a_1, can be expressed as a linear combination of the others:

$$a_1 = k_2 a_2 + \cdots + k_n a_n.$$

Then, using Property (ii),

$$\begin{aligned} D(a_1, \ldots, a_n) &= D(k_2 a_2 + \cdots + k_n a_n, a_2, \ldots, a_n) \\ &= k_2 D(a_2, a_2, \ldots, a_n) + \cdots + k_n D(a_n, a_2, \ldots, a_n). \end{aligned}$$

By property (i), all terms in the last line are zero. $\qquad \square$

Next we introduce the concept of *permutation*. A permutation is a mapping p of n objects, say the numbers $1, 2, \ldots, n$, *onto* themselves. Like all functions, permutations can be composed. Being onto, they are one-to-one and so can be inverted. Thus they form a group; these groups, except for $n = 2$, are noncommutative.

We denote $p(k)$ as p_k; it is convenient to display the action of p by a table:

1	2	$\ldots$	n
p_1	p_2	$\ldots$	p_n

Example 1. $p = \frac{1234}{2413}$. Then

$$p^2 = \frac{1234}{4321}, \qquad p^{-1} = \frac{1234}{3142},$$

$$p^3 = \frac{1234}{3142}, \qquad p^4 = \frac{1234}{1234}.$$

Next we introduce the concept of *signature* of a permutation, denoted as $\sigma(p)$. Let $x_1, \ldots, x_n$ be n variables; their *discriminant* is defined to be

$$P(x_1, \ldots, x_n) = \prod_{i<j} (x_i - x_j). \tag{5}$$

Let p be any permutation. Clearly,

$$P(p(x_1, \ldots, x_n)) = \prod_{i<j} (x_{pi} - x_{pj})$$

is either $P(x_1, \ldots x_n)$ or $-P(x_1, \ldots, x_n)$.

Definition.   The signature $\sigma(p)$ of a permutation p is defined by

$$P(p(x_1, \ldots, x_n)) = \sigma(p)P(x_1, \ldots, x_n). \tag{6}$$

Properties of signature:

(a) $\qquad\qquad\qquad \sigma(p) = +1 \text{ or } -1.$

(b) $\qquad\qquad\qquad \sigma(p_1 \circ p_2) = \sigma(p_1)\sigma(p_2).$ $\qquad\qquad$ (7)

EXERCISE 1.   Prove properties (7).

We look now at a special kind of permutation, an interchange. These are defined for any pair of indices, j, k, $j \neq k$ as follows:

$$p(i) = i \qquad \text{for } i \neq j \text{ or } k,$$
$$p(j) = k, \; p(k) = j.$$

Such a permutation is called a *transposition*. We claim that transposition has the following properties:
   (c) The signature of a transposition t is minus one:

$$\sigma(t) = -1. \tag{8}$$

   (d) Every permutation p can be written as a composition of transpositions:

$$p = t_k \circ \cdots \circ t_1. \tag{9}$$

EXERCISE 2.   Prove (c) and (d) above.

Combining (7) with (8) and (9) we get that

$$\sigma(p) = (-1)^k, \tag{10}$$

where k is the number of factors in the decomposition (9) of p.

EXERCISE 3.   Show that the decomposition (9) is not unique, but that the parity of the member k of factors is unique.

Example 2.   The permutation $p = \frac{12345}{24513}$ is the product of three transpositions $t_1 = \frac{12345}{12543}$, $t_2 = \frac{12345}{21345}$, $t_3 = \frac{12345}{42315}$.

$$p = t_3 \circ t_2 \circ t_1.$$

We return now to the function D. Its arguments a_j are column vectors

$$a_j = \begin{pmatrix} a_{1j} \\ \vdots \\ a_{nj} \end{pmatrix}, \qquad j = 1, \dots, n. \tag{11}$$

This is the same as

$$a_j = a_{1j}e_1 + \dots + a_{nj}e_n. \tag{11}'$$

Using Property (ii), multilinearity, we can write

$$\begin{aligned} D(a_1, \dots, a_n) &= D(a_{11}e_1 + \dots + a_{n1}e_n, a_2, \dots, a_n) \\ &= a_{11}D(e_1, a_2, \dots, a_n) + \dots + a_{n1}D(e_n, a_2, \dots, a_n). \end{aligned} \tag{12}$$

Next we express a_2 as a linear combination of $e_1, \dots, e_n$ and obtain a formula like (12) but containing n^2 terms. Repeating this process n times we get

$$D(a_1, \dots, a_n) = \sum_f a_{f_1 1} \, a_{f_2 2} \cdots a_{f_n n} \, D(e_{f_1}, \dots, e_{f_n}), \tag{13}$$

where the summation is over all functions f mapping $\{1, \dots, n\}$ into $\{1, \dots, n\}$. If the mapping f is not a permutation, then $f_i = f_j$ for some pair $i \neq j$ and by Property (i).

$$D(e_{f_1}, \dots, e_{f_n}) = 0. \tag{14}$$

This shows that in (13) we need sum only over those f that are permutations.

We saw earlier that each permutation can be decomposed into k transpositions (9). According to Property (iv), a single transposition of its arguments changes the value of D by a factor of (-1). Therefore k transpositions change it by the factor $(-1)^k$. Thus, using (10),

$$D(e_{p_1}, \ldots, e_{p_n}) = \sigma(p)D(e_1, \ldots, e_n) \tag{15}$$

for any permutation. Setting (14) and (15) into (13) we get, after using the normalization (4), that

$$D(a_1, \ldots, a_n) = \sum \sigma(p)a_{p_1 1} \cdots a_{p_n n}. \tag{16}$$

This is the formula for D in terms of the components of its arguments.

Formula (16) was derived using solely properties (i), (ii), and (iii) of determinants. Therefore we conclude with the following theorem.

Theorem 1. Properties (i), (ii), and (iii) uniquely determine the determinant as a function of $a_1, \ldots, a_n$.

EXERCISE 4. Show that D defined by (16) has Properties (ii), (iii) and (iv).

EXERCISE 5. Show that Property (iv) implies Property (i), unless the field K has characteristic two, that is, $1 + 1 = 0$.

Definition. Let A be an $n \times n$ matrix; denote its column vectors by $a_1, \ldots, a_n$: $A = (a_1, \ldots, a_n)$. Its determinant, denoted as det A, is

$$\det A = D(a_1, \ldots, a_n), \tag{17}$$

where D is defined by formula (16).

The determinant has properties (i)–(v) that have been derived and verified for the function D. We state now an additional important property.

Theorem 2. For all pairs of $n \times n$ matrices A and B,

$$\det(BA) = \det A \det B. \tag{18}$$

Proof. According to equation (7) of Chapter 4, the jth column of BA is $(BA)e_j$. The jth column a_j of A is Ae_j; therefore the jth column of BA is

$$(BA)e_j = BAe_j = Ba_j.$$

By definition (17),

$$\det(BA) = D(Ba_1, \ldots, Ba_n). \tag{19}$$

We assume now that det B $\neq 0$ and define the function C as follows:

$$C(a_1,\ldots,a_n) = \frac{\det(BA)}{\det B}. \tag{20}$$

Using (19) we can express C as follows:

$$C(a_1,\ldots,a_n) = \frac{D(Ba_1,\ldots,Ba_n)}{\det B}. \tag{20}'$$

We claim that the function C has Properties (i)–(iii) postulated for D.

(i) If $a_i = a_j$, $i \neq j$, then $Ba_i = Ba_j$; since D has Property (i), it follows that the right-hand side of (20)' is zero. This shows that C also has Property (i).

(ii) Since Ba_i is a linear function of a_i, and since D is a multilinear function, it follows that the right-hand side of (20)' is also a multilinear function. This shows that C is a multilinear function of $a_1,\ldots,a_n$, that is, has Property (ii).

(iii) Setting $a_i = e_i$, $i = 1,2,\ldots,n$ into formula (20)', we get

$$C(e_1,\ldots,e_n) = \frac{D(Be_1,\ldots,Be_n)}{\det B}. \tag{21}$$

Now Be_i is the ith column b_i of B, so that the right-hand side of (21) is

$$\frac{D(b_1,\ldots,b_n)}{\det B}. \tag{22}$$

By definition (17) applied to B, (22) equals 1; setting this into (21) we see that $C(e_1,\ldots,e_n) = 1$. This proves that C satisfies Property (iii).

We have shown in Theorem 1 that a function C that satisfies Properties (i)–(iii) is equal to the function D. So

$$C(a_1,\ldots,a_n) = D(a_1,\ldots,a_n) = \det A.$$

Setting this into (20) proves (18), when det B $\neq 0$.

When det B $= 0$ we argue as follows: define the matrix B(t) as

$$B(t) = B + tI.$$

Clearly, B(0) $=$ B. Formula (16) shows that $D(B(t))$ is a polynomial of degree n, and that the coefficient of t^n equals one. Therefore, $D(B(t))$ is zero for no more than n values of t; in particular $D(B(t)) \neq 0$ for all t near zero but not equal to zero. According to what we have already shown, $\det(B(t)A) = \det A \det B(t)$ for all such values of t; letting t tend to zero yields (18). $\quad\square$

Corollary 3.   An $n \times n$ matrix A is invertible iff det A $\neq$ 0.

Proof. Suppose A is not invertible; then its range is a proper subspace of $\mathbb{R}^n$. The range of A consists of all linear combinations of the columns of A; therefore the columns are linearly dependent. According to property (v), this implies that det A = 0.

Suppose, on the other hand, that A is invertible; denote its inverse by B:

$$BA = I.$$

According to Theorem 2

$$\det B \det A = \det I.$$

By property (iii), det $I = 1$; so, since $D(I) = 1$,

$$\det B \det A = 1,$$

which shows that det A $\neq$ 0.   □

The geometric meaning of the multiplicative property of determinants is this: the linear mapping B maps every simplex onto another simplex whose volume is $|\det B|$ times the volume of the original simplex. Since every open set is the union of simplices, it follows that the volume of the image under B of any open set is $|\det B|$ times the original volume.

We turn now to yet another property of determinants. We need the following lemma.

Lemma 4.   Let A be an $n \times n$ matrix whose first column is e_1:

$$A = \begin{pmatrix} 1 & \times \times \times \\ 0 & \\ \vdots & A_{11} \\ 0 & \end{pmatrix}; \qquad (23)$$

here A_{11} denotes the $(n-1) \times (n-1)$ submatrix formed by entries $a_{ij}, i > 1, j > 1$. We claim that

$$\det A = \det A_{11}. \qquad (24)$$

Proof. As first step we show that

$$\det A = \det \begin{pmatrix} 1 & 0...0 \\ 0 & A_{11} \\ 0 & \end{pmatrix}. \qquad (25)$$

For it follows from Properties (i) and (ii) that if we alter a matrix by adding a multiple of one of its columns to another, the altered matrix has the same determinant as the original. Clearly, by adding suitable multiplies of the first column of A to the others we can turn it into the matrix on the right in (25).

We regard now

$$C(A_{11}) = \det \begin{pmatrix} 1 & 0 \\ 0 & A_{11} \end{pmatrix}$$

as a function of the matrix A_{11}. Clearly it has Properties (i)–(iii). Therefore it must be equal to $\det A_{11}$. Combining this with (25) gives (24). □

EXERCISE 6.   Verify that $C(A_{11})$ has properties (i)–(iii).

Corollary 5.   Let A be a matrix whose jth column is e_i. Then

$$\det A = (-1)^{i+j} \det A_{ij}, \tag{25}'$$

where A_{ij} is the $(n-1) \times (n-1)$ matrix obtained by striking out the ith row and jth column of A; A_{ij} is called the (ij)th minor of A.

EXERCISE 7.   Deduce Corollary 5 from Lemma 4.

We deduce now the so-called Laplace expansion of a determinant according to its columns.

Theorem 6.   Let A be any $n \times n$ matrix and j any index between 1 and n. Then

$$\det A = \sum_i (-1)^{i+j} a_{ij} \det A_{ij}. \tag{26}$$

Proof. To simplify notation, we take $j = 1$. We write a_1 as a linear combination of standard unit vectors:

$$a_1 = a_{11} e_1 + \cdots + a_{n1} e_n.$$

Using multilinearity, we get

$$\det A = D(a_1, \ldots, a_n) = D(a_{11} e_1 + \cdots + a_{n1} e_n, a_2, \ldots, a_n)$$
$$= a_{11} D(e_1, a_2, \ldots, a_n) + \cdots + a_{n1} D(e_n, a_2, \ldots, a_n).$$

Using Corollary 5, we obtain (26). □

We show now how determinants can be used to express solutions of systems of equations of the form

$$Ax = u, \tag{27}$$

A an invertible $n \times n$ matrix. Write

$$x = \sum x_j e_j;$$

according to (7) of Chapter 4, $Ae_i = a_j$, the jth column of A. So (27) is equivalent to

$$\sum_j x_j a_j = u. \tag{27}'$$

We consider now the matrix A_k obtained by replacing the kth column of A by u:

$$A_k = (a_1, \ldots, a_{k-1}, u, a_{k+1}, \ldots, a_n)$$
$$= (a_1, \ldots, a_{k-1}, \sum x_j a_j, a_{k+1}, \ldots, a_n).$$

We form the determinant and use its multilinearity,

$$\det A_k = \sum_j x_j \det(a_1, \ldots, a_{k-1}, a_j, a_{k+1}, \ldots, a_n).$$

Because of Property (i) of determinants, the only nonzero term on the right is the kth, so we get

$$\det A_k = x_k \det A.$$

Since A is invertible, $\det A \neq 0$; so

$$x_k = \frac{\det A_k}{\det A}. \tag{28}$$

We use now the Laplace expansion of $\det A_k$ according to its kth column; we get

$$\det A_k = \sum_i (-1)^{i+k} \det A_{ik} u_i$$

and so, using (28),

$$x_k = \sum_i (-1)^{i+k} \frac{\det A_{ik}}{\det A} u_i. \tag{29}$$

This is called Cramer's rule for finding the solution of the system of equations (27). We now translate (29) into matrix language.

Theorem 7. The inverse matrix A^{-1} of an invertible matrix A has the form

$$(A^{-1})_{ki} = (-1)^{i+k} \frac{\det A_{ik}}{\det A}. \tag{30}$$

Proof. Since A is invertible, $\det A \neq 0$. A^{-1} acts on the vector u; see formula (1) of Chapter 4,

$$(A^{-1}u)_k = \sum_i (A^{-1})_{ki} u_i. \tag{31}$$

Using (30) in (31) and comparing it to (29) we get that

$$(A^{-1}u)_k = x_k, \qquad k = 1, \ldots, n, \tag{32}$$

that is,

$$A^{-1}u = x.$$

This shows that A^{-1} as defined by (30) is indeed the inverse of A whose action is given in (27).   □

We caution that reader that for $n > 3$, formula (30) is not a practical numerical method for inverting matrices.

EXERCISE 8.   Show that for any square matrix

$$\det A^T = \det A, \qquad A^T = \text{transpose of A}. \tag{33}$$

[*Hint*: Use formula (16) and show that for any permutation $\sigma(p) = \sigma(p^{-1})$.]

EXERCISE 9.   Given a permutation p of n objects, we define an associated so-called *permutation matrix* P as follows:

$$P_{ij} = \begin{cases} 1, & \text{if } j = p(i), \\ 0, & \text{otherwise.} \end{cases} \tag{34}$$

Show that the action of P on any vector x performs the permutation p on the components of x. Show that if p, q are two permutations and P, Q are the associated permutation matrices, then the permutation matrix associated with $p \circ q$ is the product PQ.

The determinant is an important scalar-valued function of $n \times n$ matrices. Another equally important scalar-valued function is the *trace*.

Definition.   The trace of a square matrix A, denoted as tr A, is the sum of the entries on its diagonal:

$$\text{tr } A = \sum_i a_{ii}. \tag{35}$$

Theorem 8.   (a) Trace is a linear function:

$$\text{tr } k A = k \text{ tr } A, \qquad \text{tr}(A + B) = \text{tr } A + \text{tr } B.$$

(b) Trace is "commutative"; that is,

$$\text{tr}(AB) = \text{tr}(BA) \tag{36}$$

for any pair of matrices.

Proof. Linearity is obvious from definition (35). To prove part (b), we use the rule, [see (10)′ of Chapter 4] for matrix multiplication:

$$(AB)_{ii} = \sum_k a_{ik} b_{ki}$$

and

$$(BA)_{ii} = \sum_k b_{ik} a_{ki}.$$

So

$$\text{tr}(AB) = \sum_{i,k} a_{ik} b_{ki} = \sum_{i,k} b_{ik} a_{ki} = \text{tr}(BA)$$

follows if one interchanges the names of the indices i, k.   □

We recall from the end of Chapter 3 the notion of *similarity*. The matrix A is called similar to the matrix B if there is an invertible matrix S such that

$$A = SBS^{-1}. \tag{37}$$

We recall from Theorem 8 of Chapter 3 that similarity is an equivalence relation; that is, it is the following:

(i) Reflexive: A is similar to itself.
(ii) Symmetric: if A is similar to B, B is similar to A,
(iii) Transitive: if A is similar to B, and B is similar to C, then A is similar to C.

Theorem 9. Similar matrices have the same determinat and the same trace.

Proof. Using Theorem 2, we get from (37)

$$\det A = (\det S)(\det B)(\det S^{-1}) = (\det B)(\det S)\det(S^{-1})$$
$$= \det B \, \det(SS^{-1}) = (\det B)(\det I) = \det B.$$

To show the second part we use Theorem 7(b):

$$\operatorname{tr} A = \operatorname{tr}(SBS^{-1}) = \operatorname{tr}((SB)S^{-1}) = \operatorname{tr}(S^{-1}(SB)) = \operatorname{tr} B. \qquad \square$$

At the end of Chapter 4 we remarked that any linear map T of an n-dimensional linear space X into itself can, by choosing a basis in X, be represented as an $n \times n$ matrix. Two different representations, coming from two different choices of bases, are similar. In view of Theorem 9, we can define the determinant and trace of such a linear map T as the determinant and trace of a matrix representing T.

EXERCISE 10. Let A be an $m \times n$ matrix, B an $n \times m$ matrix. Show that

$$\operatorname{tr} AB = \operatorname{tr} BA.$$

EXERCISE 11. Let A be an $n \times n$ matrix, A^T its transpose. Show that

$$\operatorname{tr} AA^T = \sum a_{ij}^2.$$

The square root of the double sum on the right is called the Euclidean, or Hilbert–Schmidt, norm of the matrix A.

In Chapter 9, Theorem 4, we shall derive an interesting connection between determinant and trace.

EXERCISE 12. Show that the determinant of the 2×2 matrix

$$\begin{pmatrix} a & b \\ c & d \end{pmatrix}$$

is $D = ad - bc$.

EXERCISE 13. Show that the determinant of an upper triangular matrix, one whose elements are zero below the main diagonal, equals the product of its elements along the diagonal.

EXERCISE 14. How many multiplications does it take to evaluate det A by using Gaussian elimination to bring it into upper triangular from?

EXERCISE 15. How many multiplications does it take to evaluate det A by formula (16)?

EXERCISE 16. Show that the determinant of a (3×3) matrix

$$A = \begin{pmatrix} a & b & c \\ d & e & f \\ g & h & i \end{pmatrix}$$

can be calculated as follows. Copy the first two columns of A as a fourth and fifth column:

$$\begin{pmatrix} a & b & c & a & b \\ d & e & f & d & e \\ g & h & i & g & h \end{pmatrix}.$$

$$\det A = aei + bfg + cdh - gec - hfa - idb.$$

Show that the sum of the products of the three entries along the dexter diagonals, minus the sum of the products of the three entries along the sinister diagonals is equal to the determinant of A.

CHAPTER 6

Spectral Theory

Spectral theory analyzes linear mappings of a space into itself by decomposing them into their basic constituents. We start by posing a problem originating in the stability of periodic motions and show how to solve it using spectral theory.

We assume that the *state of the system* under study can be described by a finite number n of parameters; these we lump into a single vector x in $\mathbb{R}^n$. Second, we assume that the *laws governing the evolution* in time of the system under study determine uniquely the state of the system at any future time if the initial state of the system is given.

Denote by x the state of the system at time $t = 0$; its state at $t = 1$ is then completely determined by x; we denote it as $F(x)$. We assume F to be a differentiable function. We assume that the laws governing the evolution of the system are the same at all times; it follows then that if the state of the system at time $t = 1$ is z, its state at time $t = 2$ is $F(z)$. More generally, F relates the state of the system at time t to its state at $t + 1$.

Assume that the motion starting at $x = 0$ is periodic with period one, that is that it returns to 0 at time $t = 1$. That means that

$$F(0) = 0, \tag{1}$$

This periodic motion is called *stable* if, starting at any point h sufficiently close to zero, the motion tends to zero as t tends to infinity.

The function F describing the motion is differentiable; therefore for small h, $F(h)$ is accurately described by a linear approximation:

$$F(h) \simeq Ah. \tag{2}$$

For purposes of this discussion we assume that F is a linear function

$$F(h) = Ah, \tag{3}$$

Linear Algebra and Its Applications, Second Edition, by Peter D. Lax
Copyright © 2007 John Wiley & Sons, Inc.

A an $n \times n$ matrix. The system starting at h will, after the elapse of N units of time, be in the position

$$A^N h. \tag{4}$$

In the next few pages we investigate such sequences, that is, of the form

$$h, Ah, \ldots, A^N h, \ldots. \tag{5}$$

First a few examples of how powers A^N of matrices behave; we choose $N = 1024$, because then A^N can be evaluated performing ten squaring operations:

Case	(a)	(b)
A	$\begin{pmatrix} 3 & 2 \\ 1 & 4 \end{pmatrix}$	$\begin{pmatrix} 5 & 6.9 \\ -3 & -4 \end{pmatrix}$
A^{1024}	$>10^{700}$	$<10^{-78}$

These numerical experiments strongly suggest that

(a) $A^N \to \infty$ as $N \to \infty$,

(b) $A^N \to 0$ as $N \to \infty$, that is, each entry of A^N tends to zero.

We turn now to a theoretical analysis of the behavior of sequences of the form (5). Suppose that a vector $h \neq 0$ has the special property with respect to the matrix A that Ah is merely a multiple of h:

$$Ah = ah, \quad \text{where } a \text{ is a scalar and } h \neq 0. \tag{6}$$

Then clearly

$$A^N h = a^N h. \tag{6$_N$}$$

In this case the behavior of the sequence (5) is as follows:

(i) If $|a| > 1$, $A^N h \to \infty$.

(ii) If $|a| < 1$, $A^N h \to 0$.

(iii) If $a = 1$, $A^N h = h$ for all N.

This simple analysis is applicable only if (6) is satisfied. A vector h satisfying (6) is called an *eigenvector* of A; a is called an *eigenvalue* of A.

How farfetched is it to assume that A has an eigenvector? We shall show that every $n \times n$ matrix over the field of *complex numbers* has an eigenvector. Choose any nonzero vector w and build the following set of $n + 1$ vectors:

$$w, Aw, A^2 w, \ldots, A^n w.$$

Since $n + 1$ vectors in the n-dimentional space $\mathbb{C}^n$ are linearly dependent, there is a nontrivial linear relation between them:

$$\sum_0^n c_j A^j w = 0,$$

not all c_j zero. We rewrite this relation as

$$p(A)w = 0, \tag{7}$$

where $p(t)$ is the polynomial

$$p(t) = \sum_0^n c_j t^j.$$

Every polynomial over the complex numbers can be written as a product of linear factors:

$$p(t) = c \prod (x - a_j), \quad c \neq 0.$$

$p(A)$ can be similarly factored and (7) rewritten as

$$c \prod (A - a_j I)w = 0.$$

This shows that the product $\prod(A - a_j I)$ maps the nonzero vector w into 0 and is therefore not invertible. According to Theorem 4 of Chapter 3, a product of invertible mappings is invertible. It follows that at least one of the matrices $A - a_j I$ is not invertible; such a matrix has a nontrivial nullspace. Denote by h any nonzero vector in the nullspace:

$$(A - aI)h = 0, \qquad a = a_j. \tag{6$'$}$$

This is our eigenvalue equation (6).

The argument above shows that every matrix A has at least one eigenvalue, but it does not show how many or how to calculate them. Here is another approach.

Equation (6)$'$ says that h belongs to the nullspace of $(A - aI)$; therefore the matrix $A - aI$ is not invertible. We saw in Corollary 3 of Chapter 5 that this can happen if and only if the determinant of the matrix $A - aI$ is zero:

$$\det(aI - A) = 0. \tag{8}$$

So equation (8) is necessary for a to be an eigenvalue of A. It is also sufficient; for if (8) is satisfied, the matrix $A - aI$ is not invertible. By Theorem 1 of Chapter 3 this noninvertible matrix has a nonzero nullvector h; (6)$'$ shows that h is an eigenvector of A. When the determinant is expressed by formula (16) of Chapter 5, (8) appears as an algebraic equation of degree n for a, where A is an $n \times n$ matrix. The left-hand side of (8) is called the *characteristic polynomial* of the matrix A and is denoted as p_A.

Example I

$$A = \begin{pmatrix} 3 & 2 \\ 1 & 4 \end{pmatrix};$$

$$\det(A - aI) = \det \begin{pmatrix} 3-a & 2 \\ 1 & 4-a \end{pmatrix} = (3-a)(4-a) - 2$$

$$= a^2 - 7a + 10 = 0.$$

This equation has two roots,

$$a_1 = 2, \qquad a_2 = 5.$$

These are eigenvalues; there is an eigenvector corresponding to each:

$$(A - a_1 I)h_1 = \begin{pmatrix} 1 & 2 \\ 1 & 2 \end{pmatrix} h_1 = 0$$

is satisfied by

$$h_1 = \begin{pmatrix} 2 \\ -1 \end{pmatrix}$$

and of course by any scalar multiple of h_1. Similarly,

$$(A - a_2 I)h_2 = \begin{pmatrix} -2 & 2 \\ 1 & -1 \end{pmatrix} h_2 = 0$$

is satisfied by

$$h_2 = \begin{pmatrix} 1 \\ 1 \end{pmatrix}$$

and of course by any multiple of h_2.

The vectors h_1 and h_2 are not multiples of each other, so they are linearly independent. Thus any vector h in $\mathbb{R}^2$ can be expressed as a linear combination of h_1 and h_2:

$$h = b_1 h_1 + b_2 h_2. \tag{9}$$

We apply A^N to (9) and use relation $(6)_N$,

$$A^N h = b_1 a_1^N h_1 + b_2 a_2^N h_2. \tag{9}_N$$

Since $a_1 = 2, a_2 = 5$, both $a_1^N = 2^N$ and $a_1^N = 5^N$ tend to infinity; since h_1 and h_2 are linearly independent, it follows that also $A^N h$ tends to infinity, unless both b_1 and b_2 are zero, in which case, by (9), $h = 0$. Thus we have shown that for $A = \begin{pmatrix} 3 & 2 \\ 1 & 4 \end{pmatrix}$ and any $h \neq 0, A^N h \to \infty$ as $N \to \infty$; that is, each component tends to infinity. This bears out our numerical result in case (a). In fact, $A^N \sim 5^N$, also borne out by the calculations.

Example 2. Here is a more interesting case. The Fibonacci sequence $f_0, f_1, \ldots$ is defined by the recurrence relation

$$f_{n+1} = f_n + f_{n-1}, \tag{10}$$

with the starting data $f_0 = 0$, $f_1 = 1$. The first ten terms of the sequence are

$$0, 1, 1, 2, 3, 5, 8, 13, 21, 34;$$

they seem to be growing rapidly. We shall construct a formula for f_n that displays its rate of growth. We start by rewriting the recurrence relation (10) in matrix–vector form:

$$\begin{pmatrix} 0 & 1 \\ 1 & 1 \end{pmatrix} \begin{pmatrix} f_{n-1} \\ f_n \end{pmatrix} = \begin{pmatrix} f_n \\ f_{n+1} \end{pmatrix} \tag{10}'$$

We deduce recursively that

$$\begin{pmatrix} f_n \\ f_{n+1} \end{pmatrix} = A^n \begin{pmatrix} f_0 \\ f_1 \end{pmatrix}, \qquad A = \begin{pmatrix} 0 & 1 \\ 1 & 1 \end{pmatrix}, \tag{11}$$

We shall represent the nth power of A in terms of its eigenvalues and eigenvectors.

$$\det(A - aI) = \det \begin{pmatrix} -a & 1 \\ 1 & 1-a \end{pmatrix} = a^2 - a - 1.$$

The zeros of the characteristic polynomial of A are

$$a_1 = \frac{1 + \sqrt{5}}{2}, \qquad a_2 = \frac{1 - \sqrt{5}}{2}.$$

Note that a_1 is positive and greater than 1, whereas a_2 is negative and in absolute value much smaller than 1.

The eigenvectors satisfy the equations

$$\begin{pmatrix} -a_1 & 1 \\ 1 & 1-a_1 \end{pmatrix} h_1 = 0, \qquad \begin{pmatrix} -a_2 & 1 \\ 1 & 1-a_2 \end{pmatrix} h_2 = 0.$$

These equations are easily solved by looking at the first component:

$$h_1 = \begin{pmatrix} 1 \\ a_1 \end{pmatrix}, \qquad h_2 = \begin{pmatrix} 1 \\ a_2 \end{pmatrix};$$

of course any scalar multiples of them are eigenvectors as well.

Next we express the initial vector $(f_0, f_1)^T = (0, 1)^T$ as a linear combination of the eigenvectors:

$$\begin{pmatrix} 0 \\ 1 \end{pmatrix} = c_1 h_1 + c_2 h_2.$$

Comparing the first component shows that $c_2 = -c_1$. The second component yields $c_1 = 1/\sqrt{5}$. So

$$\begin{pmatrix} f_0 \\ f_1 \end{pmatrix} = \frac{1}{\sqrt{5}} h_1 - \frac{1}{\sqrt{5}} h_2.$$

Set this into (11); we get

$$\begin{pmatrix} f_n \\ f_{n+1} \end{pmatrix} = A^n \frac{1}{\sqrt{5}} (h_1 - h_2) = \frac{a_1^n}{\sqrt{5}} h_1 - \frac{a_2^n}{\sqrt{5}} h_2.$$

The first component of this vector equation is

$$f_n = a_1^n/\sqrt{5} - a_2^n/\sqrt{5}.$$

Since $a_2^n/\sqrt{5}$ is less than $1/2$, and since f_n is an integer, we can put this relation in the following form:

$$f_n = \text{nearest integer to } \frac{a_1^n}{\sqrt{5}}.$$

EXERCISE 1.   Calculate f_{32}.

We return now to the general case (6), (8). The *characteristic polynomial* of the matrix A,

$$\det(aI - A) = p_A(a),$$

is a polynomial of degree n; the coefficient of the highest power a^n is 1.

According to the fundamental theorem of algebra, a polynomial of degree n with complex coefficients has n complex roots; some of the roots may be multiple. The roots of the characteristic polynomial are the eigenvalues of A. To make sure that these polynomials have a full set of roots, the spectral theory of linear maps is formulated in linear spaces over the field of complex numbers.

Theorem 1.   Eigenvectors of a matrix A corresponding to distinct eigenvalues are linearly independent.

Proof. Suppose $a_i \neq a_k$ for $i \neq k$ and

$$Ah_i = a_i h_i, \qquad h_i \neq 0. \tag{12}$$

Suppose now that there were a nontrivial linear relation among the h_i. There may be several; since all $h_i \neq 0$, all involve at least two eigenvectors. Among them there is one which involves the *least number m* of eigenvectors:

$$\sum_1^m b_j h_j = 0, \qquad b_j \neq 0, \ j = 1, \ldots, m; \tag{13}$$

here we have renumbered the h_i. Apply A to (13) and use (12); we get

$$\sum b_j A h_j = \sum b_j a_j h_j = 0. \tag{13'}$$

Multiply (13) by a_m and subtract from (13)':

$$\sum_1^m (b_j a_j - b_j a_m) h_j = 0. \tag{13''}$$

Clearly the coefficient of h_m is zero and none of the others is zero, so we have a linear relation among the h_j involving only $m - 1$ of the vectors, contrary to m being the smallest number of vectors satisfying such a relation.   □

Using Theorem 1 we deduce Theorem 2.

Theorem 2.   If the characteristic polynomial of the $n \times n$ matrix A has n distinct roots, then A has n linearly independent eigenvectors.

In this case the n eigenvectors form a basis; therefore every vector h in $\mathbb{C}^n$ can be expressed as a linear combination of the eigenvectors:

$$h = \sum_1^n b_j h_j. \tag{14}$$

Applying A^N to (13) and using $(6)_N$ we get

$$A^N h = \sum b_j a_j^N h_j. \tag{14}'$$

This formula can be used to answer the stability question raised at the beginning of this chapter:

EXERCISE 2. (a) Prove that if A has n distinct eigenvalues a_j and all of them are less than one in absolute value, then all h in $\mathbb{C}^n$,

$$A^N h \to 0 \qquad \text{as } N \to \infty,$$

that is, all components of $A^N h$ tend to zero.

(b) Prove that if all a_j are greater than one in absolute value, then for all $h \neq 0$,

$$A^N h \to \infty \qquad \text{as } N \to \infty,$$

that is, some components of $A^N h$ tend to infinity.

There are two simple and useful relations between the eigenvalues of A and the matrix A itself.

Theorem 3. Denote by $a_1, \ldots, a_n$ the eigenvalues of A, with the same multiplicity they have as roots of the characteristic equation of A. Then

$$\sum a_i = \operatorname{tr} A, \qquad \prod a_i = \det A. \tag{15}$$

Proof. We claim that the characteristic polynomial of A has the form

$$p_A(s) = s^n - (\operatorname{tr} A)s^{n-1} + \cdots + (-1)^n \det A. \tag{15}'$$

According to elementary algebra, the polynomial p_A can be factored as

$$p_A(s) = \prod_1^n (s - a_i); \tag{16}$$

this shows that the coefficient of s^{n-1} in p_A is $-\sum a_i$, and the constant term is $(-1)^n \prod a_j$. Comparing this with (15)' gives (15).

To prove (15)', we use first formula (16) in Chapter 5 for the determinant as a sum of products:

$$p_A(s) = \det(sI - A) = \det \begin{pmatrix} s - a_{11} & -a_{12} & \cdots & -a_{1n} \\ -a_{21} & s - a_{22} & & \\ \vdots & & & \vdots \\ -a_{n1} & & \cdots & s - a_{nn} \end{pmatrix}$$

$$= \sum \sigma(p) \prod (s\delta_{p_i i} - a_{p_i i}).$$

Clearly the terms of degree n and $n - 1$ in s come from the single product of the diagonal elements.

$$\prod (s - a_{ii}) = s^n - (\text{tr } A) \, s^{n-1} + \cdots .$$

This identifies the terms of order n and $(n - 1)$ in (15). The term of order zero, $p_A(0)$, is det $(-A) = (-1)^n$ det A. This proves (15)' and completes the proof of Theorem 3.   □

EXERCISE 3.   (a) Verify for the matrices discussed in Examples 1 and 2,

$$\begin{pmatrix} 3 & 2 \\ 1 & 4 \end{pmatrix} \quad \text{and} \quad \begin{pmatrix} 0 & 1 \\ 1 & 1 \end{pmatrix},$$

that the sum of the eigenvalues equals the trace, and their product is the determinant of the matrix.

Relation $(6)_N$, $A^n h = a^n h$, shows that if a is an eigenvalue of A, a^N is an eigenvalue of A^N. Now let q be any polynomial:

$$q(s) = \sum q_N s^N .$$

Multiplying $(6)_N$ by q_N and summing we get

$$q(A)h = q(a)h. \tag{17}$$

The following result is called the *spectral mapping theorem*.

Theorem 4.   (a) Let q be any polynomial, A a square matrix, a an eigenvalue of A. Then $q(a)$ is an eigenvalue of $q(A)$.
(b) Every eigenvalue of $q(A)$ is of the form $q(a)$, where a is an eigenvalue of A.

Proof. Part (a) is merely a verbalization of relation (17), which shows also that A and $q(A)$ have h as common eigenvector.
To prove (b), let b denote an eigenvalue of $q(A)$; that means that $q(A) - bI$ is not invertible. Now factor the polynomial $q(s) - b$:

$$q(s) - b = c \prod (s - r_i).$$

We may set A in place of s:

$$q(A) - bI = c \prod (A - r_i I).$$

By taking b to be an eigenvalue of $q(A)$, the left-hand side is not invertible. Therefore neither is the right-hand side. Since the right-hand side is a product, it follows that at least one of the factors $A - r_i I$ is not invertible. That means that some r_i is an eigenvalue of A. Since r_i is a root of $q(s) - b$,

$$q(r_i) = b.$$

This completes the proof of part (b).   □

If in particular we take q to be the characteristic polynomial p_A of A, we conclude that all eigenvalues of $p_A(A)$ are zero. In fact a little more is true.

Theorem 5 (Cayley–Hamilton). Every matrix A satisfies its own characteristic equation:

$$p_A(A) = 0. \tag{18}$$

Proof. If A has distinct eigenvalues, then according to Theorem 2 it has n linearly independent eigenvectors $h_j, j = 1, \ldots, n$. Using (4) we apply $p_A(A)$:

$$p_A(A)h = \sum p_A(a_j)b_j h_j = \sum 0 = 0$$

for all h, proving (18) in this case. For a proof that holds for all matrices we use the following lemma.

Lemma 6. Let P and Q be two polynomials with *matrix* coefficients

$$P(s) = \sum P_j s^j, \qquad Q(s) = \sum Q_k s^k.$$

The product $PQ = R$ is then

$$R(s) = \sum R_l s^l, \qquad R_l = \sum_{j+k=l} P_j Q_k.$$

Suppose that the matrix A commutes with the coefficients of Q; then

$$P(A)Q(A) = R(A). \tag{19}$$

The proof is self-evident.

We apply Lemma 6 to $Q(s) = sI - A$ and $P(s)$ defined as the matrix of cofactors of $Q(s)$; that is,

$$P_{ij}(s) = (-1)^{i+j} D_{ji}(s), \tag{20}$$

D_{ij} the determinant of the ijth minor of $Q(s)$. According to the formula (30) of Chapter 5,

$$P(s)Q(s) = \det Q(s)\, I = p_A(s)\, I, \tag{21}$$

where $p_A(s)$ is the characteristic polynomial of A. A commutes with the coefficients of Q; therefore by Lemma 6 we may set $s = A$ in (21). Since $Q(A) = 0$, it follows that

$$p_A(A) = 0.$$

This proves Theorem 5.   □

We are now ready to investigate matrices whose characteristic equation has multiple roots. First a few examples.

Example 3.   A $= I$,

$$p_A(s) = \det(sI - I) = (s - 1)^n;$$

1 is an n-fold zero. In this case every nonzero vector h is an eigenvector of A.

Example 4.   A $= \left(\begin{smallmatrix} 3 & 2 \\ -2 & -1 \end{smallmatrix}\right)$, tr A $= 2$, det A $= 1$; therefore by Theorem 3,

$$p_A(s) = s^2 - 2s + 1,$$

whose roots are one, with multiplicity two. The equation

$$Ah = \begin{pmatrix} 3h_1 + 2h_2 \\ -2h_1 - h_2 \end{pmatrix} = \begin{pmatrix} h_1 \\ h_2 \end{pmatrix}$$

has as solution all vectors h whose components satisfy

$$h_1 + h_2 = 0.$$

All these are multiples of A $= \left(\begin{smallmatrix} -1 \\ 1 \end{smallmatrix}\right)$. So in this case A does not have two independent eigenvectors.

We claim that if A has only one eigenvalue a and n linearly independent eigenvectors, then A $= aI$. For in this case every vector in $\mathbb{R}^n$ can be written as in (14), a linear combination of eigenvectors. Applying A to (14) and using $a_i = a$ for $i = 1, \ldots, n$ gives that

$$Ah = ah$$

for all h; then A $= aI$. We further note that every 2×2 matrix A with tr A $= 2$, det A $= 1$ has 1 as a double root of its characteristic equation. These matrices form a two-parameter family; only one member of this family, A $= I$, has two linearly independent eigenvectors. This shows that, in general, when the characteristic equation of A has multiple roots, we cannot expect A to have n linearly independent eigenvectors.

To make up for this defect one turns to *generalized eigenvectors*. In the first instance a generalized eigenvector f is defined as satisfying

$$(A - aI)^2 f = 0. \tag{22}$$

We show first that these behave almost as simply under applications of A^N as the genuine eigenvectors. We set

$$(A - aI)f = h. \tag{23}$$

Applying $(A - aI)$ to this and using (22), we get

$$(A - aI)h = 0, \tag{23}'$$

that is, h is a genuine eigenvector. We rewrite (23) and (23)$'$ as

$$Af = af + h, \qquad Ah = ah. \tag{24}$$

Applying A to the first equation of (24) and using the second equation gives

$$A^2 f = aAf + Ah = a^2 f + 2ah.$$

Repeating this N times gives

$$A^N f = a^N f + Na^{N-1}h. \tag{25}$$

EXERCISE 4.   Verify (25) by induction on N.

EXERCISE 5.   Prove that for any polynomial q,

$$q(A)f = q(a)f + q'(a)h, \tag{26}$$

where q' is the derivative of q and f satisfies (22).

Formula (25) shows that if $|a| < 1$, and f is a generalized eigenvector of A, $A^N f \to 0$.

We now generalize the notion of a generalized eigenvector.

Definition. f is a generalized eigenvector of A, with eigenvalue a, if $f \neq 0$ and

$$(A - aI)^m f = 0 \tag{27}$$

for some positive integer m.

We state now one of the principal results of linear algebra.

Theorem 7 (Spectral Theorem). Let A be an $n \times n$ matrix with complex entries. Every vector in $\mathbb{C}^n$ can be written as a sum of eigenvectors of A, genuine or generalized.

For the proof, we need the following results of algebra.

Lemma 8. Let p and q be a pair of polynomials with complex coefficients and assume that p and q have no common zero. Then there are two other polynomials a and b such that

$$ap + bq \equiv 1. \tag{28}$$

Proof. Denote by $\mathscr{I}$ all polynomials of the form $ap + bq$. Among them there is one, nonzero, of lowest degree; call it d. We claim that d divides both p and q; for suppose not; then the division algorithm yields a remainder r, say

$$r = p - md.$$

Since p and d belong to $\mathscr{I}$, so does $p - md = r$; since r has lower degree than d, this is a contradiction.

We claim that d has degree zero; for if it had degree greater than zero, it would, by the fundamental theorem of algebra, have a root. Since d divides p and q, this would be a common root of p and q. Since we have assumed the contrary, deg $d = 0$ follows; since $d \not\equiv 0, d \equiv$ const., say $\equiv 1$. This proves (28). $\qquad \square$

Lemma 9. Let p and q be as in Lemma 8, and let A be a square matrix with complex entries. Denote by N_p, N_q, and N_{pq} the null spaces of $p(A)$, $q(A)$, and $p(A)q(A)$, respectively. Then N_{pq} is the direct sum of N_p and N_q:

$$N_{pq} = N_p \oplus N_q, \tag{29}$$

by which we mean that every x in N_{pq} can be decomposed uniquely as

$$x = x_p + x_q, \qquad x_p \text{ in } N_p, \qquad x_q \text{ in } N_q. \tag{29$'$}$$

Proof. We replace the argument of the polynomials in (28) by A; we get

$$a(A)p(A) + b(A)q(A) = I. \tag{30}$$

Letting both sides act on x we obtain

$$a(A)p(A)x + b(A)q(A)x = x. \tag{31}$$

We claim that if x belongs to N_{pq}, then the first term on the left in (31) is in N_q, and the second in N_p. To see this we use the commutativity of polynomials of the same matrix:

$$q(A)a(A)p(A)x = a(A)p(A)q(A)x = 0,$$

since x belongs to the nullspace of $p(A)q(A)$. This proves that the first term on the left in (31) belongs to the nullspace of $q(A)$; analogously the second term belongs to the nullspace of $p(A)$. This shows that (31) gives the desired decomposition (29)'.

To show that the decomposition is unique, we argue as follows: If

$$x = x_p + x_q = x'_p + x'_q,$$

then

$$y = x_p - x'_p = x'_q - x_q$$

is an element that belongs to both N_p and N_q. Let (30) act on y:

$$a(A)p(A)y + b(A)q(A)y = y.$$

Both terms on the left-hand side are zero; therefore so is the right-hand side, y. This proves that $x_p = x'_p, x_q = x'_q$. $\square$

Corollary 10. Let $p_1, \ldots, p_k$ be a collection of polynomials that are pairwise without a common zero. Denote the nullspace of the product $p_1(A) \ldots p_k(A)$ by $N_{p_1 \cdots p_k}$. Then

$$N_{p_1 \cdots p_k} = N_{p_1} \oplus \cdots \oplus N_{p_k}. \tag{32}$$

EXERCISE 6. Prove (32) by induction on k.

Proof of Theorem 7. Let x be any vector; the $n + 1$ vectors $x, Ax, A^2x, \ldots A^nx$ must be linearly dependent; therefore there is a polynomial p of degree less than or equal to n such that

$$p(A)x = 0 \tag{33}$$

We factor p and rewrite this as

$$\prod (A - r_j I)^{m_j} x = 0, \tag{33'}$$

r_j the roots of p, m_j their multiplicity. When r_j is not an eigenvalue of A, $A - r_j I$ is invertible; since the factors in (33)' commute, all invertible factors can be removed. The remaining r_j in (33)' are all eigenvalues of A. Denote

$$p_j(s) = (s - r_j)^{m_j}; \tag{34}$$

then (33)' can be written as $\Pi\ p_j(A)x = 0$, that is, x belongs to $N_{p_1 \cdots p_R}$. Clearly the p_j pairwise have no common zero, so Corollary 10 applies: x can be decomposed as a sum of vectors in N_{p_j}. But by (34) and Definition (27), every x_j in N_{p_j} is a generalized eigenvector. Thus we have a decomposition of x as a sum of generalized eigenvectors, as asserted in Theorem 7.   □

We have shown earlier in Theorem 5, the Cayley–Hamiltonian Theorem, that the characteristic polynomial p_A of A satisfies $p_A(A) = 0$. We denote by $\mathcal{I} = \mathcal{I}_A$ the set of all polynomials p which satisfy $p(A) = 0$. Clearly, the sum of two polynomials in $\mathcal{I}$ belongs to $\mathcal{I}$; furthermore, if p belongs to $\mathcal{I}$, so does every multiple of p. Denote by $m = m_A$ a nonzero polynomial of smallest degree in $\mathcal{I}$; we claim that all p in $\mathcal{I}$ are multiples of m. Because, if not, then the division process

$$p = qm + r$$

gives a remainder r of lower degree than m. Clearly, $r = p - qm$ belongs to $\mathcal{I}$, contrary to the assumption that m is one of lowest degree. Except for a constant factor, which we fix so that the leading coefficient of m_A is 1, $m = m_A$ is unique. This polynomial is called the *minimal polynomial* of A.

To describe precisely the minimal polynomial we return to the definition (27) of a generalized eigenvector. We denote by $N_m = N_m(a)$ the *nullspace* of $(A - aI)^m$. The subspaces N_m consist of generalized eigenvectors; they are indexed increasingly, that is,

$$N_1 \subset N_2 \subset \cdots. \tag{35}$$

Since these are subspaces of a finite-dimensional space, they must be equal from a certain index on. We denote by $d = d(a)$ the smallest such index, that is,

$$N_d = n_{d+1} = \cdots \tag{35'}$$

but

$$N_{d-1} \neq N_d; \tag{35''}$$

$d(a)$ is called the *index* of the eigenvalue a.

EXERCISE 7. Show that A maps N_d into itself.

Theorem 11. Let A be an $n \times n$ matrix: denote its distinct eigenvalues by $a_1, \ldots, a_k$, and denote the index of a_j by d_j. We claim that the minimal polynomial m_A is

$$m_A(s) = \prod_1^h (s - a_i)^{d_i}.$$

EXERCISE 8. Prove Theorem 11.

Let us denote $N_{d_j}(a_j)$ by $N^{(j)}$; then Theorem 7, the spectral theorem, can be formulated as follows:

$$\mathbb{C}^n = N^{(1)} \oplus N^{(2)} \oplus \cdots \oplus N^{(k)}. \tag{36}$$

The dimension of $N^{(j)}$ equals the *multiplicity* of a_j as the root of the characteristic equation of A. Since our proof of this proposition uses calculus, we postpone it until Theorem 11 of Chapter 9.

A maps each subspace $N^{(j)}$ into itself; such subspaces are called *invariant* under A. We turn now to studying the action of A on each subspace; this action is completely described by the dimensions of $N_1, N_2, \ldots, N_d$ in the following sense.

Theorem 12. (i) Suppose the pair of matrices A and B are similar in the sense explained in Chapter 5 [see equation (37)],

$$A = SBS^{-1}, \tag{37}$$

S some invertible matrix. Then A and B have the same eigenvalues:

$$a_1 = b_1, \ldots, a_k = b_k; \tag{38}$$

furthermore, the nullspaces

$$N_m(a_j) = \textit{nullspace of } (A - a_j I)^m$$

and

$$M_m(a_j) = \textit{nullspace of } (B - a_j I)^m$$

have for all j and m the same dimensions:

$$\dim N_m(a_j) = \dim M_m(a_j). \tag{39}$$

(ii) Conversely, if A and B have the same eigenvalues, and if condition (39) about the nullspaces having the same dimension is satisfied, then A and B are similar.

Proof. Part (i) is obvious; for if A and B are similar, so are $A - aI$ and $B - aI$, and so is any power of them:

$$(A - aI)^m = S(B - aI)^m S^{-1}. \tag{40}$$

Since S is a 1-to-1 mapping, the nullspaces of two similar matrices have the same dimension. Relations (39) and in particular (38), follow from the observation.

The converse proposition will be proved in Appendix 15.

Theorems 4, 7, and 12 are the basic facts of the spectral theory of matrices. We wish to point out that the concepts that enter these theorems—eigenvalue, eigenvector, generalized eigenvector, index—remain meaningful for any mapping A of any finite dimensional linear space X over $\mathbb{C}$ into itself. The three theorems remain true in this abstract context and so do the proofs.

The usefulness of spectral theory in an abstract setting is shown in the following important generalization of Theorem 7.

Theorem 14. Denote by X a finite-dimensional linear space over the complex numbers, by A and B linear maps of X into itself, which commute:

$$AB = BA. \tag{41}$$

Then there is a basis in X which consists of eigenvectors and generalized eigenvectors of both A and B.

Proof. According to the Spectral Theorem, Theorem 7, equation (36), X can be decomposed as a direct sum of generalized eigenspaces of A:

$$X = N^{(1)} \oplus \cdots \oplus N^{(k)},$$

$N^{(j)}$ the nullspace of $(A - a_j I)^{d_j}$. We claim that B maps $N^{(j)}$ into $N^{(j)}$; for B is assumed to commute with A, and therefore commutes with $(A - aI)^d$:

$$B(A - aI)^d x = (A - aI)^d Bx. \tag{42}$$

If a is an eigenvalue and x belongs to $N^{(j)}$, the left-hand side of (42) is 0; therefore so is the right-hand side, which proves that Bx is in $N^{(j)}$. Now we apply the Spectral Theorem to the linear mapping B acting on $N^{(j)}$ and obtain a spectral decomposition of each $N^{(j)}$ with respect to B. This proves Theorem 14. $\qquad\square$

Corollary 15. Theorem 14 remains true if A, B are replaced by any number of pairwise commuting linear maps.

EXERCISE 9.   Prove Corollary 15.

In Chapter 3 we defined the *transpose* A' of a linear map. When A is a matrix, that is, a map $\mathbb{C}^n \to \mathbb{C}^n$, its transpose A^T is obtained by interchanging the rows and columns of A.

Theorem 16.   Every square matrix A is similar to its transpose A^T.

Proof. We have shown in Chapter 3, Theorem 6, that a mapping A of a space X into itself, and its transpose A' mapping X' into itself, have nullspaces of the same dimension. Since the transpose of $A - a\mathrm{I}$ is $A' - a\mathrm{I}'$ it follows that A and A' have the same eigenvalues, and that their eigenspaces have the same dimension.

The transpose of $(A - a\mathrm{I})^j$ is $(A' - a\mathrm{I}')^j$; therefore their nullspaces have the same dimension. We can now appeal to Theorem 12 and conclude that A and A', interpreted as matrices, are similar. $\square$

Theorem 17.   Let X be a finite-dimensional linear space over $\mathbb{C}$, A a linear mapping of X into X. Denote by X' the dual of X, $A': X' \to X'$ the transpose of A. Let a and b denote two distinct eigenvalues of A: $a \neq b$, x an eigenvector of A with eigenvalue a, l an eigenvector of A' with eigenvalue b. Then l and x annihilate each other:

$$(l, x) = 0. \tag{43}$$

Proof. The transpose of A is defined in equation (9) of Chapter 3 by requiring that for every x in X and every l in X'

$$(A'l, x) = (l, Ax).$$

If in particular we take x to be an eigenvector of A and l to be an eigenvector of A',

$$Ax = ax, \qquad A'l = bl,$$

and we deduce that

$$b(l, x) = a(l, x).$$

Since we have taken $a \neq b$, (l, x) must be zero. $\square$

Theorem 17 is useful in calculating and studying the properties of expansions of vectors x in terms of eigenvectors.

Theorem 18. Suppose the mapping A has n distinct eigenvalues $a_1, \ldots, a_n$. Denote the corresponding eigenvectors of A by $x_1, \ldots, x_n$, those of A' by $l_1, \ldots, l_n$. Then

(a) $(l_i, x_i) \neq 0, i = 1, \ldots, n$.

(b) Let

$$x = \sum k_j x_j. \tag{44}$$

be the expansion of x as a sum of eigenvectors; then

$$k_i = (l_i, x)/(l_i, x_i), \qquad i = 1, \ldots, n. \tag{45}$$

EXERCISE 10.   Prove Theorem 18.

EXERCISE 11.   Take the matrix

$$\begin{pmatrix} 0 & 1 \\ 1 & 1 \end{pmatrix}$$

from equation (10)' of Example 2.

(a) Determine the eigenvector of its transpose.

(b) Use formulas (44) and (45) to determine the expansion of the vector $(0, 1)'$ in terms of the eigenvectors of the original matrix. Show that your answer agrees with the expansion obtained in Example 2.

EXERCISE 12.   In Example 1 we have determined the eigenvalues and corresponding eigenvector of the matrix

$$\begin{pmatrix} 3 & 2 \\ 1 & 4 \end{pmatrix}$$

as $a_1 = 2, h_1 = \begin{pmatrix} 2 \\ -1 \end{pmatrix}$, and $a_2 = 5, h_2 = \begin{pmatrix} 1 \\ 1 \end{pmatrix}$.

Determine eigenvectors l_1 and l_2 of its transpose and show that

$$(l_i, h_j) = \begin{cases} 0 & \text{for } i \neq j \\ \neq 0 & \text{for } i = j \end{cases}$$

EXERCISE 13.   Show that the matrix

$$A = \begin{pmatrix} 0 & 1 & 1 \\ 1 & 0 & 1 \\ 1 & 1 & 0 \end{pmatrix}$$

has 1 as an eigenvalue. What are the other two eigenvalues?

CHAPTER 7

Euclidean Structure

In this chapter we abstract the concept of Euclidean distance. We gain no greater generality; we gain simplicity, transparency and flexibility.

We review the basic structure of Euclidean spaces. We choose a point 0 as origin in real n-dimensional Euclidean space; the *length* of any vector x in space, denoted as $\| x \|$, is defined as its *distance* to the origin.

Let us introduce a Cartesian coordinate system and denote the Cartesian coordinates of x as $x_1, \ldots, x_n$. By repeated use of the Pythagorean theorem we can express the length of x in terms of its Cartesian coordinates.

$$\| x \| = \sqrt{x_1^2 + \cdots + x_n^2}. \tag{1}$$

The *scalar product* of two vectors x and y, denoted as (x, y), is defined by

$$(x, y) = \sum x_j y_j. \tag{2}$$

Clearly, the two concepts are related; we can express the length of a vector as

$$\| x \|^2 = (x, x). \tag{2}'$$

The scalar product is commutative:

$$(x, y) = (y, x) \tag{3}$$

and bilinear:

$$\begin{aligned} (x + u, y) &= (x, y) + (u, y), \\ (x, y + v) &= (x, y) + (x, v). \end{aligned} \tag{3}'$$

Linear Algebra and Its Applications, Second Edition, by Peter D. Lax
Copyright © 2007 John Wiley & Sons, Inc.

Using these algebraic properties of scalar product we can derive the identity

$$(x - y, x - y) = (x, x) - 2(x, y) + (y, y).$$

Using (2)', we can rewrite this identity as

$$\| x - y \|^2 = \| x \|^2 - 2(x, y) + \| y \|^2. \tag{4}$$

The term on the left is the distance of x from y, squared; the first and third terms on the right are the distances of x and y from 0, squared. These three quantities have geometric meaning; therefore they have the same value in any Cartesian coordinate system. If follows therefore from (4) that also the scalar product (2) has the same value in all Cartesian coordinate systems. By choosing special coordinate axes, the first one through x, the second so that y is contained in the plane spanned by the first two axes, we can uncover the geometric meaning of (x, y).

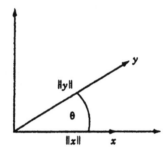

The coordinates of the vector x and y in this coordinate system are $x = (\| x \|, 0 \ldots 0)$ and $y = (\| y \| \cos \theta \ldots)$. Therefore

$$(x, y) = \| x \| \| y \| \cos \theta, \tag{5}$$

θ the angle between x and y.

The three points 0, x, y form a triangle whose sides are $a = \| x \|$, $b = \| y \|$, $c = \| x - y \|$, forming an angle θ at 0:

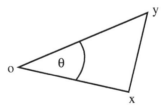

Relations (4) and (5) can be written as

$$c^2 = a^2 + b^2 - 2ab \cos \theta. \tag{4'}$$

This is the classical *law of cosine*; a special case of it, $\theta = \pi/2$, is the Pythagorean theorem.

Most texts derive formula (5) for the scalar product from the law of cosine. This is a pedagogical blunder, for most students have long forgotten the law of cosine, if they ever knew it.

We shall give now an abstract, that is axiomatic, definition of Euclidean space.

Definition. A Euclidean structure in a linear space X over the reals is furnished by a real-valued function of two vector arguments called a *scalar product* and denoted as (x, y), which has the following properties:

(i) (x, y) is a bilinear function; that is, it is a linear function of each argument when the other is kept fixed.
(ii) It is symmetric:

$$(x, y) = (y, x). \tag{6}$$

(iii) It is positive:

$$(x, x) > 0 \qquad \text{except for } x = 0. \tag{7}$$

Note that the scalar product (2) satisfies these axioms. We shall show now that, conversely, all of Euclidean geometry is contained in these simple axioms.

We define the Euclidean length (also called *norm*) of x by

$$\| x \| = (x, x)^{1/2}. \tag{8}$$

A scalar product is also called an *inner product*, or a *dot product*.

Definition. The distance of two vectors x and y in a linear space with Euclidean norm is defined as $\| x - y \|$.

Theorem 1 (Schwarz Inequality). For all x, y,

$$|(x, y)| \leq \| x \| \| y \|. \tag{9}$$

Proof. Consider the function $q(t)$ of the real variable t defined by

$$q(t) = \| x + ty \|^2. \tag{10}$$

Using the definition (8) and properties (i) and (ii) we can write

$$q(t) = \| x \|^2 + 2t(x, y) + t^2 \| y \|^2. \tag{10'}$$

Assume that $y \neq 0$ and set $t = -(x, y)/\| y \|^2$ in $(10)'$. Since (10) shows that $q(t) \geq 0$ for all t, we get that

$$\| x \|^2 - \frac{(x, y)^2}{\| y \|^2} \geq 0$$

This proves (9). For $y = 0$, (9) is trivially true. $\square$

Note that for the concrete scalar product (2), inequality (9) follows from the representation (5) of (x, y) as $\| x \| \| y \| \cos \theta$.

Theorem 2

$$\| x \| = \max(x, y), \| y \| = 1. \tag{11}$$

EXERCISE I. Prove Theorem 2.

Theorem 3 (Triangle Inequality). For all x, y

$$\| x + y \| \leq \| x \| + \| y \|. \tag{12}$$

Proof. Using the algebraic properties of scalar product, we derive, analogously to (4), the identity

$$\| x + y \|^2 = \| x \|^2 + 2(x, y) + \| y \|^2 \tag{12'}$$

and estimate the middle term by the Schwarz inequality. $\square$

Motivated by (5) we make the following definitions.

Definition. Two vectors x and y are called *orthogonal (perpendicular)*, denoted as $x \perp y$, if

$$(x, y) = 0. \tag{13}$$

From $(12)'$ we deduce the Pythagorian theorem

$$\| x + y \|^2 = \| x \|^2 + \| y \|^2 \qquad \text{if } x \perp y. \tag{13'}$$

Definition. Let X be a finite-dimensional linear space with a Eulerian structure, $x^{(1)}, \ldots, x^{(n)}$ a basis for X. This basis is called *orthonormal* with respect to a given Euclidean structure if

$$(x^{(j)}, x^{(k)}) = \begin{cases} 0, & \text{for } j \neq k, \\ 1, & \text{for } j = k. \end{cases} \tag{14}$$

Theorem 4 (Gram–Schmidt).   Given an arbitrary basis $y^{(1)}, \ldots, y^{(n)}$ in a finite-dimensional linear space equipped with a Euclidean structure, there is a related basis $x^{(1)}, \ldots, x^{(n)}$ with the following properties:

(i) $x^{(1)}, \ldots, x^{(n)}$ is an orthonormal basis.

(ii) $x^{(k)}$ is a linear combination of $y^{(1)}, \ldots, y^{(k)}$, for all k.

Proof. We proceed recursively; suppose $x^{(1)}, \ldots, x^{(k-1)}$ have already been constructed. We set

$$x^{(k)} = c \left(y^{(k)} - \sum_{1}^{k-1} c_j x^{(j)} \right).$$

Since $x^{(1)}, \ldots, x^{(k-1)}$ are already orthonormal, it is easy to see that $x^{(k)}$ defined above is orthogonal to them if we choose

$$c_l = (y^{(k)}, x^{(l)}), \qquad l = 1, \ldots, k - 1.$$

Finally we choose c so that $\| x^k \| = 1$. $\square$

Theorem 4 guarantees the existence of plenty of orthonormal bases. Given such a basis, any x can be written as

$$x = \sum_{1}^{n} a_j x^{(j)}. \tag{15}$$

Take the scalar product of (15) with $x^{(l)}$; using the orthonormality relations (14) we get

$$(x, x^{(l)}) = a_l. \tag{16}$$

Let y be any other vector in X; it can be expressed as

$$y = \sum b_k x^{(k)}.$$

Take the scalar product of y with x, using the expression (15). Then, using (14), we get

$$(x, y) = \sum \sum a_j b_k (x^{(j)}, x^{(k)}) = \sum a_j b_j. \tag{17}$$

In particular, for $y = x$ we get

$$\| x \|^2 = \sum a_j^2. \tag{17}'$$

Equation (17) shows that the mapping defined by (16),

$$x \rightarrow (a_1, \ldots, a_n),$$

carries the space X with a Euclidean structure into $\mathbb{R}^n$, and carries the scalar product of X into the standard scalar product (2) of $\mathbb{R}^n$.

Since the scalar product is bilinear, for y fixed (x, y) is a linear function of x. Conversely, we have the following theorem.

Theorem 5. Every linear function $l(x)$ on a finite-dimensional linear space X with Euclidean structure can be written in the form

$$l(x) = (x, y), \tag{18}$$

y some element of X.

Proof. Introduce an orthonormal basis $x^{(1)}, \ldots, x^{(n)}$ in X; denote the value of l on $x^{(k)}$ by

$$l(x^{(k)}) = b_k.$$

Set

$$y = \sum b_k x^{(k)}. \tag{19}$$

It follows from orthonormality that $(x^{(k)}, y) = b_k$. This shows that (18) holds for $x = x^{(k)}, k = 1, 2, \ldots, n$; but if two linear functions have the same value for all vectors that form a basis, they have the same value for all vectors x. $\qquad\square$

Corollary 5′. The mapping $l \rightarrow y$ is an isomorphism of the Euclidean space X with its dual.

Definition. Let X be a finite-dimensional linear space with Euclidean structure, Y a subspace of X. The orthogonal complement of Y, denoted as $Y^\perp$, consists of all vectors z in X that are orthogonal to every y in Y:

$$z \text{ in } Y^\perp \text{ if } (y, z) = 0 \qquad \text{for all } y \text{ in } Y.$$

Recall that in Chapter 2 we denoted by $Y^\perp$ the set of linear functionals that vanish on Y. The notation $Y^\perp$ introduced above is consistent with the previous notation when the dual of X is identified with X via (18). In particular, $Y^\perp$ is a subspace of X.

Theorem 6. For any subspace Y of X,

$$X = Y \oplus Y^\perp. \tag{20}$$

The meaning of (20) is that every x in X can be decomposed uniquely as

$$x = y + y^\perp, \qquad y \text{ in } Y, \ y^\perp \text{ orthogonal to } Y. \tag{20}'$$

Proof. We show first that a decomposition of form (20)′ is unique. Suppose we could write

$$x = z + z^\perp, \qquad z \text{ in } Y, \ z^\perp \text{ in } Y^\perp.$$

Comparing this with (20)′ gives

$$y - z = z^\perp - y^\perp.$$

It follows from this that $y - z$ belongs both to Y and to $Y^\perp$, and thus is orthogonal to itself:

$$0 = (y - z, \ z^\perp - y^\perp) = (y - z, \ y - z) = \| \, y - z \, \|^2,$$

but by positivity of norm, $y - z = 0$.

To prove that a decomposition of form (20)′ is always possible, we construct an orthonormal basis of X whose first k members lie in Y; the rest must lie in $Y^\perp$. We can construct such a basis by starting with an orthonormal basis in Y, then complete it to a basis in X, and then orthonormalize the rest of the basis by the procedure described in Theorem 3. Then x can be decomposed as in (15). We break this decomposition into two parts:

$$x = \sum_1^n a_j x^{(j)} = \sum_1^k + \sum_{k+1}^n = y + y^\perp; \tag{21}$$

clearly, y lies in Y and $y^\perp$ in $Y^\perp$.   □

In the decomposition (20)′, the component y is called the *orthogonal projection* of x into Y, denoted by

$$y = P_Y x. \tag{22}$$

Theorem 7. (i) The mapping P_Y is linear.
(ii) $P_Y^2 = P_Y$.

Proof. Let w be any vector in X, unrelated to x, and let its decomposition (20)′ be

$$w = z + z^\perp, \qquad z \text{ in } Y, z^\perp \text{ in } Y^\perp.$$

Adding this to (20)′ gives

$$x + w = (y + z) + (y^\perp + z^\perp),$$

the decomposition of $x + w$. This shows that $P_Y(x + w) = P_Y x + P_Y w$, Similarly, $P_Y(kx) = kP_Y x$.

To show that $P_Y^2 = P_Y$, we take any x and decompose it as in $(20)'$; $x = y + y^\perp$. The vector $y = P_x$ needs no further decomposition: $P_Y y = y$. □

Theorem 8. Let Y be a linear subspace of the Euclidean space X, x some vector in X. Then among all elements z of Y, the one closest in Euclidean distance to x is $P_Y x$.

Proof. Using the decomposition $(20)'$ of x we have

$$x - z = y - z + y^\perp, \qquad y = P_y x.$$

Since y and z both belong to Y, so does $y - z$.

Therefore by the Pythagorean theorem $(13)'$,

$$\| x - z \|^2 = \| y - z \|^2 + \| y^\perp \|^2;$$

clearly this is smallest when $z = y$. Since the distance between two vectors x, z is $\| x - z \|$, this proves Theorem 8. □

We turn now to linear mappings of a Euclidean space X into another Euclidean space U. Since a Euclidean space can be identified in a natural way with its own dual, the transpose of a linear map A of such a space X into U maps U into X. To indicate this distinction, and for yet another reason explained at the end of this chapter, the transpose of a map A of Euclidean X into U is called the *adjoint* of A and is denoted by A^*.

Here is the full definition of the adjoint A^* of a linear mapping A of a Euclidean space X into another Euclidean space U:

Given any u in U,

$$l(x) = (Ax, u)$$

is a linear function of x, According to Theorem 5, this linear function $l(x)$ can be represented as (x, y), y in X. Therefore for all x in X

$$(x, y) = (Ax, u). \tag{23}$$

The vector y depends on u; Since scalar products are bilinear, y depends linearly on u; we denote this dependence as $y = A^* u$, and rewrite (23) as

$$(x, A^* u) = (Ax, u). \tag{23'}$$

Note that A^* maps U into X; the parentheses on the left denote the scalar product in X, while those on the right denote the scalar product in U.

The next theorem lists the basic properties of adjointness:

Theorem 9. **(i)** If A and B are linear mappings of X into U, then

$$(A + B)^* = A^* + B^*.$$

(ii) If A is a linear map of X into U, while C is a linear map of U into V, then

$$(CA)^* = A^*C^*.$$

(iii) If A is a 1-to-1 mapping of X onto U, then

$$(A^{-1})^* = (A^*)^{-1}.$$

(iv) $(A^*)^* = A$.

Proof. **(i)** is an immediate consequence of (23)$'$; **(ii)** can be demonstrated in two steps:

$$(CAx, v) = (Ax, C^*v) = (x, A^*C^*v).$$

(iii) follows from **(ii)** applied to $A^{-1}A = I$, I the identity mapping, and the observation that $I^* = I$. **(iv)** follows if we use the symmetry of the scalar product to rewrite (23)$'$ as

$$(u, Ax) = (A^*u, x). \qquad \square$$

When we take X to be $\mathbb{R}^n$ and U to be $\mathbb{R}^m$ with their standard Euclidean structures, and interpret A and A* as *matrices*, they are *transposes* of each other.

We present now an important application of the notion of the adjoint.

There are many situations where quantities $x_1, \ldots, x_n$ cannot be measured directly, but certain linear combinations of them,

$$a_1 x_1 + \cdots + a_n x_n,$$

can. Suppose that n such linear combinations have been measured. We can put all this information in the form of a matrix equation

$$Ax = p, \tag{24}$$

where $p_1, \ldots, p_m$ are the measured values, and A is an $m \times n$ matrix. We shall examine the case where the number m of measurements *exceeds* the number n of quantities whose value is of interest to us. Such a system of equations is *overdetermined* and in general does not have a solution. This is not as alarming as it sounds, because no measurement is perfect, and therefore none of the equations is expected to hold exactly. In such a situation, we seek that vector x that comes closest to satisfying all the equations in the sense that makes $\| Ax - p \|^2$ as small as possible.

For such an x to be determined uniquely, A cannot have nonzero nullvectors. For if $Ay = 0$, and x a minimizer of $\| Ax - p \|$, then so is $x + ky$, k any number.

Theorem 10.   Let A be an $m \times n$ matrix, $m > n$, and suppose that A has only the trivial nullvector 0. The vector x that minimizes $\| Ax - p \|^2$ is the solution z of

$$A^*Az = A^*p. \tag{25}$$

Proof. We show first that equation (25) has a unique solution. Since the range of A^* is $\mathbb{R}^n$, (25) is a system of n equations for n unknowns. According to Corollary B in Chapter 3, a unique solution is guaranteed if the homogeneous equation

$$A^*Ay = 0 \tag{25'}$$

has only the trivial solution $y = 0$. To see that this is the case, take the scalar product of (25)' with y. We get, using the definition (23)' of adjointness, $0 = (A^*Ay, y) = (Ay, Ay) = \| Ay \|^2$. Since $\| \; \|$ is positive, it follows that $Ay = 0$. Since we have assumed that A has only the trivial nullspace, $y = 0$ follows.

A maps $\mathbb{R}^n$ into an n-dimensional subspace of $\mathbb{R}^m$. Suppose z is a vector in $\mathbb{R}^n$ with the following property:

$Az - p$ is orthogonal to the range of A. We claim that such a z minimizes $\| Ax - p \|^2$. To see this let x be any vector in $\mathbb{R}^n$; split it as $x = z + y$; then

$$Ax - p = A(z + y) - p = Az - p + Ay.$$

By hypothesis $Az - p$ and Ay are orthogonal; therefore by the Pythagorean theorem,

$$\| Ax - p \|^2 = \| Az - p \|^2 + \| Ay \|^2,$$

this demonstrates the minimizing property of z.

To find z, we write the condition imposed on z in the form

$$(Az - p, Ay) = 0 \qquad \text{for all } y.$$

Using the adjoint of A we can rewrite this as

$$(A^*(Az - p), y) = 0 \qquad \text{for all } y.$$

The range of A^* is $\mathbb{R}^n$, so for this condition to hold for all y, $A^*(Az - p)$ must be 0, which is equation (25) for z.   □

Theorem 11.   An orthogonal projection P_Y defined in equation (22) is its own adjoint,

$$P_Y^* = P_Y.$$

EXERCISE 2.   Prove Theorem 11.

We turn now to the following question: what mappings M of a Euclidean space into itself preserve the distance of any pair of points, that is, satisfy for all x, y,

$$\| M(x) - M(y) \| = \| x - y \| \ ? \tag{26}$$

Such a mapping is called an *isometry*. It is obvious from the definition that the composite of two isometries is an isometry. An elementary example of an isometry is *translation*:

$$M(x) = x + a,$$

a some fixed vector. Given any isometry, one can compose it with a translation and produce an isometry that maps zero to zero. Conversely, any isometry is the composite of one that maps zero to zero and a translation.

Theorem 12.   Let M be an isometric mapping of a Euclidean space into itself that maps zero to zero:

$$M(0) = 0. \tag{27}$$

(i) M is linear.

(ii) $$M^*M = I. \tag{28}$$

Conversely, if (28) is satisfied, M is an isometry.
(iii) M is invertible and its inverse is an isometry.
(iv) det $M = \pm 1$.

Proof. It follows from (26) with $y = 0$ and (27) that

$$\| M(x) \| = \| x \|. \tag{29}$$

Now let us abbreviate the action of M by $'$:

$$M(x) = x', \qquad M(y) = y'.$$

By (29),

$$\| x' \| = \| x \|, \qquad \| y' \| = \| y \|. \tag{29$'$}$$

By (26),

$$\| x' - y' \| = \| x - y \|.$$

Square and use expansion (4) on both sides:

$$\| x' \|^2 - 2(x',y') + \| y' \|^2 = \| x \|^2 - 2(x,y) + \| y \|^2.$$

Using (29)′, we conclude that

$$(x',y') = (x,y); \tag{30}$$

that is, M preserves the scalar product.

Let z be any other vector, $z' = M(z)$; then, using (4) twice we get

$$\| z' - x' - y' \|^2 = \| z' \|^2 + \| y' \|^2 + \| x' \|^2$$
$$- 2(z',x') - 2(z',y') + 2(x',y').$$

Similarly,

$$\| z - x - y \|^2 = \| z \|^2 + \| y \|^2 + \| x \|^2 - 2(z,x) - 2(z,y) + 2(x,y).$$

Using (29)′ and (30) we deduce that

$$\| z' - x' - y' \|^2 = \| z - x - y \|^2.$$

We choose now $z = x + y$; then the right-hand side above is zero; therefore so is $\| z' - x' - y' \|^2$. By positive definiteness of the norm $z' - x' - y' = 0$. This proves part (i) of Theorem 12.

To prove part (ii), we take relation (30) and use the adjointness identity (23)′:

$$(Mx, My) = (x, M^*My) = (x,y)$$

for all x and y, so

$$(x, M^*My - y) = 0.$$

Since this holds for all x, it follows that $M^*My - y$ is orthogonal to itself, and so, by positiveness of norm, that for all y,

$$M^*My - y = 0.$$

The converse follows by reversing the steps: this proves part (ii).

It follows from (29) that the nullspace of M consists of the zero vector; it follows then from Corollary (B)′ of Chapter 3 that M is invertible. That M^{-1} is an isometry is obvious. This proves (iii).

It was pointed out in equation (33) of Chapter 5 that for every matrix det $M^* =$ det M; it follows from (28) and the product rule for determinants [see (18) in Chapter 5] that $(\det M)^2 = \det I = 1$, which implies that

$$\det M = \pm 1. \tag{31}$$

This proves part (iv) of Theorem 12.   □

The geometric meaning of (iv) is that a mapping that preserves distances also preserves volume.

Definition.   A matrix that maps $\mathbb{R}^n$ onto itself isometrically is called *orthogonal*.

The orthogonal matrices of a given order form a *group* under matrix multiplication. Clearly, composites of isometries are isometric, and so, by part (iii) of Theorem 12, are their inverses.

The orthogonal matrices whose determinant is plus 1 form a subgroup, called the *special orthogonal group*. Examples of orthogonal matrices with determinant plus 1 in three-dimensional space are rotations; see Chapter 11.

EXERCISE 3.   Construct the matrix representing reflection of points in $\mathbb{R}^3$ across the plane $x_3 = 0$. Show that the determinant of this matrix is -1.

EXERCISE 4.   Let R be reflection across any plane in $\mathbb{R}^3$.

(i) Show that R is an isometry.
(ii) Show that $R^2 = I$.
(iii) Show that $R^* = R$.

We recall from Chapter 4 that the ijth entry of the matrix product AB is the scalar product of the ith row of A with the jth column of B. The ith row of M^* is the transpose of the ith column of M. Therefore the identity $M^*M = I$ characterizing orthogonal matrices can be formulated as follows:

Corollary 12′.   A matrix M is orthogonal iff its columns are pairwise orthogonal unit vectors.

EXERCISE 5.   Show that a matrix M is orthogonal iff its rows are pairwise orthogonal unit vectors.

How can we measure the *size* of a linear mapping A of one Euclidean space X into another Euclidean space U? Recall from a rigorous course on the foundations of calculus the concept of *least upper bound*, also called *supremum*, of a bounded set of real numbers, abbreviated as *sup*. Each component of Ax is a linear function of the components of x; $\| Ax \|^2$ is a quadratic function of the components of x, and therefore the set of numbers $\| Ax \|^2$, $\| x \|^2 = 1$ is a bounded set.

Definition

$$\| A \| = \sup_{\|x\|=1} \| Ax \|. \tag{32}$$

Note that $\| Ax \|$ is measured in U, $\| x \|$ in X. $\| A \|$ is called the *norm* of A.

Theorem 13. Let A be a linear mapping from the Euclidean space X into the Euclidean space U, where $\| A \|$ is its norm.

(i)
$$\| Az \| \leq \| A \| \| z \| \qquad \text{for all } z \text{ in } X. \tag{33}$$

(ii)
$$\| A \| = \sup_{\|x\|=1, \|v\|=1} (Ax, v). \tag{34}$$

Proof. **(i)** follows for unit vectors z from the definition (32) of $\| A \|$. For any $z \neq 0$, write $z = kx, x$ a unit vector; since $\| Akx \| = \| kAx \| = |k| \| Ax \|$ and $\| kx \| = |k| \| x \|$, (33) follows. For $z = 0$, (33) is obviously true.

(ii) According to Theorem 2,

$$\| u \| = \max (u, v), \qquad \| v \| = 1.$$

Set $Ax = u$ in definition (32), and we obtain (34). $\qquad \square$

EXERCISE 6. Show that $|a_{ij}| \leq \| A \|$.

Theorem 14. For A as in Theorem 13, we have the following:

(i) $\| kA \| = |k| \| A \|$ for any scalar k.
(ii) For any pair of linear mappings A and B of X into U,

$$\| A + B \| \leq \| A \| + \| B \|. \tag{35}$$

(iii) Let A be a linear mapping of X into U, and let C be a linear mapping of U into V; then

$$\| CA \| \leq \| C \| \| A \|. \tag{36}$$

(iv)
$$\| A^* \| = \| A \|. \tag{37}$$

Proof. **(i)** follows from the observation that $\| kAx \| = |k| \| Ax \|$.
(ii) By the triangle inequality (12), for all x in X we obtain

$$\| (A + B)x \| = \| Ax + Bx \| \leq \| Ax \| + \| Bx \|.$$

The supremum of the left-hand side for $\| x \| = 1$ is $\| A + B \|$. The right-hand side is a sum of two terms; the supremum of the sum is $\le$ the sum of the suprema, which is $\| A \| + \| B \|$.

(iii) By inequality (33),

$$\| CAx \| \le \| C \| \| Ax \|.$$

Combined with (33), this yields

$$\| CAx \| \le \| C \| \| A \| \| x \|.$$

Taking the supremum for all unit vectors x gives (36).

(iv) According to (23)′,

$$(Ax, v) = (x, A^*v);$$

since the scalar product is a symmetric function, we obtain

$$(Ax, v) = (A^*v, x).$$

Take the supremum of both sides for all x and v, $\| x \| = 1$, $\| v \| = 1$. According to (34), on the left-hand side we get $\| A \|$, and on the right-hand side we obtain $\| A^* \|$. □

The following result is enormously useful:

Theorem 15. Let A be a linear mapping of a finite-dimensional Euclidean space X into itself that is invertible. Denote by B another linear mapping of X into X close to A in the sense of the following inequality:

$$\| A - B \| < 1 / \| A^{-1} \|. \tag{38}$$

Then B is invertible.

Proof. Denote $A - B = C$, so that $B = A - C$. Factor B as

$$B = A(I - A^{-1}C) = A(I - S),$$

where $S = A^{-1}C$.

We have seen in Chapter 3 that the product of invertible maps is invertible; therefore it suffices to show that $I - S$ is invertible. We see that it suffices to show that the nullspace of $I - S$ is trivial. Suppose not; that is, $(I - S)x = 0$, $x \ne 0$. Then $x = Sx$; using the definition of the norm of S,

$$\| x \| = \| Sx \| \le \| S \| \| x \|.$$

Since $x \neq 0$, it follows that

$$1 \leq \| S \|. \tag{39}$$

But according to part (iii) of Theorem 14,

$$\| S \| = \| A^{-1}C \| \leq \| A^{-1} \| \| C \| < 1,$$

where in the last step we have used inequality (38): $\| C \| < 1/\| A \|^{-1}$. This contradicts (39). □

Note. In this proof we have used the finite dimensionality of X. A proof of Theorem 15 given in Chapter 15 is valid for infinite-dimensional X.

We recall now another concept from a rigorous calculus course:

Convergence. A sequence of numbers $\{a_k\}$ tends to a,

$$\lim a_k = a,$$

if $|a_k - a|$ tends to zero. Recall furthermore the notion of a Cauchy sequence of numbers $\{a_k\}$; it is a sequence for which $|a_k - a_j|$ tends to zero as j and k tend to ∞. A basic property of real numbers is that every Cauchy sequence of numbers converges to a limit.

This property of real numbers is called *completeness*.

A second basic notion about real numbers is local *compactness*: Every *bounded sequence* of real numbers contains a *convergent subsequence*.

We now show how to extend these notions and results from numbers to vectors in a finite-dimensional Euclidean space.

Definition. A sequence of vectors $\{x_k\}$ in a linear space X with Euclidean structure converges to the limit x:

$$\lim_{k \to \infty} x_k = x$$

if $\| x_k - x \|$ tends to zero as $k \to \infty$.

Theorem 16. A sequence of vectors $\{x_k\}$ in a Euclidean space X is called a *Cauchy sequence* if $\| x_k - x_j \| \to 0$ as k and $j \to \infty$.

(i) Every Cauchy sequence in a finite-dimensional Euclidean space converges to a limit.

A sequence of vectors $\{x_k\}$ in a Euclidean space X is called *bounded* if $\| x_k \| \leq R$ for all k, R some real number.

(ii) In a finite-dimensional Euclidean space every bounded sequence contains a convergent subsequence.

Proof. **(i)** Let x and y be two vectors in X, a_j and b_j their jth component; then

$$|a_j - b_j| \leq \| x - y \|.$$

Denote by $a_{k,j}$ the jth component of x_k. Since $\{x_k\}$ is a Cauchy sequence, it follows that the sequence of numbers $\{a_{k,j}\}$ also is a Cauchy sequence. Since the real numbers are complete, the $\{a_{k,j}\}$ converge to a limit a_j. Denote by x the vector whose components are $(a_1, \ldots, a_n)$. From the definition of Euclidean norm,

$$\| x_k - x \|^2 = \sum_1^n |a_{k,j} - a_j|^2, \tag{40}$$

it follows that $\lim x_k = x$.

(ii) Since $|a_{k,j}| \leq \| x_k \|$, it follows that $|a_{k,j}| \leq R$ for all k. Because the real numbers are locally compact, a subsequence of $\{a_{k,1}\}$ converges to a limit a_1.

This subsequences of $k - s$ contains a further subsubsequence such that $\{a_{k,2}\}$ converges to a limit a_2. Proceeding in this fashion we can construct a subsequence of $\{x_k\}$ for which all sequences $\{a_{k,j}\}$, converge to a limit $a_j, 1, \ldots, n$, where n is the dimension of X. Denote by x the vector whose components are $(a_1, \ldots, a_n)$. From (40) we deduce that the subsequence of $\{x_k\}$ converges to x. $\qquad \square$

It follows from part (ii) of Theorem 16 that the supremum in the definition (32) of $\| A \|$ is a maximum:

$$\| A \| = \max_{\|x\|=1} \| Ax \|. \tag{32$'$}$$

It follows from the definition of supremum that $\| A \|$ cannot be replaced by any smaller number that is an upper bound of $\| Ax \|$, $\| x \| = 1$. It follows that there is a sequence of unit vectors $\{x_k\}$, $\| x_k \| = 1$, such that

$$\lim_{k \to \infty} \| Ax_k \| = \| A \|.$$

According to Theorem 16, this sequence has a subsequence that converges to a limit x. This vector x maximizes $\| Az \|$ for all unit vectors z.

Part (ii) of Theorem 16 has a converse:

Theorem 17. Let X be a linear space with a Euclidean structure, and suppose that it is locally compact—that is, that every bounded sequence $\{x_k\}$ of vectors in X has a convergent subsequence. Then X is finite dimensional.

Proof. We shall show that if X is not finite dimensional, then it is not locally compact. Not being finite dimensional means that given any linearly independent set of vectors $y_1, \ldots, y_k$, there is a vector y_{k+1} that is not a linear combination of them. In this way we obtain an infinite sequence of vectors $y_1, y_2, \ldots$ such that every finite set

$\{y_1, \ldots, y_k\}$ is linearly independent. According to Theorem 4, we can apply the Gram–Schmidt process to construct pairwise orthogonal unit vectors $\{x_1, \ldots, x_k\}$, which are linear combinations of $y_1, \ldots, y_k$. For this infinite sequence

$$\| x_k - x_j \|^2 = \| x_k \|^2 - 2(x_k, x_j) + \| x_j \|^2 = 2$$

for all $k \neq j$. Therefore this sequence, which is bounded, contains *no* convergent subsequence. $\square$

Theorem 17 is a very useful, and therefore important criterion for a Euclidean space to be finite dimensional. In Chapter 14 we shall show how to extend it to all normed linear spaces.

In Appendix 12 we shall give an interesting application.

Definition.   A sequence $\{A_n\}$ of mappings converges to a limit A if

$$\lim \| A_n - A \| = 0.$$

EXERCISE 7.   Show that $\{A_n\}$ converges to A iff for all x, $A_n x$ converges to Ax.

Note. The result in Exercise 7 does not hold in infinite-dimensional spaces.

We conclude this chapter by a brief discussion of *complex* Euclidean structure. In the concrete definition of complex Euclidean space, definition (2) of the scalar product in $\mathbb{R}^n$ has to be replaced in $\mathbb{C}^n$ by

$$(x, y) = \sum x_i \bar{y}_i, \tag{41}$$

where the bar $\bar{}$ denotes the *complex conjugate*. The definition of the *adjoint* of a matrix is as in (23)', but in the complex case has a slightly different interpretation. Writing

$$A = (a_{ij}), \qquad (Ax)_i = \sum_j a_{ij} x_j$$

and using the (41) definition of scalar product we can write

$$(Ax, u) = \sum_i \left(\sum_j a_{ij} x_j \right) \bar{u}_i.$$

This can be rewritten as

$$\sum_j x_j \left(\overline{\sum \bar{a}_{ij} u_i} \right).$$

which shows that $(Ax, u) = (x, A^* u)$, where

$$(A^* u)_j = \sum_i \bar{a}_{ij} u_i;$$

that is, the adjoint A^* of the matrix A is the *complex conjugate* of the *transpose* of A. We now define the abstract notion of a complex Euclidean space.

Definition.   A complex Euclidean structure in a linear space X over the complex numbers is furnished by a complex valued function of two vector arguments, called a *scalar product* and denoted as (x, y), with these properties:

(i) (x, y) is a linear function of x for y fixed.

(ii) Conjugate symmetry: for all x, y,

$$\overline{(x, y)} = (y, x). \tag{42}$$

Note that conjugate symmetry implies that (x, x) is real for all x.

(iii) Positivity:

$$(x, x) > 0 \qquad \text{for all } x \neq 0.$$

The theory of complex Euclidean spaces is analogous to that for real ones, with a few changes where necessary. For example, it follows from (i) and (ii) that for x fixed, (x, y) is a *skew linear* function of y, that is, additive in y and satisfying for any complex number k,

$$(x, ky) = \bar{k}(x, y). \tag{43}$$

Instead of repeating the theory, we indicate those places where a slight change is needed. In the complex case identity $(12)'$ is

$$\begin{aligned} \| x + y \|^2 &= \| x \|^2 + (x, y) + (y, x) + \| y \|^2 \\ &= \| x \|^2 + 2\text{Re}(x, y) + \| y \|^2, \end{aligned} \tag{44}$$

where Re k denotes the real part of the complex number k.

EXERCISE 8.   Prove the Schwarz inequality for complex linear spaces with a Euclidean structure.

EXERCISE 9.   Prove the complex analogues of Theorems 6, 7, and 8.

We define the adjoint A^* of a linear map A of an abstract complex Euclidean space into itself by relation $(23)'$ as before:

$$(x, A^*u) = (Ax, u).$$

EXERCISE 10.   Prove the complex analogue of Theorem 9.

We define isometric maps of a complex Euclidean space as in the real case:

$$\| Mx \| = \| x \|.$$

Definition. A linear map of a complex Euclidean space into itself that is isometric is called *unitary*.

EXERCISE 11. Show that a unitary map M satisfies the relations

$$M^*M = 1 \tag{45}$$

and, conversely, that every map M that satisfies (45) is unitary.

EXERCISE 12. Show that if M is unitary, so is M^{-1} and M^*.

EXERCISE 13. Show that the unitary maps form a group under multiplication.

EXERCISE 14. Show that for a unitary map M, $|\det M| = 1$.

EXERCISE 15. Let X be the space of continuous complex-valued functions on $[-1, 1]$ and define the scalar product in X by

$$(f, g) = \int_{-1}^{1} f(s)\bar{g}(s)ds.$$

Let $m(s)$ be a continuous function of absolute value 1: $|m(s)| = 1, -1 \le s \le 1$.
Define M to be multiplication by m:

$$(Mf)(s) = m(s)f(s).$$

Show that M is unitary.

We give now a simple but useful lower bound for the norm of a matrix mapping a complex Euclidean space X *into itself*. The definition of the norm of such a matrix is the same as in the real case, given by equation (32)':

$$\| A \| = \max_{\|x\|=1} \| Ax \|.$$

Let A be any square matrix with complex entries, h one of its eigenvectors, chosen to have length 1, and a the eigenvalue:

$$Ah = ah, \qquad \| h \| = 1.$$

Then

$$\| Ah \| = \| ah \| = |a|.$$

Since $\| A \|$ is the maximum of $\| Ax \|$ for all unit vectors x, it follows that $\| A \| \geq |a|$. This is true for every eigenvalue; therefore

$$\| A \| \geq \max_i |a_i|, \tag{46}$$

where the a_i range over all eigenvalues of A.

Definition. The *spectral radius* $r(A)$ of a linear mapping A of a linear space into itself is

$$r(A) = \max |a_j|, \tag{47}$$

where the a_j range over all eigenvalues of A. So (46) can be restated as follows:

$$\| A \| \geq r(A). \tag{48}$$

Recall that the eigenvalues of the powers of A are the powers of the eigenvalues of A:

$$A^j h = a^j h.$$

Applying (48) to A^j, we conclude that

$$\| A^j \| \geq r(A)^j.$$

Taking the jth root gives

$$\| A^j \|^{1/j} \geq r(A). \tag{48}_j$$

Theorem 18. As j tends to ∞, $(48)_j$ tends to be an equality; that is,

$$\lim_{j \to \infty} \| A^j \|^{1/j} = r(A).$$

A proof will be furnished in Appendix 10.

We shall give now a simple and useful upper bound for the norm of a real $m \times n$ matrix

$$A = (a_{ij}),$$

mapping $\mathbb{R}^n$ into $\mathbb{R}^m$. For any x in $\mathbb{R}^n$, set $Ax = y$, y in $\mathbb{R}^m$. The components of y are expressed in terms of the components of x as follows:

$$y_i = \sum_j a_{ij} x_j.$$

Estimate the right-hand side using the Schwarz inequality:

$$y_i^2 = \left(\sum_j a_{ij} x_j \right)^2 \le \left(\sum_j a_{ij}^2 \right) \left(\sum x_j^2 \right).$$

adding all these inequalities, $i = 1, \ldots, m$, we get

$$\sum_i y_i^2 \le \left(\sum_{i,j} a_{ij}^2 \right) \sum_j x_j^2. \tag{49}$$

Using the definition of norm in Euclidean space [see equation (1)], we can rewrite inequality (49) as

$$\| y \|^2 \le \left(\sum_{i,j} a_{ij}^2 \right) \| x \|^2.$$

Take the square root of this inequality; since $y = Ax$, we can write it as

$$\| Ax \| \le \left(\sum a_{ij}^2 \right)^{1/2} \| x \|. \tag{50}$$

The definition of the norm $\| A \|$ of the matrix A is

$$\sup \| Ax \|, \qquad \| x \| = 1.$$

It follows from (50) that

$$\| A \| \le \left(\sum_{i,j} a_{ij}^2 \right)^{1/2}; \tag{51}$$

this is the upper bound for $\| A \|$ we set out to prove.

EXERCISE 16.   Prove the following analogue of (51) for matrices with complex entries:

$$\| A \| \le \left(\sum_{i,j} |a_{ij}|^2 \right)^{1/2} \tag{51}'$$

EXERCISE 17.   Show that

$$\sum_{i,j} |a_{ij}|^2 = \text{tr } AA^*. \tag{52}$$

EXERCISE 18. Show that

$$\operatorname{tr} AA^* = \operatorname{tr} A^* A.$$

EXERCISE 19. Find an upper bound and a lower bound for the norm of the 2×2 matrix

$$A = \begin{pmatrix} 1 & 2 \\ 0 & 3 \end{pmatrix}.$$

The quantity $\left(\sum_{i,j} |a_{ij}|^2 \right)^{1/2}$ is called the *Hilbert–Schmidt norm* of the matrix A. Let T denote a 3×3 matrix, its columns x, y, and z:

$$T = (x, y, z).$$

The determinant of T is, for x and y fixed, a linear function of z:

$$\det (x, y, z) = l(z). \tag{53}$$

According to Theorem 5, every linear function can be represented as a scalar product:

$$l(z) = (w, z), \tag{54}$$

where w is some vector depending on x and y:

$$w = w(x, y).$$

Combining (53) and (54) gives

$$\det (x, y, z) = (w(x, y), z). \tag{55}$$

We formulate the properties of the dependence of w on x and y as a series of exercises:

EXERCISE 20. **(i)** w is a *bilinear* function of x and y. Therefore we write w as a *product* of x and y, denoted as

$$w = x \times y,$$

and called the *cross product*.
(ii) Show that the cross product is antisymmetric:

$$y \times x = -x \times y.$$

(iii) Show that $x \times y$ is orthogonal to both x and y.

(iv) Let R be a rotation in $\mathbb{R}^3$; show that

$$(Rx) \times (Ry) = R(x \times y).$$

(v) Show that

$$\| x \times y \| = \pm \| x \| \| y \| \sin \theta,$$

where θ is the angle between x and y.

(vi) Show that

$$\begin{pmatrix} 1 \\ 0 \\ 0 \end{pmatrix} \times \begin{pmatrix} 0 \\ 1 \\ 0 \end{pmatrix} = \begin{pmatrix} 0 \\ 0 \\ 1 \end{pmatrix}.$$

(vii) Using Exercise 16 in Chapter 5, show that

$$\begin{pmatrix} a \\ b \\ c \end{pmatrix} \times \begin{pmatrix} d \\ e \\ f \end{pmatrix} = \begin{pmatrix} bf - ce \\ cd - af \\ ae - bd \end{pmatrix}.$$

EXERCISE 21. Show that in a Euclidean space every pair of vector satisfies

$$\| u + v \|^2 + \| u - v \|^2 = 2 \| u \|^2 + 2 \| v \|^2. \tag{56}$$

CHAPTER 8

Spectral Theory of Self-Adjoint Mappings of a Euclidean Space into Itself

In this chapter we shall study mappings A of Euclidean spaces into themselves that are self-adjoint—that is, are their own adjoints:

$$A^* = A.$$

When A acts on a real Euclidean space, any matrix representing it in an orthonormal system of coordinates is symmetric, that is,

$$A_{ij} = A_{ji}.$$

Such mappings are therefore also called *symmetric*. When A acts on a complex Euclidean space, its matrix representations are *conjugate symmetric*;

$$A_{ij} = \bar{A}_{ji}.$$

Such mappings are also called *Hermitean*. We saw in Theorem 11 of Chapter 7 that orthogonal projections are self-adjoint. Below we describe another large class of self-adjoint matrices. In Chapter 11 we shall see that matrices that describe the motion of mechanical systems are self-adjoint.

Definition. Let M be an arbitrary linear mapping in a Euclidean space, We define its self-adjoint part as

$$M_s = \frac{M + M^*}{2}, \tag{1}$$

Linear Algebra and Its Applications, Second Edition, by Peter D. Lax

EXERCISE I. Show that

$$\text{Re}(x, Mx) = (x, M_s x). \tag{2}$$

Let $(x_1, \ldots, x_n) = f(x)$ be a real-valued twice-differentiable function of n real variables $x_1, \ldots, x_n$ written as a single vector variable x. The Taylor approximation to f at a up to second order reads

$$f(a + y) = f(a) + l(y) + \frac{1}{2}q(y) + ||y||^2 \epsilon(||y||), \tag{3}$$

where $\epsilon(d)$ denotes some function that tends to 0 as $d \to 0$, $l(y)$ is a linear function of y, and $q(y)$ is a quadratic function. A linear function has the form (see Theorem 5 of Chapter 7)

$$l(y) = (y, g); \tag{4}$$

g is the *gradient* of f at a; according to Taylor's theorem

$$g_j = \left.\frac{\partial f}{\partial x_j}\right|_{x=a}. \tag{5}$$

The quadratic function q has the form

$$q(y) = \sum_{i,j} h_{ij} y_i y_j. \tag{6}$$

The *matrix* (h_{ij}) is called the *Hessian* H of f; according to Taylor's theorem,

$$h_{ij} = \left.\frac{\partial^2}{\partial x_j \partial x_i} f\right|_{x=a}. \tag{7}$$

Employing matrix notation and the Euclidean scalar product, we can write q, given by (4), in the form

$$q(y) = (y, Hy). \tag{8}$$

The matrix H is *self-adjoint*, that is, $H^* = H$:

$$h_{ij} = h_{ji}; \tag{9}$$

this follows from definition (5), and the fact that the mixed partials of a twice-differentiable function are equal.

Suppose now that a is a critical point of the function f, that is where $\text{grad}\, f = g$ is zero. Around such a point Taylor's formula (3) shows that the behavior of f is governed by the quadratic term. Now the behavior of functions near critical points is of fundamental importance for dynamical systems, as well as in geometry; this is what gives quadratic functions such an important place in mathematics and makes the analysis of symmetric matrices such a central topic in linear algebra.

To study a quadratic function it is often useful to introduce new variables:

$$Ly = z, \tag{10}$$

where L, is some invertible matrix, in terms of which q has a simpler form.

Theorem 1. (a) Given a real quadratic form (6) it is possible to change variables as in (10) so that in terms of the new variables, z, q is diagonal, that is, of the form

$$q(L^{-1}z) = \sum_{1}^{n} d_i z_i^2. \tag{11}$$

(b) There are many ways to introduce new variables which diagonalize q; however, the number of positive, negative, and zero-diagonal terms d_i appearing in (11) is the same in all of them.

Proof. Part (a) is entirely elementary and constructive. Suppose that one of the diagonal elements of q is nonzero, say $h_{11} \neq 0$. We then group together all terms containing y_1:

$$q(y) = h_{11}y_1^2 + \sum_{2}^{n} h_{1j}y_1y_j + \sum_{2}^{n} h_{ij}y_iy_j.$$

Since H is symmetric, $h_{j1} = h_{1j}$; so we can write q as

$$h_{11}\left(y_1 + h_{11}^{-1}\sum_{2}^{n} h_{1j}y_j\right)^2 - h_{11}^{-1}\left(\sum_{2}^{n} h_{1j}y_j\right)^2.$$

Set

$$y_1 + h_{11}^{-1}\sum_{2}^{n} h_{1j}y_j = z_1. \tag{12}$$

We can then write

$$q(y) = h_{11}z_1^2 + q_2(y), \tag{13}$$

where q_2 depends only on $y_2, \ldots, y_n$.

If all diagonal terms of q are zero but there is some nonzero off-diagonal term, say $h_{12} = h_{21} \neq 0$, then we introduce $y_1 + y_2$ and $y_1 - y_2$ as new variables, which produces a nonzero diagonal term. If all diagonal and off-diagonal terms are zero, then $q(y) \equiv 0$ and there is nothing to prove.

We now apply induction on the number of variables n; using (13) shows that if the quadratic function q_2 in $(n - 1)$ variables can be written in form (11), then so can q itself. Since $y_2, \dots, y_n$ are related by an invertible matrix to $z_2, \dots, z_n$, it follows from (12) that the full set y is related to z by an invertible matrix. $\square$

EXERCISE 2.   We have described above an algorithm for diagonalizing q; implement it as a computer program.

We turn now to part (b); denote by p_+, p_-, and p_0 the number of terms in (11) that are positive, negative, and zero, respectively. We shall look at the behavior of q on subspaces S of $\mathbb{R}^n$. We say that q is *positive* on the subspace S if

$$q(u) > 0 \qquad \text{for every } u \text{ in } S, \qquad u \neq 0. \tag{14}$$

Lemma 2.   The dimension of the largest subspace of $\mathbb{R}^n$ on which q is positive is p_+:

$$p_+ = \max \dim S, \qquad q \text{ positive on } S. \tag{15}$$

Similarly,

$$p_- = \max \dim S, \qquad q \text{ negative on } S. \tag{15}'$$

Proof. We shall use representation (11) for q in terms of the coordinates $z_1, \dots, z_n$; suppose we label them so that $d_1, \dots, d_p$ are positive, $p = p_+$, the rest nonpositive. Define the subspace S_+ to consist of all vectors for which $z_{p+1} = \dots = z_n = 0$. Clearly $\dim S_+ = p_+$, and equally clearly, q is positive on S_+. This proves that p_+ is less than or equal to the right-hand side of (15). We claim that the equality holds. Let S be any subspace whose dimension exceeds p_1. For any vector u in S, define P_u as the vector whose p_+ components are the same as the first p_+ components of u, and the rest of the components are zero. The dimension p_+ of the target space of this map is smaller than the dimension of the domain space S. Therefore, according to Corollary A of Theorem 2, Chapter 3, there is a nonzero vector y in the nullspace of P. By definition of P, the first p_+ of the z-components of this vector y are zero. But then it follows from (11) that $q(y) \leq 0$; this shows that q is not positive on S. This proves (15); the proof of (15)' is analogous. $\square$

Lemma 2 shows that the numbers p_- and p_+ can be defined in terms of the quadratic form q itself, intrinsically, and are therefore independent of the special choice of variables that puts q in form (11). Since $p_+ + p_- + p_0 = n$, this proves part (b) of Theorem 1. $\square$

Part (b) of Theorem 1 is called the *law of inertia*.

EXERCISE 3.   Prove that

$$p_+ + p_0 = \max \dim S, \qquad q \geq 0 \text{ on } S$$

and

$$p_- + p_0 = \max \dim S, \qquad q \leq 0 \text{ on } S.$$

Using form (6) of q we can reinterpret Theorem 1 in matrix terms. It is convenient for this purpose to express y in terms of z, rather than the other way around as in (10). So we multiply (10) by L^{-1}, obtaining

$$y = Mz, \tag{16}$$

where M abbreviates L^{-1}. Setting (16) into (8) gives, using the adjoint of M,

$$q(y) = (y, Hy) = (Mz, HMz) = (z, M^*HMz). \tag{17}$$

Clearly, q in terms of z is of form (11) iff M^*HM is a diagonal matrix. So part (*a*) of Theorem 1 can be put in the following form:

Theorem 3.   Given any real self-adjoint matrix H, there is a real invertible matrix M such that

$$M^*HM = D, \tag{18}$$

D a diagonal matrix.

For many applications it is of utmost importance to change variables so that the Euclidean length of the old and the new variables is the same:

$$\| y \|^2 = \| z \|^2.$$

For the matrix M in (16) this means that M is an isometry. According to (28) of Chapter 7, this is the case iff M is orthogonal, that is, satisfies

$$M^*M = I. \tag{19}$$

It is one of the basic theorems of linear algebra, nay, of mathematics itself, that given a real-valued quadratic form q, it is possible to diagonalize it by an *isometric* change of variables. In matrix language, given a real symmetric matrix H, there is a real invertible matrix M such that *both* (18) *and* (19) *hold*.

We shall give two proofs of this important result. The first is based on the spectral theory of general matrices presented in Chapter 6, specialized to self-adjoint mappings in *complex Euclidean* space.

We recall from Chapter 7 that the *adjoint* H* of a linear map H of a complex Euclidean space X into itself is defined by requiring that

$$(Hx, y) = (x, H^*y) \tag{20}$$

hold for all pairs of vectors x,y. Here the bracket (,) is the conjugate-symmetric scalar product introducd at the end of Chapter 7. A linear map H is called *self-adjoint* if

$$H^* = H.$$

For H self-adjoint, (20) becomes

$$(Hx, y) = (x, Hy). \tag{20$'$}$$

Theorem 4.   A self-adjoint map H of complex Euclidean space X into itself has real eigenvalues and a set of eigenvectors that form an orthonormal basis of X.

Proof. According to the principal result of spectral theory, Theorem 7 of Chapter 6, the eigenvectors and generalized eigenvectors of H span X. To deduce Theorem 4 from Theorem 7, we have to show that a self-adjoint mapping H has the following additional properties:

(**a**) H *has only real eigenvalues.*
(**b**) H *has no generalized eigenvectors, only genuine ones.*
(**c**) *Eigenvectors of* H *corresponding to different eigenvalues are orthogonal.*

(**a**) If $a + ib$ is an eigenvalue of H, then ib is an eigenvalue of $H - aI$, also self-adjoint. Therefore, it suffices to show that a self-adjoint H cannot have a purely imaginary eigenvalue ib. Suppose it did, with eigenvector z:

$$Hz = ibz.$$

Take the scalar product of both sides with z:

$$(Hz, z) = (ibz, z) = ib(z, z). \tag{21}$$

Setting both x and y equal to z in (20)$'$, we get

$$(Hz, z) = (z, Hz). \tag{21$'$}$$

Since the scalar product is conjugate symmetric, we conclude that the two sides of (21)$'$ are conjugates. Since they are equal, the left-hand side of (21) is real. Therefore so is the right-hand side; since (z, z) is positive, this can be only if $b = 0$, as asserted in (a).

(**b**) A generalized eigenvector z satisfies

$$H^d z = 0; \tag{22}$$

here we have taken the eigenvalue to be zero, by replacing H with $H - aI$. We want to show that then z is a genuine eigenvector:

$$Hz = 0. \tag{22}'$$

We take first the case $d = 2$:

$$H^2 z = 0; \tag{23}$$

we take the scalar product of both sides with z:

$$(H^2 z, z) = 0. \tag{23}'$$

Using (20)′ with $x = Hz, y = z$, we get

$$(H^2 z, z) = (Hz, Hz) = \| Hz \|^2;$$

using (23)′, we conclude that $\| Hz \| = 0$, which, by positivity, holds only when $Hz = 0$.

We do now an induction on d; we rewrite (22) as

$$H^2 H^{d-2} z = 0.$$

Abbreviating $H^{d-2} z$ as w, we rewrite this as $H^2 w = 0$; this implies, as we have already shown, that $Hw = 0$. Using the definition of w this can be written as

$$H^{d-1} z = 0.$$

This completes the inductive step and proves (b).

(**c**) Consider two eigenvalues a and b of H, $a \neq b$:

$$Hx = ax, \qquad Hy = by.$$

We form the scalar product of the first relation with y and of the second with x; since b is real we get

$$(Hx, y) = a(x, y), \qquad (x, Hy) = b(x, y).$$

By (20)' the left-hand sides are equal; therefore so are the right-hand sides. But for $a \neq b$ this can be only if $(x, y) = 0$. This completes the proof of (c).   □

Definition.   The set of eigenvalues of H is called the *spectrum* of H.

We show now that Theorem 4 has the consequence that real quadratic forms can be diagonalized by real isometric transformation. Using the matrix formulation given in Theorem 3, we state the result as follows.

Theorem 4'.   Given any real self-adjoint matrix H, there is an orthogonal matrix M such that

$$M^*HM = D, \tag{24}$$

D a diagonal matrix whose entries are the eigenvalues of H. M satisfies $M^*M = I$.

Proof. The eigenvectors f of H satisfy

$$Hf = af. \tag{25}$$

H is a real matrix, and according to (a), the eigenvalue a is real. It follows from (25) that the real and imaginary parts of f also are eigenvectors. It follows from this easily that we may choose an orthonormal basis consisting of real eigenvectors in each eigenspace N_a. Since by (c), eigenvectors belonging to distinct eigenvalues are orthogonal, we have an orthonormal basis of X consisting of real eigenvectors f_j of H. Every vector y in X can be expressed as a linear combination of these eigenvectors:

$$y = \sum z_j f_j. \tag{25'}$$

For y real, the z_j are real. We denote the vector with components z_j as z: $z = (z_1, \ldots, z_n)$. Since the $\{f_j\}$ form an orthonormal basis,

$$\| y \|^2 = \sum z_j^2 = \| z \|^2. \tag{26}$$

Letting H act on (25)', we get, using (25), that

$$Hy = \sum z_j a_j f_j. \tag{25''}$$

Setting (25) and (25)' into (6) we can express the quadratic form q as

$$q(y) = (y, Hy) = \sum a_j z_j^2. \tag{26'}$$

This shows that the introduction of the new variables z diagonalizes the quadratic form q. Relation (26) says that the new vector has the same length as the old.

Denote by M the relation of z to y:

$$y = Mz.$$

Set this into (26)′; we get

$$q(y) = (y, Hy) = (Mz, HMz) = (z, M^*HMz).$$

Using (26)′, we conclude that $M^*HM = D$, as claimed in (24). This completes the proof of Theorem 4′, $\square$

Multiply (24) by M on the left and M^* on the right. Since MM^* also equals I for an isometry M, we get

$$H = MDM^*. \tag{24'}$$

EXERCISE 4.   Show that the columns of M are the eigenvectors of H.

We restate now Theorem 4, the spectral theorem for self-adjoint maps, in a slightly different language. Theorem 4 asserts that the whole space X can be decomposed as the direct sum of *pairwise orthogonal* eigenspaces:

$$X = N^{(1)} \oplus \cdots \oplus N^{(k)}, \tag{27}$$

where $N^{(j)}$ consists of eigenvectors of H with real eigenvalues $a_j, a_j \neq a_i$ for $j \neq i$. That means that each x in X can be decomposed uniquely as the sum

$$x = x^{(1)} + \cdots + x^{(k)}, \tag{27'}$$

where $x^{(j)}$ belongs to $N^{(j)}$. Since $N^{(j)}$ consists of eigenvectors, applying H to (27)′ gives

$$Hx = a_1 x^{(1)} + \cdots + a_k x^{(k)}. \tag{28}$$

Each $x^{(j)}$ occurring in (27)′ is a function of x; we denote this dependence as

$$x^{(j)} = P_j(x).$$

Since the $N^{(j)}$ are linear subspaces of X, it follows that $x^{(j)}$ depends linearly on x, that is, the P_j are linear mappings. We can rewrite (27)′ and (28) as follows:

$$I = \sum_j P_j, \tag{29}$$

$$H = \sum_j a_j P_j. \tag{30}$$

Claim: The operators P_j have the following properties:

(a) $$P_j P_k = 0 \qquad \text{for } j \neq k, P_j^2 = P_j. \tag{31}$$

(b) Each P_j is self-adjoint:

$$P_j^* = P_j. \tag{32}$$

Proof. **(a)** Relations (31) are immediate consequences of the definition of P_j. **(b)** Using the expansion (27)′ for x and the analogous one for y we get

$$(P_j x, y) = (x^{(j)}, y) = \left(x^{(j)}, \sum_i y^{(i)} \right) = \sum_i (x^{(j)}, y^{(i)}) = (x^{(j)}, y^{(j)}).$$

where in the last step we have used the orthogonality of $N^{(j)}$ to $x^{(i)}$ for $j \neq i$. Similarly we can show that

$$(x, P_j y) = (x^{(j)}, y^{(j)}).$$

Putting the two together shows that

$$(P_j x, y) = (x, P_j y).$$

According to (20), this expresses the self-adjointness of P_j. This proves (32).   □

We recall from Chapter 7 that a self-adjoint operator P which satisfies $P^2 = P$ is an *orthogonal projection*. A decomposition of the form (29), where the P_j satisfy (31), is called a *resolution of the identity*. H in form (30) gives the *spectral resolution of* H.

We can now restate Theorem 4 as

Theorem 5. Let X be a complex Euclidean space, H: $X \to X$ a self-adjoint linear map. Then there is a resolution of the identity, in the sense of (29), (31), and (32) that gives a spectral resolution (30) of H.

The restated form of the spectral theorem is very useful for defining functions of self-adjoint operators. We remark that its greatest importance is as the model for the infinite-dimensional version.

Squaring relation (30) and using properties (31) of the P_j we get

$$H^2 = \sum a_j^2 P_j.$$

By induction, for any natural number m,

$$H^m = \sum a_j^m P_j.$$

It follows that for any polynomial p,

$$p(H) = \sum p(a_j)P_j. \tag{33}$$

Let $f(a)$ be any real valued function defined on the spectrum of H. We define $f(H)$ by formula (33):

$$f(H) = \sum f(a_j)P_j. \tag{33}'$$

An example:

$$e^{Ht} = \sum e^{a_j}P_j.$$

We shall say more about this in Chapter 9.

We present a series of no-cost extensions of Theorem 5.

Theorem 6.   Suppose H and K are a pair of self-adjoint matrices that commute:

$$H^* = H, \qquad K^* = K, \qquad HK = KH.$$

Then they have a common spectral resolution, that is, there exist orthogonal projections satisfying (29), (31), and (32) so that (30) holds, as well as

$$\sum b_j P_j = K. \tag{30}'$$

Proof. Denote by N one of the eigenspaces of H; then for every x in N

$$Hx = ax,$$

Applying K, we get

$$KHx = aKx.$$

Since H and K commute, we can rewrite this as

$$HKx = aKx,$$

which shows that Kx is an eigenvector of H. So K maps N into itself. The restriction of K to N is self-adjoint. We now apply spectral resolution of K over N; combining all these resolutions gives the joint spectral resolution of H and K.   □

This result can be generalized to any finite collection of pairwise commuting self-adjoint mappings.

Definition.   A linear mapping A of Euclidean space into itself is called *anti-self-adjoint* if

$$A^* = -A.$$

It follows from the definition of adjoint and the property of conjugate symmetry of the scalar product that for any linear map M of a complex Euclidean space into itself,

$$(iM)^* = -iM^*. \tag{34}$$

In particular, if A is anti-self-adjoint, iA is self-adjoint, and Theorem 4 applies. This yields Theorem 7.

Theorem 7.   Let A be an anti-self-adjoint mapping of a complex Euclidean space into itself. Then

(a) The eigenvalues of A are purely imaginary.
(b) We can choose an orthonormal basis consisting of eigenvectors of A.

We introduce now a class of maps that includes self-adjoint, anti-self-adjoint, and unitary maps as special cases.

Definition.   A mapping N of a complex Euclidean space into itself is called *normal* if it commutes with its adjoint:

$$NN^* = N^*N.$$

Theorem 8.   A normal map N has an orthonormal basis consisting of eigenvectors.

Proof. If N and N* commute, so do

$$H = \frac{N + N^*}{2} \quad \text{and} \quad A = \frac{N - N^*}{2}. \tag{35}$$

Clearly, H is adjoint and A is anti-self-adjoint. According to Theorem 6 applied to H and $K = iA$, they have a common spectral resolution, so that there is an orthonormal basis consisting of common eigenvectors of both H and A. But since by (35),

$$N = H + A, \tag{35'}$$

it follows that these are also eigenvectors of N as well as of N*.   □

Here is an application of Theorem 8.

Theorem 9. Let U be a unitary map of a complex Euclidean space into itself, that is, an isometric linear map.

(a) There is an orthonormal basis consisting of genuine eigenvectors of U.

(b) The eigenvalues of U are complex numbers of absolute value $= 1$.

Proof. According to equation (42) of Chapter 7, an isometric map U satisfies $U^*U = 1$. This relation says that U^* is a left inverse for U. We have shown in Chapter 3 (see Corollary B of Theorem 1 there) that a mapping that has a left inverse is invertible, and its left inverse is also its right inverse: $UU^* = 1$. These relations show that U commutes with U^*; thus U is normal and Theorem 8 applies, proving part (a). To prove part (b), let f be an eigenvector of U, with eigenvalue $u : Uf = uf$. It follows that $\| Uf \| = \| uf \| = |u| \, \| f \|$. Since U is isometric, $|u| = 1$. □

Our first proof of the spectral resolution of self-adjoint mappings is based on the spectral resolution of general linear mappings. This necessitates the application of the fundamental theorem of algebra on the existence of complex roots, which then are shown to be real. The question is inescapable: Is it possible to prove the spectral resolution of self-adjoint mappings without resorting to the fundamental theorem of algebra? The answer is "Yes." The new proof, given below, is in every respect superior to the first proof. Not only does it avoid the fundamental theorem of algebra, but in the case of real symmetric mappings it avoids the use of complex numbers. It gives a variational characterization of eigenvalues that is very useful in estimating the location of eigenvalues; this will be exploited systematically in Chapter 10. Most important, the new proof can be carried over to infinite-dimensional spaces.

Second Proof of Theorem 4. We start by *assuming* that X has an orthonormal basis of eigenvectors of H. We use the representations (26) and (26)' to write

$$\frac{(x, Hx)}{(x, x)} = \frac{\sum a_i z_i^2}{\sum z_i^2}. \tag{36}$$

We arrange the a_i in increasing order:

$$a_1 \leq a_2 \leq \cdots \leq a_n. \tag{36}'$$

It is clear from (36)' that choosing $z_1 \neq 0$ and all the other $z_i, i = 2, \ldots, n = 0$, makes (36) as small as possible. So

$$a_1 = \min_{s \neq 0} \frac{(x, Hx)}{(x, x)}. \tag{37}$$

Similarly,

$$a_n = \min_{x \neq 0} \frac{(x, Hx)}{(x, x)}. \tag{37'}$$

The minimum and maximum, respectively, are taken on at points $x = f$ that are eigenvectors of H with eigenvalues a_1 and a_n, respectively.

We shall show now, *without* using the representation (36), that the minimum problem (37) has a solution and that this solution is an eigenvector of H. From this we shall deduce, by induction, that H has a full set of eigenvectors.

The quotient (36) is called the *Rayleigh quotient of* H and is abbreviated by $R = R_H$. The numerator is abbreviated, see (6), as q; we shall denote the denominator by p,

$$R(x) = \frac{q(x)}{p(x)} = \frac{(x, Hx)}{(x, x)}.$$

Since H is self-adjoint, by (21)′ R is real-valued; furthermore, R is a homogeneous function of x of degree zero, that is, for every scalar k,

$$R(kx) = R(x).$$

Therefore in seeking its maximum or minimum, it suffices to confine the search to the unit sphere $\| x \| = 1$. In Chapter 7, Theorem 15, we have shown that in a finite-dimensional Euclidean space X, every sequence of vectors on the unit sphere has a convergent subsequence. It follows that $R(x)$ takes on its minimum at some point of the unit sphere; call this point f. Let g be any other vector and t be a real variable; $R(f + tg)$ is the quotient of two quadratic functions of t.

Using the self-adjointness of H and the conjugate symmetry of the scalar product we can express $R(f + tg)$ as

$$R(f + tg) = \frac{(f, Hf) + 2t\mathrm{Re}(g, Hf) + t^2(g, Hg)}{(f, f) + 2t\mathrm{Re}(g, f) + t^2(g, g)} = \frac{q(t)}{p(t)}. \tag{38}$$

Since R achieves its minimum at f, $R(f + tg)$ achieves its minimum at $t = 0$; by calculus its derivative there is zero:

$$\frac{d}{dt} R(f + tg)\Big|_{t=0} = \dot{R} = \frac{\dot{q}p - q\dot{p}}{p^2} = 0.$$

Since $\| f \| = 1, p = 1$; denoting $R(f) = \min R$ by a, we can rewrite the above as

$$\dot{R} = \dot{q} - a\dot{p} = 0. \tag{38'}$$

Using (38), we get readily

$$\dot{q}(f + tg)|_{t=0} = 2\mathrm{Re}(g, Hf),$$
$$\dot{p}(f + tg)|_{t=0} = 2\mathrm{Re}(g, f).$$

Setting this into (38)′ yields

$$2\mathrm{Re}(g, Hf - af) = 0.$$

Replacing g by ig we deduce that for all g in X,

$$2(g, Hf - af) = 0. \tag{39}$$

A vector orthogonal to all vectors g is zero; since (39) holds for all g, it follows that

$$Hf - af = 0, \tag{39'}$$

that is, f is an eigenvector and a is an eigenvalue of H.

We prove now by induction on the dimension n of X that H has a complete set of n orthogonal eigenvectors in X. We consider the orthogonal complement X_1, of f, that is, all x such that

$$(x, f) = 0. \tag{39''}$$

Clearly, $\dim X_1 = \dim X - 1$. We claim that H maps the space X_1 into itself; that is, if $x \in X_1$, then $(Hx, f) = 0$. By self-adjointness and (39)″,

$$(Hx, f) = (x, Hf) = (x, af) = a(x, f) = 0.$$

H restricted to X_1 is self-adjoint: since $\dim X_1 = n - 1$, induction on the dimension of the underlying space shows that H has a full set of eigenvectors on X. These together with f give a full set of n orthogonal eigenvectors of H on X. Instead of arguing by induction we can argue by recursion; we can pose the same minimum problem in X_1 that we have previously posed in the whole space, to minimize

$$\frac{(x, Hx)}{(x, x)}$$

among all nonzero vectors in X_1. Again this minimum value is taken on by some vector $x = f_2$ in X_1, and f_2 is an eigenvector of H. The corresponding eigenvalue is a_2:

$$Hf_2 = a_2 f_2,$$

where a_2 is the second smallest eigenvalue of H. In this fashion we produce successively a full set of eigenvectors. Notice that the jth eigenvector goes with the jth eigenvalue arranged in increasing order.   □

In the argument sketched above, the successive eigenvalues, arranged in increasing order, are calculated through a sequence of restricted minimum problems. We give now a characterization of the jth eigenvalue that makes no reference to the eigenvectors belonging to the previous eigenvalues. This characterization is due to E. Fischer.

Theorem 10.   Let H be a real symmetric linear map of a real Euclidean space X of finite dimension. Denote the eigenvalues of H, arranged in increasing order, by $a_1, \ldots, a_n$. Then

$$a_j = \min_{\dim S = j} \ \max_{x \text{ in } S. \, x \neq 0} \frac{(x, Hx)}{(x, x)}, \tag{40}$$

S linear subspaces of X.

Note. (40) is called the *minmax principle.*

Proof. We shall show that for any linear subspace S of X of dim $S = j$.

$$\max_{x \text{ in } S} \frac{(x, Hx)}{(x, x)} \geq a_j. \tag{41}$$

To prove this it suffices to display a single vector $x \neq 0$ in S for which

$$\frac{(x, Hx)}{(x, x)} \geq a_j. \tag{42}$$

Such an x is one that satisfies the $j - 1$ linear conditions

$$(x, f_i) = 0, \qquad i = 1, \ldots, j - 1, \tag{43}$$

where f_i is the ith eigenvector of H. It follows from Corollary A of Theorem 1 in Chapter 3 that every subspace S of dimension j has a nonzero vector x satisfying $j - 1$ linear conditions (43). The expansion (25) of such an x in terms of the eigenvectors of H contains no contribution from the first $j - 1$ eigenvectors; that is, in (36), $z_i = 0$ for $i < j$. It follows then from (36) that for such x, (42) holds. This completes the proof of (41).

To complete the proof of Theorem 10 we have to exhibit a single subspace S of dimension j such that

$$a_j \geq \frac{(x, Hx)}{(x, x)} \tag{44}$$

holds for all x in S. Such a subspace is the space spanned by $f_1, \ldots, f_j$. Every x in this space is of form $\sum_1^j z_i f_i$; since $a_i \leq a_j$ for $i \leq j$, inequality (44) follows from (36). □

The calculations and arguments presented above show an important property of the Rayleigh quotient:

(i) Every eigenvector h of H is a *critical point* of R_H; that is, the first derivatives of $R_H(x)$ are zero when x is an eigenvector of H. Conversely, the eigenvectors are the only critical points of $R_H(x)$.

(ii) The value of the Rayleigh quotient at an eigenvector f is the corresponding eigenvalue of H:

$$R_H(f) = a \qquad \text{when} \qquad Hf = af.$$

This observation has the following important consequence:
Suppose g is an approximation of an eigenvector f within a deviation of ϵ:

$$\| g - f \| \leq \epsilon. \tag{45}$$

Then $R_H(g)$ is an approximation of the eigenvalue a within $o(\epsilon^2)$:

$$|R_H(g) - a| \leq 0(\epsilon^2). \tag{45'}$$

This result is a direct consequence of the Taylor approximation of the function $R_H(x)$ near the point $x = f$.

The estimate (45)' is very useful for devising numerical methods to calculate the eigenvalues of matrices.

We now give a useful extension of the variational characterization of the eigenvalues of a self-adjoint mapping. In a Euclidean space X, real or complex, we consider two self-adjoint mappings, H and M; we assume that the second one, M, is *positive*.

Definition. A self-adjoint mapping M of a Euclidean space X into itself is called *positive* if for all nonzero x in X

$$(x, Mx) > 0.$$

It follows from the definition and properties of scalar product that the identity I is positive. There are many others; these will be studied systematically in Chapter 10.

We now form a generalization of the Rayleigh quotient:

$$R_{H.M}(x) = \frac{(x, Hx)}{(x, Mx)}. \tag{46}$$

Note that when $M = I$, we are back at the old Rayleigh quotient. We now pose for the generalized Rayleigh quotient the same minimum problem that we posed before for the original Rayleigh quotient: Minimize $R_{H,M}(x)$, that is, find a nonzero vector x that solves

$$\min \frac{(x, Hx)}{(x, Mx)}. \tag{47}$$

EXERCISE 5.   (a) Show that the minimum problem (47) has a nonzero solution f.
(b) Show that a solution f of the minimum problem (47) satisfies the equation

$$Hf = bMf, \tag{48}$$

where the scalar b is the value of the minimum (47).
(c) Show that the constrained minimum problem

$$\min_{(y.Mf)=0} \frac{(y, Hy)}{(y, My)} \tag{47'}$$

has a nonzero solution g.
(d) Show that a solution g of the minimum problem (47)′ satisfies the equation

$$Hg = cMg, \tag{48'}$$

where the scalar c is the value of the minimum (47)′.

Theorem 11.   Let X be a finite-dimensional Euclidean space, let H and M be two self-adjoint mappings of X into itself, and let M be positive. Then there exists a basis $f_1, \ldots, f_n$ of X where each f_i satisfies an equation of the form

$$Hf_i = b_i Mf_i, \qquad b_j \text{ real} \tag{49}$$

and

$$(f_i, Mf_j) = 0 \qquad \text{for } i \neq j.$$

EXERCISE 6.   Prove Theorem 11.

EXERCISE 7.   Characterize the numbers b_i in Theorem 11 by a minimax principle similar to (40).
The following useful result is an immediate consequence of Theorem 11.

Theorem 11′.   Let H and M be self-adjoint, M positive. Then all the eigenvalues of $M^{-1}H$ are real. If H is positive, all eigenvalues of $M^{-1}H$ are positive.

EXERCISE 8. Prove Theorem 11'.

EXERCISE 9. Give an example to show that Theorem 11' is false if M is not positive.

We recall from formula (32)' of Chapter 7 the definition of the norm of a linear mapping A of a Euclidean space X into itself.

$$\| A \| = \max \| Ax \|, \| x \| = 1.$$

When the mapping is normal, that is, commutes with its adjoint, we can express its norm as follows.

Theorem 12. Suppose N is a normal mapping of a Euclidean space X into itself. Then

$$\| N \| = \max |n_j|, \tag{50}$$

where the n_j are the eigenvalues of N.

EXERCISE 10. Prove Theorem 12. (*Hint*: Use Theorem 8.)

EXERCISE 11. We define the cyclic shift mapping S, acting on vectors in $\mathbb{C}^n$, by $S(a_1, a_2, \ldots, a_n) = (a_n, a_1, \ldots, a_{n-1})$.

(a) Prove that S is an isometry in the Euclidean norm.
(b) Determine the eigenvalues and eigenvectors of S.
(c) Verify that the eigenvectors are orthogonal.

Remark. The expansion of a vector v in terms of the eigenvectors of S is called the *finite Fourier transform of v*. See Appendix 9.

Theorem 13. Let A be a linear mapping of a finite-dimensional Euclidean space X into another finite-dimensional Euclidean space U. The norm $\| A \|$ of A equals the square root of the largest eigenvalue of A^*A.

Proof. $\| Ax \|^2 = (Ax, Ax) = (x, A^*Ax)$. According to the Schwarz inequality, the right-hand side is $\leq \| x \| \| A^*Ax \|$. It follows that for unit vectors x, $\| x \| = 1$,

$$\| Ax \|^2 \leq \| A^*Ax \|. \tag{51}$$

A^*A is a self-adjoint mapping; according to formula (37)', we have

$$\max_{\|x\|=1} \| A^*Ax \| = a_{max},$$

where a_{max} is the largest eigenvalue of A^*A. Combining this with (50), we conclude that $\| A \|^2 \le a_{max}$. To show that equality holds, we note that for the eigenvector f of A^*A, $A^*Af = a_{max}f$ and so in the Schwarz inequality which gave (51), the sign of equality holds. $\square$

EXERCISE 12.   (i) What is the norm of the matrix

$$A = \begin{pmatrix} 1 & 2 \\ 0 & 3 \end{pmatrix}$$

in the standard Euclidean structure?

(ii) Compare the value of $\| A \|$ with the upper and lower bounds of $\| A \|$ asked for in Exercise 19 of Chapter 7.

EXERCISE 13.   What is the norm of the matrix

$$\begin{pmatrix} 1 & 0 & -1 \\ 2 & 3 & 0 \end{pmatrix}$$

in the standard Euclidean structures of $\mathbb{R}^2$ and $\mathbb{R}^3$.

CHAPTER 9

Calculus of Vector- and Matrix-Valued Functions

In Section 1 of this chapter we develop the calculus of vector- and matrix-valued functions. There are two ways of going about it: by representing vectors and matrices in terms of their components and entries with respect to some basis and using the calculus of number-valued functions or by redoing the theory in the context of linear spaces. Here we opt for the second approach, because of its simplicity and because it is the conceptual way to think about the subject; but we reserve the right to go to components when necessary.

In what follows, the field of scalars is the real or complex numbers. In Chapter 7 we defined the length of vectors and the norm of matrices; see (1) and (32). This made it possible to define convergence of sequences as follows.

(i) A sequence x_k of vectors in $\mathbb{R}^n$ converges to the vector x if

$$\lim_{k \to \infty} ||x_k - x|| = 0.$$

(ii) A sequence A_k of $n \times n$ matrices converges to A if

$$\lim_{k \to \infty} ||A_k - A|| = 0.$$

We could have defined convergence of sequences of vectors and matrices, without introducing the notion of size, by requiring that each component of x_k tend to the corresponding component of x and, in the case of matrices, that each entry of A_k tend to the corresponding entry of A. But using the notion of size introduces a simplification in notation and thinking, and is an aid in proof. There is more about size in Chapter 14 and 15.

---

1. THE CALCULUS OF VECTOR- AND MATRIX-VALUED FUNCTIONS

Let $x(t)$ be a vector-valued function of the real variable t, defined, say, for t in $(0, 1)$. We say that $x(t)$ is *continuous* at t_0 if

$$\lim_{t \to t_0} \|x(t) - x(t_0)\| = 0. \tag{1}$$

We say that x is *differentiable* at t_0, with derivative $\dot{x}(t_0)$, if

$$\lim_{h \to 0} \left\| \frac{x(t_0 + h) - x(t_0)}{h} - \dot{x}(t_0) \right\| = 0. \tag{1}'$$

Here we have abbreviated the derivative by a dot:

$$\dot{x}(t) = \frac{d}{dt} x(t).$$

The notion of continuity and differentiability of matrix-valued functions is defined similarly.

The *fundamental lemma* of differentiation holds for vector- and matrix-valued functions.

Theorem 1. If $\dot{x}(t) = 0$ for all t in $(0, 1)$, then $x(t)$ is constant.

EXERCISE I. Prove the fundamental lemma for vector valued functions. (*Hint:* Show that for every vector y, $(x(t), y)$ is constant.)

We turn to the **rules of differentiation**. *Linearity.* (**i**) The sum of two differentiable functions is differentiable, and

$$\frac{d}{dt}(x + y) = \frac{d}{dt} x + \frac{d}{dt} y.$$

(**ii**) The constant multiple of a differentiable function is differentiable, and

$$\frac{d}{dt}(kx(t)) = k \frac{d}{dt} x(t).$$

Similarly for matrix-valued differentiable functions,

(**iii**) $$\frac{d}{dt}(A(t) + B(t)) = \frac{d}{dt} A(t) + \frac{d}{dt} B(t).$$

(iv) If A is independent of t, then we have

$$\frac{d}{dt}AB(t) = A\frac{d}{dt}B(t).$$

The proof is the same as in scalar calculus.

For vector- and matrix-valued functions there is a further manifestation of the linearity of the derivative: Suppose that l is a fixed linear function defined on $\mathbb{R}^n$ and that $x(t)$ is a differentiable vector-valued function. Then $l(x(t))$ is a differentiable function, and

$$\frac{d}{dt}l(x(t)) = l\left(\frac{d}{dt}x(t)\right). \tag{2}$$

The same result applies to linear functions of matrices. In particular the trace, defined by (35) in Chapter 5, is such a linear function. So we have, for every differentiable matrix function $A(t)$, that

$$\frac{d}{dt}\mathrm{tr}(A(t)) = \mathrm{tr}\left(\frac{d}{dt}A(t)\right). \tag{2$'$}$$

The rule (sometimes called the Leibniz rule) for differentiating a *product* is the same as in elementary calculus. Here, however, we have at least five kinds of products and therefore five versions of rules.

Product Rules

(i) The product of a scalar function and a vector function:

$$\frac{d}{dt}[k(t)x(t)] = \left(\frac{dk}{dt}\right)x(t) + k(t)\frac{d}{dt}x(t).$$

(ii) The product of a matrix function times a vector function:

$$\frac{d}{dt}[A(t)x(t)] = \left(\frac{d}{dt}A(t)\right)x(t) + A(t)\frac{d}{dt}x(t).$$

(iii) The product of two matrix-valued functions:

$$\frac{d}{dt}[A(t)B(t)] = \left[\frac{d}{dt}A(t)\right]B(t) + A(t)\left[\frac{d}{dt}B(t)\right].$$

(iv) The product of a scalar-valued and a matrix-valued function:

$$\frac{d}{dt}[k(t)A(t)] = \left[\frac{dk}{dt}\right]A(t) + k(t)\frac{d}{dt}A(t).$$

(v) The scalar product of two vector functions:

$$\frac{d}{dt}(y(t), x(t)) = \left(\frac{d}{dt}y(t), x(t)\right) + \left(y(t), \frac{d}{dt}x(t)\right).$$

The proof of all these is the same as in the case of ordinary numerical functions.
The rule for differentiating the inverse of a matrix function resembles the calculus rule for differentiating the reciprocal of a function, with one subtle twist.

Theorem 2. Let A(t) be a matrix-valued function, differentiable and invertible. Then $A^{-1}(t)$ also is differentiable, and

$$\frac{d}{dt}A^{-1} = -A^{-1}\left(\frac{d}{dt}A\right)A^{-1}. \tag{3}$$

Proof. The following identity is easily verified:

$$A^{-1}(t+h) - A^{-1}(t) = A^{-1}(t+h)[A(t) - A(t+h)]A^{-1}(t).$$

Dividing both sides by h and letting $h \to 0$ yields (3). ☐

EXERCISE 2. Derive formula (3) using product rule (iii).

The chain rule of calculus says that if f and a are scalar-valued differentiable functions, so is their composite, $f(a(t))$, and

$$\frac{d}{dt}f(a(t)) = f'(a)\frac{da}{dt}, \tag{4}$$

where f' is the derivative of f. We show that the chain rule *fails* for matrix-valued functions. Take $f(a) = a^2$; by the product rule,

$$\frac{d}{dt}A^2 = A\frac{d}{dt}A + \left(\frac{d}{dt}A\right)A,$$

certainly *not* the same as (4). More generally, we claim that for any positive integer power k,

$$\frac{d}{dt}A^k = \dot{A}A^{k-1} + A\dot{A}A^{k-2} + \cdots + A^{k-1}\dot{A}. \tag{5}$$

This is easily proved by induction: We write

$$A^k = AA^{k-1}$$

and apply the product rule

$$\frac{d}{dt}A^k = \dot{A}A^{k-1} + A\frac{d}{dt}A^{k-1}.$$

Theorem 3. Let p be any polynomial, let $A(t)$ be a square matrix-valued function that is differentiable; denote the derivative of A with respect to t as $\dot{A}$.

(a) If for a particular value of t the matrices $A(t)$ and $\dot{A}(t)$ commute, then the chain rule in form (4) holds as t:

$$\frac{d}{dt}p(A) = p'(A)\dot{A}. \tag{6}$$

(b) Even if $A(t)$ and $\dot{A}(t)$ do not commute, a trace of the chain rule remains:

$$\frac{d}{dt}\operatorname{tr}p(A) = \operatorname{tr}(p'(A)\dot{A}). \tag{6'}$$

Proof. Suppose A and $\dot{A}$ commute; then (5) can be rewritten as

$$\frac{d}{dt}A^k = kA^{k-1}\dot{A}.$$

This is formula (6) for $p(s) = s^k$; since all polynomials are linear combinations of powers, using the linearity of differentiation we deduce (6) for all polynomials.

For noncommuting A and $\dot{A}$ we take the trace of (5). According to Theorem 6 of Chapter 5, trace is commutative:

$$\operatorname{tr}(A^j\dot{A}A^{k-j-1}) = \operatorname{tr}(A^{k-j-1}A^j\dot{A}) = \operatorname{tr}(A^{k-1}\dot{A}).$$

So we deduce that

$$\operatorname{tr}\frac{d}{dt}A^k = k\operatorname{tr}(A^{k-1}\dot{A}).$$

Since trace and differentiation commute [see (2)'], we deduce formula (6)' for $p(s) = s^k$. The extension to arbitrary polynomials goes as before. $\square$

We extend now the product rule to multilinear functions $M(a_1, \ldots, a_k)$. Suppose $x_1, \ldots, x_k$ are differentiable vector functions. Then $M(x_1, \ldots, x_k)$ is differentiable, and

$$\frac{d}{dt}M(x_1, \ldots, x_k) = M(\dot{x}_1, x_2, \ldots, x_k) + \cdots + M(x_1, \ldots, x_{k-1}, \dot{x}_k). \tag{7}$$

The proof is straightforward: since M is multilinear,

$$
\begin{aligned}
M(x_1(t+h), &\ldots, x_k(t+h)) - M(x_1(t), \ldots, x_k(t)) \\
&= M(x_1(t+h) - x_1(t), x_2(t+h), \ldots, x_k(t+h)) \\
&\quad + M(x_1(t), x_2(t+h) - x_2(t), x_3(t+h), \ldots, x_k(t+h)) \\
&\quad + \cdots + M(x_1(t), \ldots, x_{k-1}(t), x_k(t+h) - x_k(t)).
\end{aligned}
$$

Dividing by h and letting h tend to zero gives (7).

The most important application of (7) is to the function D, the determinant, defined in Chapter 5:

$$
\frac{d}{dt} D(x_1, \ldots, x_n) = D(\dot{x}_1, x_2, \ldots, x_n) + \cdots + D(x_1, \ldots, x_{n-1}, \dot{x}_n). \tag{8}
$$

We now show how to recast this formula to involve a matrix X itself, not its columns. We start with the case when $X(0) = I$, that is, $x_j(0) = e_j$. In this case the determinants on the right in (8) are easily evaluated at $t = 0$:

$$
\begin{aligned}
D(\dot{x}_1(0), e_2, \ldots, e_n) &= \dot{x}_{11}(0) \\
D(e_1, \dot{x}_2(0), e_3, \ldots, e_n) &= \dot{x}_{22}(0) \\
&\vdots \\
D(e_1, \ldots, e_{n-1}, \dot{x}_n(0)) &= \dot{x}_{nn}(0).
\end{aligned}
$$

Setting this into (8) we deduce that if $X(t)$ is a differentiable matrix-valued function and $X(0) = I$, then

$$
\frac{d}{dt} \det X(t)\big|_{t=0} = \operatorname{tr}\dot{X}(0). \tag{8$'$}
$$

Suppose $Y(t)$ is a differentiable square matrix-valued function, which is *invertible*. We define $X(t)$ as $Y(0)^{-1}Y(t)$, and write

$$
Y(t) = Y(0)X(t); \tag{9}
$$

clearly, $X(0) = I$, so formula (8)$'$ is applicable. Taking the determinant of (9), we get by the product rule for determinants that

$$
\det Y(t) = \det Y(0) \det X(t). \tag{9$'$}
$$

Setting (9) and (9)$'$ into (8)$'$, we get

$$
[\det Y(0)]^{-1} \frac{d}{dt} \det Y(t)\big|_{t=0} = \operatorname{tr}[Y^{-1}(0)\dot{Y}(0)].
$$

We can rewrite this as

$$\frac{d}{dt} \log \det \mathbf{Y}(t)|_{t=0} = \text{tr}[\mathbf{Y}^{-1}(t)\dot{\mathbf{Y}}(t)]_{t=0}.$$

Since now there is nothing special about $t = 0$, this relation holds for all t:

Theorem 4.   Let $\mathbf{Y}(t)$ be a differentiable square matrix-valued function. Then for those values of t for which $\mathbf{Y}(t)$ is invertible,

$$\frac{d}{dt} \log \det \mathbf{Y} = \text{tr}\left(\mathbf{Y}^{-1}\frac{d}{dt}\mathbf{Y}\right). \tag{10}$$

The importance of this result lies in the connection it establishes between determinant and trace.

So far we have defined $f(\mathbf{A})$ for matrix arguments when f is a polynomial. We show now an example of a nonpolynomial f for which $f(\mathbf{A})$ can be defined. We take $f(s) = e^s$, defined by the Taylor series

$$e^s = \sum_0^{\infty} \frac{s^k}{k!}. \tag{11}$$

We claim that the Taylor series also serves to define $e^{\mathbf{A}}$ for any square matrix $\mathbf{A}$:

$$e^{\mathbf{A}} = \sum_0^{\infty} \frac{\mathbf{A}^k}{k!}. \tag{11'}$$

The proof of convergence is the same as in the scalar case; it boils down to showing that the difference of the partial sums tends to zero. That is, denote by $e_m(\mathbf{A})$ the mth partial sum:

$$e_m(\mathbf{A}) = \sum_0^m \frac{\mathbf{A}^k}{k!}; \tag{12}$$

then

$$e_m(\mathbf{A}) - e_l(\mathbf{A}) = \sum_{j+1}^m \frac{\mathbf{A}^k}{k!}. \tag{13}$$

Using the multiplicative and additive inequalities for the norm of matrices developed in Chapter 7, Theorem 14, we deduce that

$$||e_m(\mathbf{A}) - e_l(\mathbf{A})|| \le \sum_{j+1}^m \frac{||\mathbf{A}||^k}{k!}. \tag{13'}$$

We are now back in the scalar case, and therefore can estimate the right-hand side and assert that as l and m tend to infinity, the right-hand side of (13) tends to zero, uniformly for all matrices whose norm $\|A\|$ is less than any preassigned constant.

The matrix exponential function has some but not all properties of the scalar exponential function.

Theorem 5.   **(a)** If A and B are commuting square matrices,

$$e^{A+B} = e^A e^B.$$

(b) If A and B do not commute, then in general

$$e^{A+B} \neq e^A e^B.$$

(c) If $A(t)$ depends differentiably on t, so does $e^{A(t)}$.
(d) If for a particular value of t, $A(t)$ and $\dot{A}(t)$ commute, then $(d/dt)e^A = e^A \dot{A}$.
(e) If A is anti-self-adjoint, $A^* = -A$, then e^A is unitary.

Proof. Part (a) follows from the definition (11)′ of e^{A+B}, after $(A+B)^k$ is expressed as $\sum \binom{k}{j} A^j B^{k-j}$, valid for *commuting* variables.

That commutativity is used essentially in the proof of part (a) makes part (b) plausible. We shall not make the statement more precise; we content ourselves with giving a single example:

$$A = \begin{pmatrix} 0 & 1 \\ 0 & 0 \end{pmatrix}, \qquad B = \begin{pmatrix} 0 & 0 \\ 1 & 0 \end{pmatrix}.$$

It is easy to see that $A^2 = 0$, $B^2 = 0$, so by definition (11)′,

$$e^A = I + A = \begin{pmatrix} 1 & 1 \\ 0 & 1 \end{pmatrix}, \qquad e^B = I + B = \begin{pmatrix} 1 & 0 \\ 1 & 1 \end{pmatrix}.$$

A brief calculation shows that

$$e^A e^B = \begin{pmatrix} 2 & 1 \\ 1 & 1 \end{pmatrix}, \qquad e^B e^A = \begin{pmatrix} 1 & 1 \\ 1 & 2 \end{pmatrix};$$

since these products are different, at least one must differ from e^{A+B}; actually, both do.

EXERCISE 3.   Calculate

$$e^{A+B} = \exp \begin{pmatrix} 0 & 1 \\ 1 & 0 \end{pmatrix}.$$

To prove (c) we rely on the following matrix analogue of an important property of differentiation: Let $\{E_m(t)\}$ be a sequence of differentiable matrix-valued functions defined on an interval, with these properties:

(i) $E_m(t)$ converges uniformly to a limit function $E(t)$.
(ii) The derivatives $\dot{E}_m(t)$ converge uniformly to a limit function $F(t)$.

Conclusion: E is differentiable, and $\dot{E} = F$.

EXERCISE 4.   Prove the proposition stated in the Conclusion.

We apply the same principle to $E_m(t) = e_m(A(t))$. We have already shown that $E_m(t)$ tends uniformly to $e^{A(t)}$; a similar argument shows that $\dot{E}_m(t)$ converges.

EXERCISE 5.   Carry out the details of the argument that $\dot{E}_m(t)$ converges.

Part (d) of Theorem 5 follows from the explicit formula for $(d/dt)e^{A(t)}$, obtained by differentiating the series $(11)'$ termwise.

To prove part (e) we start with the definition $(11)'$ of e^A. Since forming the adjoint is a linear and continuous operation, we can take the adjoint of the infinite series in $(11)'$ term by term:

$$(e^A)^* = \sum_0^\infty \left(\frac{A^k}{k!}\right)^* = \sum \frac{(A^*)^k}{k!} = e^{A^*} = e^{-A}.$$

It follows, using part (a), that

$$(e^A)^* e^A = e^{-A}e^A = e^0 = I.$$

According to formula (45) of Chapter 7, this shows that e^A is unitary.   □

EXERCISE 6.   Apply formula (10) to $Y(t) = e^{At}$ and show that

$$\det e^A = e A.$$

EXERCISE 7.   Prove that all eigenvalues of e^A are of the form e^a, a an eigenvalue of A. *Hint*: Use Theorem 4 of Chapter 6, along with Theorem 6 below.

We remind the reader that for *self-adjoint* matrices H we have already in Chapter 8 defined $f(H)$ for a broad class of functions; see formula $(33)'$.

2. SIMPLE EIGENVALUES OF A MATRIX

In this section we shall study the manner in which the eigenvalues of a matrix depend on the matrix. We take the field of scalars to be $\mathbb{C}$.

Theorem 6. The eigenvalues depend continuously on the matrix in the following sense: If $\{A_m\}$ is a convergent sequence of square matrices, in the sense that all entries of A_m converge to the corresponding entry of A, then the set of eigenvalues of A_m converges to the set of eigenvalues of A. That is, for every $\epsilon > 0$ there is a k such that all eigenvalues of A_m are, for $m > k$, contained in discs of radius ϵ centered at the eigenvalues of A.

Proof. The eigenvalues of A_m are the roots of the characteristic polynomial $p_m(s) = \det(sI - A_m)$. Since A_m tends to A, all entries of A_m tend to the corresponding entries of A; from this it follows that the coefficients of p_m tend to the coefficients of p. Since the roots of polynomials depend continuously on the coefficients, Theorem 6 follows.   □

Next we investigate the differentiability of the dependence of the eigenvalues on the matrix. There are several ways of formulating such a result, for example, in the following theorem.

Theorem 7. Let $A(t)$ be a differentiable square matrix-valued function of the real variable t. Suppose that $A(0)$ has an eigenvalue a_0 of multiplicity one, in the sense that a_0 is a simple root of the characteristic polynomial of $A(0)$. Then for t small enough, $A(t)$ has an eigenvalue $a(t)$ that depends differentiably on t, and which equals a_0 at zero, that is, $a(0) = a_0$.

Proof. The characteristic polynomial of $A(t)$ is

$$\det(sI - A(t)) = p(s, t),$$

a polynomial of degree n in s whose coefficients are differentiable functions of t. The assumption that a_0 is a simple root of $A(0)$ means that

$$p(a_0, 0) = 0, \qquad \frac{\partial}{\partial s} p(s, 0)|_{s=a_0} \neq 0.$$

According to the implicit function theorem, under these conditions the equation $p(s, t) = 0$ has a solution $s = a(t)$ in a neighborhood of $t = 0$ that depends differentiably on t.   □

Next we show that under the same conditions as in Theorem 7, the eigenvector pertaining to the eigenvalue $a(t)$ can be chosen to depend differentiably on t. We say "can be chosen" because an eigenvector is determined only up to a scalar factor; by inserting a scalar factor $k(t)$ that is a nondifferentiable function of t we could, with malice aforethought, spoil differentiability (and even continuity).

Theorem 8. Let $A(t)$ be a differentiable matrix-valued function of t, $a(t)$ an eigenvalue of $A(t)$ of multiplicity one. Then we can choose an eigenvector $h(t)$ of $A(t)$ pertaining to the eigenvalue $a(t)$ to depend differentiably on t.

Proof. We need the following lemma.   ☐

Lemma 9.   Let A be an $n \times n$ matrix, p its characteristic polynomial, a some simple root of p. Then at least one of the $(n-1) \times (n-1)$ principal minors of $A - aI$ has nonzero determinant, where the ith principal minor is the matrix remaining when the ith row and ith column of A are removed.

Proof. We may, at the cost of subtracting aI from A, take the eigenvalue to be zero. The condition that 0 is a simple root of $p(s)$ means that $p(0) = 0$; $(dp/ds)(0) \neq 0$. To compute the derivative of p we denote by $c_1, \dots, c_n$ the columns of A, and by $e_1, \dots, e_n$ the unit vectors. Then

$$sI - A = (se_1 - c_1, se_2 - c_2, \dots, se_n - c_n).$$

Now we use formula (8) for the derivative of a determinant:

$$\frac{dp}{ds}(0) = \frac{d}{ds}\det(sI - A)|_{s=0}$$
$$= \det(e_1, -c_2, \dots, -c_n) + \cdots + \det(-c_1, -c_2, \dots, c_{n-1}, e_n).$$

Using Lemma 2 of Chapter 5 for the determinants on the right-hand side we see that $(dp/ds)(0)$ is $(-1)^{n-1}$ times the sum of the determinants of the $(n-1) \times (n-1)$ principal minors. Since $(dp/ds)(0) \neq 0$, at least one of the determinants of these principal minors is nonzero.   ☐

Let A be a matrix as in Lemma 9 and take the eigenvalue a to be zero. Then one of the principal $(n-1) \times (n-1)$ minors of A, say the ith, has nonzero determinant. We claim that the ith component of an eigenvector h of A pertaining to the eigenvalue a is nonzero. Suppose it were denote by $h^{(i)}$ the vector obtained from h by omitting the ith component, and by A_{ji} the ith principal minor of A. Then $h^{(i)}$ satisfies

$$A_{ii}h^{(i)} = 0. \tag{14}$$

Since A_{ii} has determinant not equal to 0, A_{ii} is, according to Theorem 5 of Chapter 5, invertible. But then according to (14), $h^{(i)} = 0$. If the ith component were zero, that would make $h = 0$, a contradiction, since an eigenvector is not equal to 0. Having shown that the ith component of h is not equal to 0, we set it equal to 1 as a way of normalizing h. For the remaining components we have now an inhomogeneous system of equations:

$$A_{ii}h^{(i)} = c^{(i)}, \tag{14}'$$

where $c^{(i)}$ is -1 times the ith column of A, with the ith component removed. So

$$h^{(i)} = A_{ii}^{-1}c^{(i)}. \tag{15}$$

The matrix $A(0)$ and the eigenvalue $a(0)$ of Theorem 8 satisfy the hypothesis of Lemma 9. Then a matrix $A_{ii}(0)$ is invertible; since $A(t)$ depends continuously on t, it follows from Theorem 6 that $A_{ii}(t) - a(t)I$ is invertible for t small; for such small values of t we set the ith component of $h(t)$ equal to 1, and determine the rest of h by formula (15):

$$h^i(t) = A_{ii}^{-1}(t)\, c^i(t). \tag{16}$$

Since all terms on the right depend differentiably on t, so does $h^i(t)$. This concludes the proof of Theorem 8. □

We now extend Lemma 9 to the case when the characteristic polynomial has multiple roots and prove the following results.

Lemma 10.   Let A be an $n \times n$ matrix, p its characteristic polynomial. Let a be some root of p of multiplicity k. Then the nullspace of $(A - aI)$ is at most k-dimensional.

Proof. We may, without loss of generality, take $a = 0$. That 0 is a root of multiplicity k means that

$$p(0) = \cdots = \frac{d^{k-1}}{ds^{k-1}}p(0) = 0, \qquad \frac{d^k}{ds^k}p(0) \neq 0.$$

Proceeding as in the proof of Lemma 9, that is, differentiating k times $\det(sI - A)$, we can express the kth derivative of p at 0 as a sum of determinants of principal minors of order $(n - k) \times (n - k)$. Since the kth derivative is not equal to 0, it follows that at least one of these determinants is nonzero, say the minor obtained by removing from A the ith rows and columns, $i = 1, \ldots, k$. Denote this minor as $A^{(k)}$. We claim that the nullspace N of A contains no vector other than zero whose first k components are all zero. For, suppose h is such a vector; denote by $h^{(k)}$ the vector obtained from h by removing the first k components. Since $Ah = 0$, this shortened vector satisfies the equation

$$A^{(k)} h^{(k)} = 0. \tag{17}$$

Since $\det A^{(k)} \neq 0$, $A^{(k)}$ is invertible; therefore it follows from (17) that $h^{(k)} = 0$. Since the components that were removed are zero, it follows that $h = 0$, a contradiction.

It follows now that $\dim N \leq k$; for, if the dimension of N were greater than k, it would follow from Corollary A of Theorem 1 in Chapter 3 that the k linear conditions $h_1 = 0, \ldots, h_k = 0$ are satisfied by some nonzero vector h in N. Having just shown that no nonzero vector h in N satisfies these conditions, we conclude that $\dim N \leq k$. □

Lemma 10 can be used to prove Theorem 11, announced in Chapter 6.

Theorem 11. Let A be an $n \times n$ matrix, p its characteristic polynomial, a some root of p of multiplicity k. The dimension of the space of generalized eigenvectors of A pertaining to the eigenvalue a is k.

Proof. We saw in Chapter 6 that the space of generalized eigenvectors is the nullspace of $(A - aI)^d$, where d is the index of the eigenvalue a. We take $a = 0$. The characteristic polynomial p_d of A^d can be expressed in terms of the characteristic polynomial p of A as follows:

$$sI - A^d = \prod_0^{d-1} (s^{1/d}I - \omega^j A),$$

where ω is a primitive dth root of unity. Taking determinants and using the multiplicative property of determinants we get

$$p_d(s) = \det(sI - A^d) = \prod_0^{d-1} \det(s^{1/d}I - \omega^j A)$$

$$= \pm \prod_0^{d-1} \det(\omega^{-j}s^{1/d}I - A) = \pm \prod_0^{d-1} p(\omega^{-j}s^{1/d}). \tag{18}$$

Since $a = 0$ is a root of p of multiplicity k, it follows that

$$p(s) \sim \text{const. } s^k$$

as s tends to zero. It follows from (18) that as s tends to zero,

$$p_d(s) \sim \text{const. } s^k;$$

therefore p_d also has a root of multiplicity k at 0. It follows then from Lemma 10 that the nullspace of A^d is *at most* k dimensional.

To show that equality holds, we argue as follows. Denote the roots of p as $a_1, \ldots, a_j$ and their multiplicities as $k_1, \ldots, k_j$. Since p is a polynomial of degree n, according to the fundamental theorem of algebra,

$$\sum k_i = n. \tag{19}$$

Denote by N_i the space of generalized eigenvectors of A pertaining to the eigenvalue a_i. According to Theorem 7, the spectral theorem, of Chapter 6, every vector can be decomposed as a sum of generalized eigenvectors: $\mathbb{C}^n = N_1 \oplus \cdots \oplus N_j$. It follows that

$$n = \sum \dim N_i. \tag{20}$$

N_i is the nullspace of $(A - a_i I)^{d_i}$; we have already shown that

$$\dim N_i \le k_i. \tag{21}$$

Setting this into (20), we obtain

$$n \le \sum k_i.$$

Comparing this with (19), we conclude that in all inequalities (21) the sign of equality holds.   □

We show next how to actually calculate the derivative of the eigenvalue $a(t)$ and the eigenvector $h(t)$ of a matrix function $A(t)$ when $a(t)$ is a simple root of the characteristic polynomial of $A(t)$. We start with the eigenvector equation

$$Ah = ah. \tag{22}$$

We have seen in Chapter 5 that the transpose A^T of a matrix A has the same determinant as A. It follows that A and A^T have the same characteristic polynomial. Therefore if a is an eigenvalue of A, it is also an eigenvalue of A^T:

$$A^T l = al. \tag{22$'$}$$

Since a is a simple root of the characteristic polynomial of A^T, by Theorem 11 the space of eigenvectors satisfying (22)$'$ is one dimensional, and there are no generalized eigenvectors.

Now differentiate (22) with respect to t:

$$\dot{A}h + A\dot{h} = \dot{a}h + a\dot{h}. \tag{23}$$

Let l act on (23):

$$(l, \dot{A}h) + (l, A\dot{h}) = \dot{a}(l, h) + a(l, \dot{h}). \tag{23$'$}$$

We use now the definition of the transpose, equation (9) of Chapter 3, to rewrite the second term on the left as $(A^T l, \dot{h})$. Using equation (22)$'$, we can rewrite this as $a(l, \dot{h})$, the same as the second term on the right; after cancellation we are left with

$$(l, \dot{A}h) = \dot{a}(l, h). \tag{24}$$

We claim that $(l, h) \ne 0$, so that (24) can be used to determine $\dot{a}$. Suppose on the contrary that $(l, h) = 0$; we claim that then the equation

$$(A^T - a I)m = l \tag{25}$$

would have a solution m. To see this we appeal to Theorem 2' of Chapter 3, according to which the range of $T = A^T - aI$ consists of those vectors which are annihilated by the vectors in the nullspace of $T^T = A - aI$. These are the eigenvectors of A and are multiples of h. Therefore if $(l, h) = 0$, l would satisfy the criterion of belonging to the range of $A^T - aI$, and equation (25) would have a solution m. This m would be a generalized eigenvector of A^T, contrary to the fact that there aren't any.

Having determined $\dot{a}$ from equation (24), we determine $\dot{h}$ from equation (23), which we rearrange as

$$(A - aI)\dot{h} = (\dot{a} - \dot{A})h. \tag{26}$$

Appealing once more to Theorem 2' of Chapter 3 we note that (26) has a solution $\dot{h}$ if the right-hand side is annihilated by the nullspace of $A^T - aI$. That nullspace consists of multiples of l, and equation (24) is precisely the requirement that it annihilate the right-hand side of (26). Note that equation (26) does not determine $\dot{h}$ uniquely, only up to a multiple of h. That is as it should be, since the eigenvectors $h(t)$ are determined only up to a scalar factor that can be taken as an arbitrary differentiable function of t.

3. MULTIPLE EIGENVALUES

We are now ready to treat multiple eigenvalues. The occurence of generalized eigenvectors is hard to avoid for general matrices and even harder to analyze. For this reason we shall discuss only self-adjoint matrices, because they have no generalized eigenvectors. Even in the self-adjoint case we need additional assumptions to be able to conclude that the eigenvectors of A depend continuously on a parameter t when $A(t)$ is a differentiable function of t. Here is a simple 2×2 example:

$$A = \begin{pmatrix} b & c \\ c & d \end{pmatrix},$$

b, c, d functions of t, so that $c(0) = 0, b(0) = d(0) = 1$. That makes $A(0) = I$, which has 1 as double eigenvalue.

The eigenvalues a of A are the roots of its characteristics polynomial.

$$a = \frac{b + d \pm \sqrt{(b - d)^2 + 4c^2}}{2}.$$

Denote the eigenvector h as $\binom{x}{y}$. The first component of the eigenvalue equation $Ah = ah$ is $bx + cy = ax$, from which

$$\frac{y}{x} = \frac{a - b}{c}.$$

Using the abbreviation $(d - b)/c = k$, we can express

$$\frac{y}{x} = \frac{a - b}{c} = \frac{k + \sqrt{k^2 + 4}}{2}.$$

We choose $k(t) = \sin(t^{-1}), c(t) = \exp(-|t|^{-1})$, and set $b \equiv 1, d = 1 + ck$. Clearly the entries of $A(t)$ are C^∞ functions, yet y/x is discontinuous as $t \to 0$.

Theorem 12 describes an additional condition under which the eigenvectors vary continuously, To arrive at these conditions we shall reverse the procedure employed for matrices with simple eigenvalues: we shall first compute the derivatives of eigenvalues and eigenvectors and prove afterwards that they are differentiable under the additional condition.

Let $A(t)$ be a differentiable function of the real variable t, whose values are selfadjoint matrices. $A^* = A$. Suppose that at $t = 0$, $A(0)$ has a_0 as eigenvalue of multiplicity $k > 1$, that is, a_0 is a k-fold root of the characteristic equation of $A(0)$. According to Theorem 11, the dimension of the generalized eigenspace of $A(0)$ pertaining to the eigenvalue a_0 is k. Since $A(0)$ is self-adjoint, it has no generalized eigenvectors; so the eigenvectors $A(0)h = a_0 h$ form a k-dimensional space which we denote as N.

We take now eigenvectors $h(t)$ and eigenvalues $a(t)$ of $A(t)$, $a(0) = a_0$, presumed to depend differentiably on t. Then the derivatives of h and a satisfy equation (23); set $t = 0$:

$$\dot{A}h + A\dot{h} = \dot{a}h + a\dot{h}. \tag{27}$$

We recall now from Chapter 8 the projection operators entering the spectral resolution; see equations (29), (30), (31), and (32). We denote by P the orthogonal projection onto the eigenspace N of A with eigenvalue $a = a_0$. Since the eigenvectors of A are orthogonal, it follows [see equations (29)–(32)] that

$$PA = aP. \tag{28}$$

Furthermore, eigenvectors h in N satisfy

$$Ph = h. \tag{28}'$$

Now apply P to both sides of (27):

$$P\dot{A}h + PA\dot{h} = \dot{a}Ph + aP\dot{h}.$$

Using (28) and (28)', we get

$$P\dot{A}Ph + aP\dot{h} = \dot{a}h + aP\dot{h}.$$

The second terms on the right- and left-hand sides are equal, so after cancellation we get

$$P\dot{A}Ph = \dot{a}h. \tag{29}$$

Since $A(t)$ is self-adjoint, so is $\dot{A}$; and since P is self-adjoint, so is $P\dot{A}P$. Clearly, $P\dot{A}P$ maps N into itself; equation (29) says that $\dot{a}(0)$ must be one of the eigenvalues of $P\dot{A}P$ on N, and $h(0)$ must be an eigenvector.

Theorem 12. Let $A(t)$ be a differentiable function of the real variable t whose values are self-adjoint matrices. Suppose that at $t = 0$, $A(0)$ has an eigenvalue a_0 of multiplicity $k > 1$. Denote by N the eigenspace of $A(0)$ with eigenvalue a_0, and by P the orthogonal projection onto N. Assume that the self-adjoint mapping $P\dot{A}(0)P$ of N into N has k *distinct* eigenvalues $d_i, i = 1, \ldots, k$. Denote by w_i corresponding normalized eigenvectors. Then for t small enough, $A(t)$ has k eigenvalues $a_j(t), j = 1, \ldots, k$, near a_0, with the following properties:

   (i) $a_i(t)$ depend differentiably on t and tend to a_0 as $t \to 0$.
   (ii) For $t \neq 0$, the $a_j(t)$ are distinct.
   (iii) The corresponding eigenvector $h_j(t)$:

$$A(t)h_j(t) = a_j(t)h_j(t), \tag{30}$$

   can be so normalized that $h_j(t)$ tends to w_j as $t \to 0$.

Proof. For t small enough the characteristic polynomial of $A(t)$ differs little from that of $A(0)$. By hypothesis, the latter has a k-fold root at a_0; it follows that the former have exactly k roots that approach a_0 as $t \to 0$. These roots are the eigenvalues $a_j(t)$ of $A(t)$. According to Theorem 4 of Chapter 8, the corresponsing eigenvectors $h_j(t)$ can be chosen to form an orthonormal set. $\square$

Lemma 13. As $t \to 0$, the distance of each of the normalized eigenvectors $h_j(t)$ from the eigenspace N tends to zero.

Proof. Using the orthogonal projection P onto N, we can reformulate the conclusion as follows:

$$\lim_{t \to 0} \|(I - P)h_j(t)\| = 0, \qquad j = 1, \ldots, k. \tag{31}$$

To show this, we use the fact that as $t \to 0$, $A(t) \to A(0)$ and $a_j(t) \to a_0$; since $\|h_j(t)\| = 1$, we deduce from equation (30) that

$$A(0)h_j(t) = a_0h_j(t) + \epsilon(t), \tag{32}$$

where $\epsilon(t)$ denotes a vector function that tends to zero as $t \to 0$. Since N consists of eigenvectors of $A(0)$, and P projects any vector onto N.

$$A(0)Ph_j(t) = a_0 Ph_j(t). \tag{32}'$$

We subtract $(32)'$ from (32) and get

$$A(0)(I - P)h_j(t) = a_0(I - P)h_j(t) + \epsilon(t). \tag{33}$$

Now suppose (31) were false; then there would be a positive number d and a sequence of $t \to 0$ such that $\|(I - P)h_j(t)\| > d$. We have shown in Chapter 7 that there is a subsequence of t for which $(I - P)h_j(t)$ tends to a limit h; this limit has norm $\geq d$. It follows from (33) that this limit satisfies

$$A(0)h = a_0 h. \tag{33}'$$

This shows that h belongs to the eigenspace N.
On the other hand, each of the vectors $(I - P)h_j(t)$ is orthogonal to N; therefore so is their limit h. But since N contains h, we have arrived at a contradiction. Therefore (31) is true. $\qquad \square$

We proceed now to prove the continuity of $h_j(t)$ and the differentiability of $a_j(t)$. Subtract $(32)'$ from (30) and divide by t; after the usual Leibniz-ish rearrangement we get

$$\frac{A(t) - A(0)}{t}h(t) + A(0)\frac{h(t) - Ph(t)}{t} = \frac{a(t) - a(0)}{t}h(t) + a(0)\frac{h(t) - Ph(t)}{t}.$$

We have dropped the subscript j to avoid clutter. We apply P to both sides; according to relation (28) $PA(0) = aP$. Since $P^2 = P$ we see that the second terms on the two sides are zero. So we get

$$P\frac{A(t) - A(0)}{t}h(t) = \frac{a(t) - a(0)}{t}Ph(t). \tag{34}$$

Since A was assumed to be differentiable,

$$\frac{A(t) - A(0)}{t} = \dot{A}(0) + \epsilon(t);$$

and by (31), $h(t) = Ph(t) + \epsilon(t)$. Setting these into (34) we get, using $P^2 = P$, that

$$P\dot{A}(0)P\, Ph(t) = \frac{a(t) - a(0)}{t}Ph(t) + \epsilon(t). \tag{35}$$

By assumption, the self-adjoint mapping $P\dot{A}(0)P$ has k distinct eigenvalues d_i on N, with corresponding eigenvectors w_i;

$$P\dot{A}(0)Pw_i = d_iw_i, \qquad i = 1,\ldots,k.$$

We expand $Ph(t)$ in terms of these eigenvectors:

$$Ph(t) = \sum x_iw_i, \tag{36}$$

where x_i are functions of t, and set it into (35):

$$\sum x_i\left(d_i - \frac{a(t) - a(0)}{t}\right)w_i = \epsilon(t). \tag{35$'$}$$

Since the $\{w_i\}$ form an orthonormal basis for N, we can express the norm of the left-hand side of (36) in terms of components:

$$\|Ph(t)\|^2 = \sum |x_i|^2.$$

According to (31), $\|Ph(t) - h(t)\|$ tends to zero. Since $\|h(t)\|^2 = 1$, we deduce that

$$\|Ph(t)\|^2 = \sum |x_i(t)|^2 = 1 - \epsilon(t), \tag{37}$$

where $\epsilon(t)$ denotes a scalar function that tends to zero. We deduce from (35)$'$ that

$$\sum \left|d_i - \frac{a(t) - a(0)}{t}\right|^2 |x_i(t)|^2 = \epsilon(t). \tag{37$'$}$$

Combining (37) and (37)$'$ we deduce that for each t small enough there is an index j such that

$$\begin{aligned}
\text{(i)} \qquad & \left|d_j - \frac{a(t) - a(0)}{t}\right| \le \epsilon(t), \\
\text{(ii)} \qquad & |x_i(t)| \le \epsilon(t) \qquad \text{for } i \ne j, \\
\text{(iii)} \qquad & |x_j(t)| = 1 - \epsilon(t).
\end{aligned} \tag{38}$$

Since $x_i(t)$ are continuous functions of t for $t \ne 0$, it follows from (38) that the index j is independent of t for t small enough.

The normalization $\|h(t)\| = 1$ of the eigenvectors still leaves open a factor of absolute value 1; we choose this factor so that not only $|x_j|$ but x_j itself is near 1:

$$x_j = 1 - \epsilon(t). \tag{38$'$}$$

Now we can combine (31), (36), (38)$_{(ii)}$, and (38)$'$ to conclude that

$$\|h(t) - w_j\| \le \epsilon(t). \tag{39}$$

We recall now that the eigenvector $h(t)$ itself was one of a set of k orthonormal eigenvectors. We claim that distinct eigenvectors $h_j(t)$ are assigned to distinct vectors w_j; for, clearly two orthogonal unit vectors cannot both differ by less than ϵ from the same vector w_j.

Inequality (39) shows that $h_j(t)$, properly normalized, tends to w_j as $t \to 0$. Inequality (38)$_{(i)}$ shows that $a_j(t)$ is differentiable at $t = 0$ and that its derivative is d_j. It follows that for t small but not equal to 0, $A(t)$ has simple eigenvalues near a_0. This concludes the proof of Theorem 12. $\quad\square$

4. ANALYTIC MATRIX-VALUED FUNCTIONS

There are further results about differentiability of eigenvectors, the existence of higher derivatives, but since these are even more tedious than Theorem 12 we shall not pursue them, except for one observation, due to Rellich. Suppose $A(t)$ is an analytic function of t:

$$A(t) = \sum_0^\infty A_i t^i, \tag{40}$$

where each A_i is a self-adjoint matrix. Then also the characteristic polynomial of $A(t)$ is analytic in t. The characteristic equation

$$p(s, t) = 0$$

defines s as a function of t. Near a value of t where the roots of p are simple, the roots $a(t)$ are regular analytic functions of t; near a multiple root the roots have an algebraic singularity and can be expressed as power series in a fractional power of t:

$$a(t) = \sum_0^\infty r_i t^{i/k}. \tag{40$'$}$$

On the other hand, we know from Theorem 4 of Chapter 8 that for real t, the matrix $A(t)$ is self-adjoint and therefore all its eigenvalues are real. Since fractional powers of t have complex values for real t, we can deduce that in (40)$'$ only integer powers of t occur, that is, that *the eigenvalues $a(t)$ are regular analytic functions of t.*

5. AVOIDANCE OF CROSSING

The discussion at the end of this chapter indicates that multiple eigenvalues of a matrix function $A(t)$ have to be handled with care, even when the values of the function are self-adjoint matrices. This brings up the question, How likely is it that $A(t)$ will have multiple eigenvalues for some values of t? The answer is, "Not very likely"; before making this precise, we describe a numerical experiment.

Choose a value of n, and then pick at random two real, symmetric $n \times n$ matrices B and M. Define $A(t)$ to be

$$A(t) = B + tM. \tag{41}$$

Calculate numerically the eigenvalues of $A(t)$ at a sufficiently dense set of values of t. The following behavior emerges: as t approaches certain values of t, a pair of adjacent eigenvalues $a_1(t)$ and $a_2(t)$ appear to be on a collision course; yet at the last minute they turn aside:

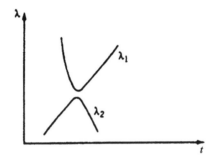

This phenomenon, called *avoidance of crossing*, was discovered by physicists in the early days of quantum mechanics. The explanation of avoidance of crossing was given by Wigner and von Neumann; it hinges on the size of the set of real, symmetric matrices which have multiple eigenvalues, called *degenerate* in the physics literature.

The set of all real, symmetric $n \times n$ matrices forms a linear space of dimension $N = n(n+1)/2$. There is another way of parametrizing these matrices, namely by their eigenvectors and eigenvalues. We recall from Chapter 8 that the eigenvalues are real, and in case they are distinct, the eigenvectors are orthogonal; we shall choose them to have length 1. The first eigenvector, corresponding to the largest eigenvalue, depends on $n - 1$ parameters; the second one, constrained to be orthogonal to the first eigenvector, depends on $n - 2$ parameters, and so on, all the way to the $(n - 1)$st eigenvector that depends on one parameter. The last eigenvector is then determined, up to a factor plus or minus 1. The total number of these parameters is $(n - 1) + (n - 2) + \cdots + 1 = n(n - 1)/2$; to these we add the n eigenvalues, for a total of $n(n - 1)/2 + n = n(n + 1)/2 = N$ parameters, as before.

We turn now to the degenerate matrices, which have two equal eigenvalues, the rest distinct from it and each other. The first eigenvector, corresponding to the largest of the simple eigenvalues, depends on $n - 1$ parameters, the next one on $n - 2$ parameters, and so on, all the way down to the last simple eigenvector that depends on two parameters. The remaining eigenspace is then uniquely determined. The total number of these parameters is $(n - 1) + \cdots + 2 = (n(n - 1))/2 - 1$: to these we add the $n-1$ distinct eigenvalues, for a total of $(n(n-1))/2) - 1 + n - 1 = (n(n+1))/2) - 2 = N - 2$.

This explains the avoidance of crossing: a line or curve lying in N-dimensional space will in general avoid intersecting a surface depending on $N-2$ parameters.

EXERCISE 8.　(a) Show that the set of all complex, self-adjoint $n \times n$ matrices forms $N = n^2$-dimensional linear space over the reals,

(b) Show that the set of complex, self-adjoint $n \times n$ matrices that have one double and $n-2$ simple eigenvalues can be described in terms of $N-3$ real parameters.

EXERCISE 9.　Choose in (41) at random two self-adjoint 10×10 matrices M and B. Using available software (MATLAB, MAPLE, etc.) calculate and graph at suitable intervals the 10 eigenvalues of $B + tM$ as functions of t over some t-segment.

The graph of the eigenvalues of such a one-parameter family of 12×12 self-adjoint matrices ornaments the cover of this volume; they were computed by David Muraki.

CHAPTER 10

Matrix Inequalities

In this chapter we study self-adjoint mappings of a Euclidean space into itself that are positive. In Section 1 we state and prove the basic properties of positive mappings and properties of the relation $A < B$. In Section 2 we derive some inequalities for the determinant of positive matrices. In Section 3 we study the dependence of the eigenvalues on the matrix in light of the partial order $A < B$. In Section 4 we show how to decompose arbitrary mappings of Euclidean space into itself as a product of self-adjoint and unitary maps.

1. POSITIVITY

We recall from Chapter 8 the definition of a positive mapping:

Definition. A self-adjoint linear mapping H from a real or complex Euclidean space into itself is called *positive* if

$$(x, Hx) > 0 \qquad \text{for all } x \neq 0. \tag{1}$$

Positivity of H is denoted as $H > O$ or $O < H$.

We call a self-adjoint map K *nonnegative* if the associated quadratic form is

$$(x, Kx) \geq 0 \qquad \text{for all } x. \tag{2}$$

Nonnegativity of K is denoted as $K \geq O$ or $O \leq K$.

The basic properties of positive maps are contained in the following theorem.

Linear Algebra and Its Applications, Second Edition, by Peter D. Lax
Copyright © 2007 John Wiley & Sons, Inc.

Theorem 1

  (i) The identity I is positive.

  (ii) If M and N are positive, so is their sum M + N, as well as aM for any positive number a.

  (iii) If H is positive and Q is invertible, then

$$Q^*HQ > O. \tag{3}$$

  (iv) H is positive iff all its eigenvalues are positive.

  (v) Every positive mapping is invertible.

  (vi) Every positive mapping has a positive square root, uniquely determined.

  (vii) The set of all positive maps is an open subset of the space of all self-adjoint maps.

  (viii) The boundary points of the set of all positive maps are nonnegative maps that are not positive.

Proof. Part (i) is a consequence of the positivity of the scalar product; part (ii) is obvious. For part (iii) we write the quadratic form associated with Q^*HQ as

$$(x, Q^*HQx) = (Qx, HQx) = (y, Hy), \tag{3}'$$

where $y = Qx$. Since Q is invertible, if $x \neq 0$, $y \neq 0$, and so by (1) the right-hand side of $(3)'$ is positive.

To prove (iv), let h be an eigenvector of H, a the eigenvalue $Hh = ah$. Taking the scalar product with h we get

$$(h, Hh) = a(h, h);$$

clearly, this is positive only if $a > 0$. This shows that the eigenvalues of a positive mapping are positive.

To show the converse, we appeal to Theorem 4 of Chapter 8, according to which every self-adjoint mapping H has an orthonormal basis of eigenvectors. Denote these by h_j and the corresponding eigenvalues by a_j:

$$Hh_j = a_j h_j. \tag{4}$$

Any vector x can be expressed as a linear combination of the h_j:

$$x = \sum x_j h_j. \tag{4}'$$

Since the h_j are eigenfunctions,

$$Hx = \sum x_j a_j h_j. \tag{4}''$$

Since the h_j form an orthonormal basis,

$$(x, x) = \sum |x_j|^2, \qquad (x, Hx) = \sum a_j |x_j|^2. \qquad (5)$$

It follows from (5) that if all a_j are positive, H is positive.

We deduce from (5) the following sharpening of inequality (1): for a positive mapping H,

$$(x, Hx) \geq a\| x \|^2, \qquad \text{for all } x, \qquad (5)'$$

where a is the smallest eigenvalue of H.

(v) Every noninvertible map has a nullvector, which is an eigenvector with eigenvalue zero. Since by (iv) a positive H has all positive eigenvalues, H is invertible.

(vi) We use the existence of an orthonormal basis formed by eigenvectors of H, H positive. With x expanded as in (4)', we define $\sqrt{H}$ by

$$\sqrt{H}x = \sum x_j \sqrt{a_j} h_j, \qquad (6)$$

where $\sqrt{a_j}$ denotes the positive square root of a_j. Comparing this with the expansion (4)'' of H itself we can verify that $(\sqrt{H})^2 = H$. Clearly, $\sqrt{H}$ as defined by (6) has positive eigenvalues, and so by (iv) is positive.

(vii) Let H be any positive mapping, and N any self-adjoint mapping whose distance from H is less than a,

$$\| N - H \| < a,$$

where a is the smallest eigenvalue of H. We claim that N is invertible. Denote $N - H$ by M; the assumption is that $\| M \| < a$. This means that for all nonzero x in X,

$$\| Mx \| < a\| x \|.$$

By the Schwarz inequality, for $x \neq 0$,

$$|(x, Mx)| \leq \| x \| \| Mx \| < a\| x \|^2.$$

Using this and (5)', we see that for $x \neq 0$,

$$(x, Nx) = (x, (H + M)x) = (x, Hx) + (x, Mx) > a\| x \|^2 - a\| x \|^2 = 0.$$

This shows that $H + M = N$ is positive.

(viii) By definition of boundary, every mapping K on the boundary is the limit of mappings $H_n > 0$:

$$\lim_{n \to \infty} H_n = K.$$

It follows from the Schwarz inequality that for every x,

$$\lim_{n\to\infty} (x, H_n x) = (x, K x).$$

Since each H_n is positive, and the limit of positive numbers is nonnegative, it follows that $K \geq 0$. K cannot be positive, for then by part (vii) it would not be on the boundary. □

EXERCISE 1. How many square roots are there of a positive mapping?

Characterizations analogous to parts of Theorem 1 hold for nonnegative mappings:

EXERICSE 2. Formulate and prove properties of nonnegative mappings similar to parts (i), (ii), (iii), (iv), and (vi) of Theorem 1.

Based on the notion of positivity we can define a *partial order* among self-adjoint mappings of a given Euclidean space into itself.

Definition. Let M and N be two selfadjoint mappings of a Euclidean space into itself. We say that M is less than N, denoted as

$$M < N \quad \text{or} \quad N > M, \tag{7}$$

if $N - M$ is positive:

$$O < N - M. \tag{7}'$$

The relation $M \leq N$ is defined analogously.
The following properties are easy consequences of Theorem 1.

Additive Property. If $M_1 < N_1$ and $M_2 < N_2$ then

$$M_1 + M_2 < N_1 + N_2. \tag{8}$$

Transitive Property. If $L < M$ and $M < N$, then $L < N$.
Multiplicative Property. If $M < N$ and Q is invertible, then

$$Q^* M Q < Q^* N Q. \tag{9}$$

The partial ordering defined in (7) and (7)' for self-adjoint maps shares some—but not all—other properties of the natural ordering of real numbers. For instance, the reciprocal property holds.

Theorem 2. Let M and N denote positive mappings that satisfy

$$O < M < N. \tag{10}$$

Then

$$M^{-1} > N^{-1}. \tag{10}'$$

First Proof. We start with the case when $N = I$. By definition, $M < I$ means that $I - M$ is positive. According to part (iv) of Theorem 1, that means that the eigenvalues of $I - M$ are positive, that is, that the eigenvalues of M are less than 1. Since M is positive, the eigenvalues of M lie between 0 and 1. The eigenvalues of M^{-1} are reciprocals of those of M; therefore the eigenvalues of M^{-1} are greater than 1. That makes the eigenvalues of $M^{-1} - I$ positive; so by part (iv) of Theorem 1, $M^{-1} - I$ is positive, which makes $M^{-1} > I$.

We turn now to any N satisfying (10); according to part (vi) of Theorem 1, we can factor $N = R^2$, $R > O$. According to part (v) of Theorem 1, R is invertible; we use now property (9), with $Q = R$, to deduce from (10) that

$$0 < R^{-1}MR^{-1} < R^{-1}NR^{-1} = I.$$

From what we have already shown, it follows from the equation that the inverse of $R^{-1}MR^{-1}$ is greater than I:

$$RM^{-1}R > I.$$

We use once more property (9), with $Q = R^{-1}$, to deduce that

$$M^{-1} > R^{-1}IR^{-1} = R^{-2} = N^{-1}. \qquad \square$$

Second Proof. We shall use the following generally useful calculus lemma.

Lemma 3. Let $A(t)$ be a differentiable function of the real variable whose values are self-adjoint mappings; the derivative $(d/dt)A$ is then also self-adjoint. Suppose that $(d/dt)A$ is positive; then $A(t)$ is an increasing function, that is,

$$A(s) < A(t) \quad \text{when } s < t. \tag{11}$$

Proof. Let x be any nonzero vector, independent of t. Then by the assumption that the derivative of A is positive, we obtain

$$\frac{d}{dt}(x, Ax) = \left(x, \frac{d}{dt}Ax\right) > 0.$$

So by ordinary calculus, $(x, A(t)x)$ is an increasing function of t:

$$(x, A(s)x) < (x, A(t)x) \qquad \text{for } s < t.$$

This implies that $A(t) - A(s) > O$, which is the meaning of (11). $\qquad \square$

Let $A(t)$ be as in Lemma 3, and in addition suppose that $A(t)$ is invertible; we claim that $A^{-1}(t)$ is a decreasing function of t. To see this we differentiate A^{-1}, using Theorem 2 of Chapter 9:

$$\frac{d}{dt}A^{-1} = A^{-1}\frac{dA}{dt}A^{-1}.$$

We have assumed that dA/dt is positive, so it follows from part (iii) of Theorem 1 that so is $A^{-1}(dA/dt)A^{-1}$. This shows that the derivative of $A^{-1}(t)$ is negative. It follows then from Lemma 3 that $A^{-1}(t)$ is decreasing.

We now define

$$A(t) = M + t(N - M), \qquad 0 \le t \le 1. \tag{12}$$

Clearly, $dA/dt = N - M$, positive by assumption (10). It further follows from assumption (10) that for $0 \le t \le 1$,

$$A(t) = (1 - t)M + tN$$

is the sum of two positive operators and therefore itself positive. By part (v) of Theorem 1 we conclude that $A(t)$ is invertible. We can assert now, as shown above, that $A(t)$ is a decreasing function:

$$A^{-1}(0) > A^{-1}(1).$$

Since $A(0) = M$, $A(1) = N$, this is inequality (10)'. This concludes the second proof of Theorem 2. □

The product of two self-adjoint mappings is not, in general, self-adjoint. We introduce the *symmetrized product S* of two self-adjoint mappings A and B as

$$S = AB + BA. \tag{13}$$

The quadratic form associated with the symmetrized product is

$$(x, Sx) = (x, ABx) + (x, BAx) = (Ax, Bx) + (Bx, Ax). \tag{14}$$

In the real case

$$(x, Sx) = 2(Ax, Bx). \tag{14}'$$

This formula shows that the symmetrized product of two positive mappings need not be positive; the conditions $(x, Ax) > 0$ and $(x, Bx) > 0$ mean that the pairs of vectors x, Ax and x, Bx make an angle less than $\pi/2$. But these restrictions do not prevent the vectors Ax, Bx from making an angle greater than $\pi/2$, which would render (14)' negative.

EXERCISE 3. Construct two real, positive 2×2 matrices whose symmetrized product is *not* positive.

In view of the Exercise 3 the following result is somewhat surprising.

Theorem 4. Let A and B denote two self-adjoint maps with the following properties:

(i) A is positive.
(ii) The symmetrized product $S = AB + BA$ is positive.
Then B is positive.

Proof. Define $B(t)$ as $B(t) = B + tA$. We claim that for $t \geq 0$ the symmetrized product of A and $B(t)$ is positive. For

$$S(t) = AB(t) + B(t)A = AB + BA + 2tA^2 = S + 2tA^2;$$

since S and $2tA^2$ are positive, their sum is positive. We further claim that for t large enough positive, $B(t)$ is positive. For

$$(x, B(t)x) = (x, Bx) + t(x, Ax); \tag{15}$$

A was assumed positive, so by $(5)'$,

$$(x, Ax) \geq a \| x \|^2, \qquad a > 0.$$

On the other hand, by the Schwarz inequality

$$|(x, Bx)| \leq \| x \| \| Bx \| \leq \| B \| \| x \|^2.$$

Putting these inequalities together with (15), we get

$$(x, B(t)x) \geq (ta - \| B \|) \| x \|^2;$$

clearly this shows that $B(t)$ is positive when $ta > \| B \|$.

Since $B(t)$ depends continuously on t, if $B = B(0)$ were not positive, there would be some nonnegative value t_0 between 0 and $\| B \|/a$, such that $B(t_0)$ lies on the *boundary* of the set of positive mappings. According to part (viii) of Theorem 1, a mapping on the boundary is nonnegative but not positive. Such a mapping $B(t_0)$ has nonnegative eigenvalues, at least one of which is zero.

So there is a nonzero vector y such that $B(t_0)y = 0$. Setting $x = y$ in (14) with $B = B(t_0)$, we obtain

$$(y, S(t_0)y) = (Ay, B(t_0)y) + (B(t_0)y, Ay) = 0;$$

this is contrary to the positivity of $S(t_0)$; therefore B is positive. □

In Section 4 we offer a second proof of Theorem 4.

An interesting consequence of Theorem 4 is the following theorem.

Theorem 5.   Let M and N denote positive mappings that satisfy

$$O < M < N; \tag{16}$$

then

$$\sqrt{M} < \sqrt{N}, \tag{16$'$}$$

where $\sqrt{\ }$ denotes the positive square root.

Proof. Define the function $A(t)$ as in (12):

$$A(t) = M + t(N - M).$$

We have shown that $A(t)$ is positive when $0 \le t \le 1$; so we can define

$$R(t) = \sqrt{A(t)}, \qquad 0 \le t \le 1, \tag{17}$$

where $\sqrt{\ }$ is the positive square root. It is not hard to show that $R(t)$, the square root of a differentiable positive function, is differentiable. We square (17), obtaining $R^2 = A$; differentiating with respect to t, we obtain

$$\dot{R}R + R\dot{R} = \dot{A}, \tag{18}$$

where the dot denotes the derivative with respect to t. Recalling the definition (13) of symmetrized product we can paraphrase (18) as follows: The symmetrized product of $\dot{R}$ and is $\dot{A}$.

By hypothesis (16), $\dot{A} = N - M$ is positive: by construction, so is R. Therefore using Theorem 4 we conclude that $\dot{R}$ is positive on the interval [0, 1]. It follows then from Lemma 3 that $R(t)$ is an increasing function of t; in particular,

$$R(0) < R(1).$$

Since $R(0) = \sqrt{A(0)} = \sqrt{M}$, $R(1) = \sqrt{A(1)} = \sqrt{N}$,   inequality $(16)'$
follows.   □

EXERCISE 4.   Show that if $0 < M < N$, then (a) $M^{1/4} < N^{1/4}$. (b) $M^{1/m} < N^{1/m}$, m a power of 2. (c) $\log M \le \log N$.

Fractional powers and logarithm are defined by the functional calculus in Chapter 8. (*Hint:* $\log M = \lim_{m \to \infty} m[M^{1/m} - I]$.)

EXERCISE 5.   Construct a pair of mappings $0 < M < N$ such that M^2 is *not* less than N^2. (*Hint:* Use Exericse 3.)

There is a common theme in Theorems 2 and 5 and Exercises 4 and 5 that can be expressed by the concept of *monotone matrix function*.

Definition.   A real-valued function $f(s)$ defined for $s > 0$ is called a *monotone matrix function* if all pairs of self-adjoint mappings M, N satisfying

$$O < M < N$$

also satisfy

$$f(M) < f(N),$$

where $f(M), f(N)$ are defined by the functional calculus of Chapter 8.

According to Theorems 2 and 5, and Exercise 4, the functions $f(s) = -1/s$, $s^{1/m}$, $\log s$ are monotone matrix functions. Exercise 5 says $f(s) = s^2$ is not.

Positive multiples, sums, and limits of monotone matrix functions are mmf's. Thus

$$-\sum \frac{m_j}{s + t_j}, \qquad m_j > 0, \qquad t_j > 0$$

are mmf's, as is

$$f(s) = as + b - \int_0^\infty \frac{dm(t)}{s + t}, \tag{19}$$

where a is positive, b is real, and $m(t)$ is a nonnegative measure for which the integral (19) converges.

Carl Loewner has proved the following beautiful theorem.

Theorem.   Every monotone matrix function can be written in the form (19).

At first glance, this result seems useless, because how can one recognize that a function $f(s)$ defined on $\mathbb{R}_+$ is of form (19)? There is, however, a surprisingly simple criterion:

Every function f of form (19) can be extended as an analytic function in the upper half-plane, and has a positive imaginary part there.

EXERCISE 6.   Verify that (19) defines $f(z)$ for a complex argument z as an analytic function, as well as that $\mathrm{Im} f(z) > 0$ for $\mathrm{Im} x > 0$.

Conversely, a classical theorem of Herglotz and F. Riesz says that every function analytic in the upper half-plane whose imaginary part is positive there, and which is real on the positive real axis, is of form (19). For a proof, consult the author's text entitled *Functional Analysis*.

The functions $-1/s$, $s^{1/m}$, $m > 1$, $\log s$ have positive imaginary parts in the upper half-plane; the function s^2 does not.

Having talked so much about positive mappings, it is time to present some examples. Below we describe a method for constructing positive matrices, in fact all of them.

Definition.   Let $f_1, \ldots, f_m$ be an ordered set of vectors in a Euclidean space. The matrix G with entries.

$$G_{ij} = (f_j, f_i) \tag{20}$$

is called the *Gram matrix* of the set of vectors.

Theorem 6.   **(i)** Every Gram matrix is nonnegative.
 (ii) The Gram matrix of a set of linearly independent vectors is positive.
(iii) Every positive matrix can be represented as a Gram matrix.

Proof. The quadratic form associated with a Gram matrix can be expressed as follows:

$$(x, Gx) = \sum_{i,j} x_i \bar{G}_{ij} \bar{x}_j = \sum (f_i, f_j) x_i \bar{x}_j$$
$$= \left(\sum_i x_i f_i, \sum x_j f_i \right) = \left\| \sum x_i f_i \right\|^2 \tag{20'}$$

Parts (i) and (ii) follow immediately from (20)'. To prove part (iii), let $(H_{ij}) = H$ be positive. Define for vectors x and y in $\mathbb{C}^n$ the *nonstandard* scalar product $(,)_H$ defined as

$$(x, y)_H = (x, Hy),$$

where (,) is the standard scalar product. The Gram matrix of the unit vectors $f_i = e_i$ is

$$(e_i, e_j)_H = (e_i, He_j) = h_{ij}. \qquad \square$$

EXAMPLE. Take the Euclidean space to consist of real-valued functions on the interval $[0, 1]$, with the scalar product

$$(f, g) = \int_0^1 f(t)g(t)dt.$$

Choose $f_j = t^{j-1}, j = 1, \ldots, n$. The associated Gram matrix is

$$G_{ij} = \frac{1}{i+j-1}. \tag{21}$$

EXERCISE 7. Given m positive numbers $r_1, \ldots, r_m$, show that the matrix

$$G_{ij} = \frac{1}{r_i + r_j + 1} \tag{22}$$

is positive.

Example. Take as scalar product

$$(f, g) = \int_0^{2\pi} f(\theta)\bar{g}(\theta)w(\theta)d\theta,$$

where w is some given positive real function. Choose $f_j = e^{ij\theta}, j = -n, \ldots, n$. The associated $(2n+1) \times (2n+1)$ Gram matrix is $G_{kj} = c_{k-j}$, where

$$c_p = \int w(\theta)e^{-ip\theta}d\theta.$$

We conclude this section with a curious result due to I. Schur.

Theorem 7. Let $A = (A_{ij})$ and $B = (B_{ij})$ denote positive matrices. Then $M = (M_{ij})$, whose entries are the products of the entries of A and B,

$$M_{ij} = A_{ij}B_{ij} \tag{23}$$

also is a positive matrix.

In Appendix 4 we shall give a one-line proof of Theorem 7 using tensor products.

2. THE DETERMINANT OF POSITIVE MATRICES

Theorem 8.   The determinant of every positive matrix is positive.

Proof. According to Theorem 3 of Chapter 6, the determinant of a matrix is the product of its eigenvalues. According to Theorem 1 of this chapter, the eigenvalues of a positive matrix are positive. Then so is their product.   □

Theorem 9.   Let A and B denote real, self-adjoint, positive $n \times n$ matrices. Then for all t between 0 and 1,

$$\det(tA + (1 - t)B) \geq (\det A)^t (\det B)^{1-t}. \tag{24}$$

Proof. Take the algorithm of both sides. Since log is a monotonic function, we get the equivalent inequality: for all t in [0, 1],

$$\log \det(tA + (1 - t)B) \geq t \log \det A + (1 - t) \log \det B. \tag{24$'$}$$

We recall the concept of a *concave* function of a single variable: A function $f(x)$ is called *concave* if its graph between two points lies above the chord connecting those points. Analytically, this means that for all t in [0, 1],

$$f(ta + (1 - t)b) \geq tf(a) + (1 - t)f(b).$$

Clearly, (24)$'$ can be interpreted as asserting that the function log det H is concave on the set of positive matrices. Note that it follows from Theorem 1 that for A and B positive, $tA + (1 - t)B$ is positive when $0 \leq t \leq 1$. According to a criterion we learn in calculus, a function whose *second derivative* is negative is concave. For example, the function $\log t$, defined for t positive, has second derivative $-1/t^2$, and so it is concave. To prove (24)$'$, we shall calculate the second derivative of the function $f(t) = \log \det(tA + (1 - t)B)$ and verify that it is negative. We use formula (10) of Theorem 4 in Chapter 9, valid for matrix valued functions $Y(t)$ that are differentiable and invertible:

$$\frac{d}{dt} \log \det Y = \operatorname{tr}(Y^{-1}\dot{Y}). \tag{25}$$

In our case, $Y(t) = B + t(A - B)$; its derivative is $\dot{Y} = A - B$, independent of t. So, differentiating (25) with respect to t, we get

$$\frac{d^2}{dt^2} \log \det Y = \operatorname{tr}(-Y^{-1}\dot{Y}Y^{-1}\dot{Y}) = -\operatorname{tr}(Y^{-1}\dot{Y})^2. \tag{25$'$}$$

Here we have used the product rule, and rules (2)$'$ and (3) from Chapter 9 concerning the differentiation of the trace and the reciprocal of matrix functions.

According to Theorem 3 of Chapter 6, the trace of a matrix is the sum of its eigenvalues; and according to Theorem 4 of Chapter 6, the eigenvalues of the square of a matrix T are the square of the eigenvalues of T. Therefore

$$\operatorname{tr}(Y^{-1}\dot{Y})^2 = \sum a_j^2, \tag{26}$$

where a_j are the eigenvalues of $Y^{-1}\dot{Y}$. According to Theorem 11′ in Chapter 8, the eigenvalues a_j of the product $Y^{-1}\dot{Y}$ of a positive matrix Y^{-1} and a self-adjoint matrix $\dot{Y}$ are real. It follows that (26) is positive; setting this into (25)′, we conclude that the second derivative of $\log \det Y(t)$ is negative. □

Second Proof. Define C as $B^{-1}A$; by Theorem 11′ of Chapter 8, the product C of two positive matrices has positive eigenvalues c_j. Now rewrite the left-hand side of (24) as

$$\det B(tB^{-1}A + (1-t)I) = \det B \det(tC + (1-t)I).$$

Divide both sides of (24) by $\det B$; the resulting right-hand side can be rewritten as

$$(\det A)^t (\det B)^{-t} = (\det C)^t.$$

What is to be shown is that

$$\det(tC + (1-t)I) \geq (\det C)^t.$$

Expressing the determinants as the product of eigenvalues gives

$$\prod (tc_j + 1 - t) \geq \prod c_j^t.$$

We claim that for all t between 0 and 1 each factor on the left is greater than the corresponding factor on the right:

$$tc + (1-t) \geq c^t.$$

This is true because c^t is a convex function of t and equality holds when $t = 0$ or $t = 1$. □

Next we give a useful estimate for the determinant of a positive matrix.

Theorem 10. The determinant of a positive matrix H does not exceed the product of its diagonal elements:

$$\det H \leq \prod h_{ii}. \tag{27}$$

Proof. Since H is positive, so are its diagonal entries. Define $d_i = 1/\sqrt{h_{ii}}$, and denote by D the diagonal matrix with diagonal entries d_i. Define the matrix B by

$$B = DHD.$$

Clearly, B is symmetric and positive and its diagonal entries are all 1's. By the multiplicative property of determinants,

$$\det B = \det H \det D^2 = \frac{\det H}{\prod h_{ii}}. \tag{28}$$

So (27) is the same as $\det B \leq 1$. To show this, denote the eigenvalues of B by $b_1, \ldots, b_n$, positive quantities since B is a positive matrix. By the arithmetic–geometric mean inequality

$$\prod b_i \leq \left(\sum b_i/n\right)^n.$$

We can rewrite this as

$$\det B \leq \left(\frac{\operatorname{tr} B}{n}\right)^n. \tag{29}$$

Since the diagonal entries of B are all 1's, $\operatorname{tr} B = n$, so $\det B \leq 1$ follows.   □

Theorem 10 has this consequence.

Theorem 11. Let T be any $n \times n$ matrix whose columns are $c_1, c_2, \ldots, c_n$. Then the determinant of T is in absolute value not greater than the product of the length of its columns:

$$|\det T| \leq \prod \| c_j \|. \tag{30}$$

Proof. Define $H = T^*T$; its diagonal elements are

$$h_{ii} = \sum_j t_{ij}^* t_{ji} = \sum_j \bar{t}_{ji} t_{ji} = \sum_j |t_{ji}|^2 = \| c_i \|^2.$$

According to Theorem 1, T^*T is positive, except when T is noninvertible, in which case $\det T = 0$, so there is nothing to prove. We appeal now to Theorem 10 and deduce that

$$\det H \leq \prod \| c_i \|^2.$$

Since the determinant is multiplicative, and since $\det T^* = \overline{\det T}$,

$$\det H = \det T^* \det T = |\det T|^2.$$

Combining the last two and taking its square root we obtain inequality (30) of Theorem 11.   □

Inequality (30) is due to Hadamard and is useful in applications. In the real case it has an obvious geometrical meaning: among all parallelepipeds with given side lengths $\| c_j \|$, the one with the largest volume is rectangular.

We return to Theorem 9 about determinants; the first proof we gave for it used the differential calculus. We present now a proof based on integral calculus. This proof works for real, symmetric matrices; it is based on an integral formula for the determinant of real positive matrices.

Theorem 12. Let H be an $n \times n$ real, symmetric, positive matrix. Then

$$\frac{\pi^{n/2}}{\sqrt{\det H}} = \int_{\mathbb{R}^n} e^{-(x, Hx)} dx. \tag{31}$$

Proof. It follows from inequality (5)' that the integral (31) converges. To evaluate it, we appeal to the spectral theorem for self-adjoint mappings, see Theorem 4' of Chapter 8, and introduce new coordinates

$$x = My, \tag{32}$$

M an orthogonal matrix so chosen that the quadratic form is diagonalized:

$$(x, Hx) = (My, HMy) = (y, M^* HMy) = \sum a_j y_j^2. \tag{33}$$

The a_j are the eigenvalues of H. We substitute (33) into (31); since the matrix M is an isometry, it preserves volume as well: $|\det M| = 1$. In terms of the new variables the integrand is a product of functions of single variables, so we can rewrite the right side of (31) as a product of one-dimensional integrals:

$$\int e^{-\Sigma a_j y_j^2} dy = \int \prod e^{-a_j y_j^2} dy = \prod \int e^{-a_j y_j^2} dy_j. \tag{34}$$

The change of variable $\sqrt{a} y = z$ turns each of the integrals on the right in (34) into

$$\int e^{-z^2} \frac{dz}{\sqrt{a_j}}.$$

According to a result of calculus

$$\int_{-\infty}^{\infty} e^{-z^2} dz = \sqrt{\pi}, \tag{35}$$

so that the right-hand side of (34) equals

$$\frac{\pi^{n/2}}{\prod \sqrt{a_j}} = \frac{\pi^{n/2}}{(\prod a_j)^{1/2}}. \tag{34}$$

According to formula (15), Theorem 3 in Chapter 6 the determinant of H is the product of its eigenvalues; so formula (31) of Theorem 12 follows from (34) and (34)′.   □

EXERCISE 8.   Look up a proof of the calculus result (35).

Proof of Theorem 9. We take in formula (35), $H = tA + (1 - t)B$, where A, B are arbitrary real, positive matrices:

$$\frac{\pi^{n/2}}{\sqrt{\det(tA + (1 - t)B)}} = \int_{\mathbb{R}^n} e^{-(x,(tA+(1-t)B)x)} dx$$

$$= \int_{\mathbb{R}^n} e^{-t(x,Ax)} e^{-(1-t)(x,Bx)} dx. \tag{36}$$

We appeal now to Hölder's inequality:

$$\int fg \, dx \le \left(\int f^p dx \right)^{1/p} \left(\int g^q dx \right)^{1/q},$$

where p, q are real, positive numbers such that

$$\frac{1}{p} + \frac{1}{q} = 1.$$

We take

$$f(x) = e^{-t(x,Ax)}, \qquad g(x) = e^{-(1-t)(x,Bx)},$$

and choose $p = 1/t$, $q = 1/(1 - t)$; we deduce that the integral on the right in (36) is not greater than

$$\left(\int_{\mathbb{R}^n} e^{-(x,Ax)} dx \right)^t \left(\int e^{-(x,Bx)} dx \right)^{1-t}.$$

Using formula (31) to express these integrals we get

$$\left(\frac{\pi^{n/2}}{\sqrt{\det A}}\right)^t \left(\frac{\pi^{n/2}}{\sqrt{\det B}}\right)^{1-t} = \frac{\pi^{n/2}}{\sqrt{(\det A)^t (\det B)^{1-t}}}.$$

Since this is an upper bound for (36), inequality (24) follows. □

Formula (31) also can be used to give another proof of Theorem 10.

Proof. In the integral on the right in (31) we write the vector variable x as $x = ue_1 + z$, where u is the first component of x and z the rest of them. Then

$$(x, Hx) = h_{11}u^2 + 2ul(z) + (z, H_{11}z),$$

where $l(z)$ is some linear function of z. Setting this into (31) gives

$$\frac{\pi^{n/2}}{\sqrt{\det H}} = \int \int e^{-h_{11}u^2 - 2ul - (z, H_{11}z)} du \, dz. \tag{37}$$

Changing the variable u to $-u$ transforms the above integral into

$$\int \int e^{-h_{11}u^2 + 2ul - (z, H_{11}z)} du \, dz.$$

Adding and dividing by 2 gives

$$\int \int e^{-h_{11}u^2 - (z, H_{11}z)} \frac{c + c^{-1}}{2} du \, dz, \tag{37'}$$

where c abbreviates e^{2ul}. Since c is positive,

$$\frac{c + c^{-1}}{2} \geq 1.$$

Therefore (37)' is bounded from below by

$$\int \int e^{-h_{11}u^2} e^{-(z, H_{11}z)} du \, dz.$$

The integrand is now the product of a function of u and of z, and so is the product of two integrals, both of which can be evaluated by (31):

$$\frac{\pi^{1/2}}{\sqrt{h_{11}}} \frac{\pi^{(n-1)/2}}{\sqrt{\det H_{11}}}$$

Since this is a lower bound for the right-hand side of (37), we obtain that $\det H \leq h_{11} \det H_{11}$. Inequality (27) follows by induction on the size of H. □

3. EIGENVALUES

In this section we present a number of interesting and useful results on eigenvalues.

Lemma 13. Let A be a self-adjoint map of a Euclidean space U into itself. We denote by $p_+(A)$ the number of positive eigenvalues of A, and denote by $p_-(A)$ the number of its negative eigenvalues.

$p_+(A)$ = maximum dimension of subspace S of U such that (Au, u) is positive on S.

$p_-(A)$ = maximum dimension of subspace S of U such (Au, u) is negative on S.

Proof. This follows from the minmax characterization of the eigenvalues of A; see Theorem 10, as well as Lemma 2 of Chapter 8. □

Theorem 14. Let U and A be as in Lemma 13, and let V be a subspace of U whose dimension is one less than the dimension of U:

$$\dim V = \dim U - 1.$$

Denote by P orthogonal projection onto V. Then PAP is a self-adjoint map of U into U that maps V into V; we denote by B the restriction of PAP to V. We claim that

$$p_+(A) - 1 \leq p_+(B) \leq p_+(A), \tag{38}_+$$

and

$$p_-(A) - 1 \leq p_-(B) \leq p_-(A). \tag{38}_-$$

Proof. Let T denote a subspace of V of dimension $p_+(B)$ on which B is positive:

$$(Bv, v) > 0, \quad v \text{ in } T, \quad v \neq 0.$$

By definition of B, we can write this as

$$0 < (PAPv, v) = (APv, Pv).$$

Since v belongs to T, a subspace of V, $Pv = v$. So we conclude that A is positive on T; this proves that

$$p_+(B) \le p_+(A).$$

To estimate $p_+(B)$ from below, we choose a subspace S of U, of dimension $p_+(A)$ on which A is positive:

$$(Au, u) > 0, \qquad u \text{ in S}, \qquad u \ne 0.$$

Denote the intersection of S and V by T:

$$T = S \cap V.$$

We claim that the dimension of T is at most one less than the dimension of S:

$$\dim S - 1 \le \dim T.$$

If S is a subspace of V, then $T = S$ and $\dim T = \dim S$. If not, choose a basis in S: $\{s_1, \ldots, s_k\}$. At least one of these, say s_1, does not belong to V; this means that s_1 has a nonzero component orthogonal to V. Then we can choose scalars $a_2, \ldots, a_k$ such that

$$s_2 - a_2 s_1, \ldots, s_k - a_k s_1$$

belong to V. They are linearly independent, since $s_1, \ldots, s_k$ are linearly independent. It follows that

$$\dim S - 1 \le \dim T,$$

as asserted.

We claim that B is positive on T. Take any $v \ne 0$ in T:

$$(Bv, v) = (PAPv, v) = (APv, Pv) = (Av, v),$$

since v belongs to V. Since v also belongs to S, $(Av, v) > 0$.

Since $p_+(B)$ is defined as the dimension of the largest subspace on which B is positive, and since $\dim T \ge \dim S - 1$, $p_+(B) \ge p_+(A) - 1$ follows. This completes the proof of $(38)_+$; $(38)_-$ can be proved similarly. $\square$

An immediate consequence of Theorem 14 is the following theorem.

Theorem 15. Let U, V, A, and B be as in Theorem 14. Denote the eigenvalues of A as $a_1, \ldots, a_n$, and denote those of B as $b_1, \ldots, b_{n-1}$. The eigenvalues of B separate the eigenvalues of A:

$$a_1 \le b_1 \le a_2 \le \cdots b_{n-1} \le a_n. \tag{39}$$

Proof. Apply Theorem 14 to $A - c$ and $B - c$. We conclude that the number of b_i less than c is not greater than the number of a_i less than c, and at most one less. We claim that $a_i \leq b_{ij}$ if not, we could choose $b_i < c < a_i$ and obtain a contradiction. We can show analogously that $b_i \leq a_{i+1}$. This proves that $a_i \leq b_i \leq a_{i+1}$, as asserted in (39). $\square$

Take U to be $\mathbb{R}^n$ with the standard Euclidean structure, and take A to be any $n \times n$ self-adjoint matrix. Fix i to be some natural number between 1 and n, and take V to consist of all vectors whose ith component is zero. Theorem 14 says that the eigenvalues of the ith principal minor of A separate the eigenvalues of A.

EXERCISE 9.   Extend Theorem 14 to the case when $\dim V = \dim U - m$, where m is greater than 1.

The following result is of fundamental interest in mathematical physics; see, for example, Theorem 4 of Chapter 11.

Theorem 16.   Let M and N denote self-adjoint $k \times k$ matrices satisfying

$$M < N. \tag{40}$$

Denote the eigenvalues of M, arranged in increasing order, by $m_1 \leq \cdots \leq m_k$, and those of N by $n_1 \leq \cdots \leq n_k$. We claim that

$$m_j < n_j, \quad j = 1, \ldots, k. \tag{41}$$

First Proof. We appeal to the minmax principle, Theorem 10 in Chapter 8, formula (40), according to which

$$m_j = \min_{\dim S = j} \max_{x \text{ in } S} \frac{(x, Mx)}{(x, x)}, \tag{42}_m$$

$$n_j = \min_{\dim S = j} \max_{x \text{ in } S} \frac{(x, Nx)}{(x, x)}. \tag{42}_n$$

Denote by T the subspace of dimension j for which the minimum in $(42)_n$ is reached, and denote by y the vector in T where $(x, Mx)/(x, x)$ achieves its maximum; we take y to be normalized as $\| y \| = 1$. Then by $(42)_m$,

$$m_j \leq (y, My),$$

while from $(42)_n$,

$$(y, Ny) \leq n_j.$$

Since the meaning of (40) is that $(y, My) < (y, Ny)$ for all $y \neq 0$, (41) follows.   □

If the hypothesis (40) is weakened to $M \leq N$, the weakened conclusion $m_j \leq n_j$ can be reached by the same argument.

Second Proof. We connect M and N by a straight line:

$$A(t) = M + t(N - M); \tag{43}$$

we also use calculus, as we have done so profitably in Section 1. Assuming for a moment that the eigenvalues of $A(t)$ are distinct, we use Theorem 7 of Chapter 9 to conclude that the eigenvalues of $A(t)$ depend differentiably on t, and we use formula (24) of that chapter for the value of the derivative. Since A is self-adjoint, we can identify in this formula the eigenvector l of A^T with the eigenvector h of A itself. Normalizing h so that $\| h \| = 1$, we have the following version of (24), Chapter 9, for the derivative of the eigenvalue a in $Ah = ah$:

$$\frac{da}{dt} = \left(h, \frac{dA}{dt} h \right). \tag{43$'$}$$

For $A(t)$ in (43), $dA/dt = N - M$ is positive according to hypothesis (41); therefore the right-hand side of (43)$'$ is positive. This proves that da/dt is positive, and therefore $a(t)$ is an increasing function of t; in particular, $a(0) < a(1)$. Since $A(0) = M$, $A(1) = N$, this proves (41) in case $A(t)$ has distinct eigenvalues for all t in [0, 1].

In case $A(t)$ has multiple eigenvalues for a finite set of t, the above argument shows that each $a_j(t)$ is increasing between two such values of t; that is enough to draw the conclusion (41). Or we can make use of the observation made at the end of Chapter 9 that the degenerate matrices form a variety of codimension 2 and can be avoided by changing M by a small amount and passing to the limit.   □

The following result is very useful.

Theorem 17.   Let M and N be self-adjoint $k \times k$ matrices m_j and n_j their eigenvalues arrayed in increasing order. Then

$$|n_j - m_j| \leq \| M - N \|. \tag{44}$$

Proof. Denote $\| M - N \|$ by d. It is easy to see that

$$N - dI \leq M \leq N + dI. \tag{44$'$}$$

Inequality (44) follows from (44)$'$ and (41).   □

EXERCISE 10. Prove inequality (44)$'$.

Wielandt and Hoffman have proved the following interesting result.

Theorem 18. Let M, N be self-adjoint $k \times k$ matrices and m_j and n_j their eigenvalues arranged in increasing order. Then

$$\sum (n_j - m_j)^2 \leq \| N - M \|_2^2, \tag{45}$$

where $\| N - M \|_2$ is the Hilbert–Schmidt norm defined by

$$\| C \|_2^2 = \sum |c_{ij}|^2. \tag{46}$$

Proof. The Hilbert–Schmidt norm of any matrix can be expressed as a trace:

$$\| C \|_2^2 = \operatorname{tr} C^* C. \tag{46$'$}$$

For C self-adjoint,

$$\| C \|_2^2 = \operatorname{tr} C^2. \tag{46$''$}$$

Using (46)$''$ we can rewrite inequality (45) as

$$\sum (n_j - m_j)^2 \leq \operatorname{tr}(N - M)^2.$$

Expanding both sides and using the linearity and commutativity of trace gives

$$\sum n_j^2 - 2n_j m_j + m_j^2 \leq \operatorname{tr} N^2 - 2\operatorname{tr}(NM) + \operatorname{tr} M^2. \tag{47}$$

According to Theorem 3 of Chapter 6, the trace of N^2 is the sum of the eigenvalues of N^2. According to the spectral mapping theorem, the eigenvalues of N^2 are n_j^2. Therefore

$$\sum n_j^2 = \operatorname{tr} N^2, \qquad \sum m_j^2 = \operatorname{tr} M^2;$$

so inequality (47) can be restated as

$$\sum n_j m_j \geq \operatorname{tr}(NM). \tag{47$'$}$$

To prove this we fix M and consider all self-adjoint matrices N whose eigenvalues are $n_1, \ldots, n_k$. The set of such matrices N forms a bounded set in the space of all self-adjoint matrices. By compactness, there is among these that matrix N that renders the right-hand side of (47)′ largest. According to calculus, the maximizing matrix N_{max} has the following property: if $N(t)$ is a differentiable function whose values are self-adjoint matrices with eigenvalues $n_1, \ldots, n_k$, and $N(0) = N_{max}$, then

$$\frac{d}{dt} \text{tr}(N(t)M)\Big|_{t=0} = 0. \tag{48}$$

Let A denote any anti-self-adjoint matrix; according to Theorem 5, part (e), Chapter 9, e^{At} is unitary for any real values of t. Now define

$$N(t) = e^{At} N_{max} e^{-At}. \tag{49}$$

Clearly, $N(t)$ is self-adjoint and has the same eigenvalues as N_{max}. According to part (d) of Theorem 5, Chapter 9,

$$\frac{d}{dt} e^{At} = Ae^{At} = e^{At}A.$$

Using the rules of differentiation developed in Chapter 9, we get, upon differentiating (49), that

$$\frac{d}{dt} N(t) = e^{At}(AN_{max} - N_{max}A)e^{-At}.$$

Setting this into (48) gives at $t = 0$

$$\frac{d}{dt} \text{tr}(N(t)M)\Big|_{t=0} = \text{tr}\left(\frac{dN}{dt}M\right)\Big|_{t=0} = \text{tr}(AN_{max}M - N_{max}AM) = 0.$$

Using the commutativity of trace, we can rewrite this as

$$\text{tr}(A(N_{max}M - MN_{max})) = 0. \tag{48'}$$

The commutator of two self-adjoint matrices N_{max} and M is anti-self-adjoint, so we may choose

$$A = N_{max}M - MN_{max}. \tag{50}$$

Setting this into (48)′ reveals that $\text{tr}\,A^2 = 0$; since by (46)′, for anti-self-adjoint A,

$$\text{tr}\,A^2 = -\sum |a_{ij}|^2,$$

we deduce that $A = 0$, so according to (50) the matrices N_{max} and M commute. Such matrices can be diagonalized simultaneously; the diagonal entries are n_j and

m_j, in some order. The trace of $N_{max}M$ can therefore be computed in this representation as

$$\sum n_{p_j} m_j, \tag{51}$$

where $p_j, j = 1, \ldots, k$ is some permutation of $1, \ldots, k$. It is not hard to show, and is left as an exercise to the reader, that the sum (51) is largest when the n_j are arranged in the same order as the m_j, that is, increasingly. This proves inequality (47)' for N_{max} and hence for all N. □

EXERCISE 11.   Show that (51) is largest when n_i and m_j are arranged in the same order.

The next result is useful in many problems of physics.

Theorem 19.   Denote by $e_{min}(H)$ the smallest eigenvalue of a self-adjoint mapping H in a Euclidean space. We claim that e_{min} is a *concave* function of H, that is, that for $0 \leq t \leq 1$,

$$e_{min}(tL + (1-t)M) \geq te_{min}(L) + (1-t)e_{min}(M) \tag{52}$$

for any pair of self-adjoint maps L and M. Similarly, $e_{max}(H)$ is a *convex* function of H; for $0 \leq t \leq 1$,

$$e_{max}(tL + (1-t)M) \leq te_{max}(L) + (1-t)e_{max}(M). \tag{52}'$$

Proof. We have shown in Chapter 8, equation (37), that the smallest eigenvalue of a mapping can be characterized as a minimum:

$$e_{min}(H) = \min_{\|x\|=1} (x, Hx). \tag{53}$$

Let y be a unit vector where (x, Hx), with $H = tL + (1-t)M$ reaches its minimum. Then

$$
\begin{aligned}
e_{min}(tL + (1-t)M) &= t(y, Ly) + (1-t)(y, My) \\
&\geq t \min_{\|x\|=1} (x, Lx) + (1-t) \min_{\|x\|=1} (x, Mx) \\
&= te_{min}(L) + (1-t)e_{min}(M).
\end{aligned}
$$

This proves (52). Since $-e_{max}(A) = e_{min}(-A)$, the convexity of $e_{max}(A)$ follows. □

Note that the main thrust of the argument above is that any function characterized as the minimum of a collection of *linear* functions is concave.

4. REPRESENTATION OF ARBITRARY MAPPINGS

Every linear mapping Z of a complex Euclidean space into itself can be decomposed, uniquely, as a sum of a self-adjoint mapping and an anti-self-adjoint one:

$$Z = H + A, \tag{54}$$

where

$$H^* = H, \qquad A^* = -A. \tag{54'}$$

Clearly, if (54) and (54)′ hold, $Z^* = H^* + A^* = H - A$, so H and A are given by

$$H = \frac{Z + Z^*}{2}, \qquad A = \frac{Z^* - Z}{2}.$$

H is called the self-adjoint part of Z, A the anti-self-adjoint part.

Theorem 20.   Suppose the self-adjoint part Z is positive:

$$Z + Z^* > 0.$$

Then the eigenvalues of Z have positive real part.

Proof. Using the conjugate symmetry of scalar product in a complex Euclidean space, and the definition of adjoint, we have the following identity for any vector h:

$$2\,\mathrm{Re}(Zh, h) = (Zh, h) + \overline{(Zh, h)} = (Zh, h) + (h, Zh) = (Zh, h) + (Z^*h, h)$$
$$= ((Z + Z^*)h, h).$$

Since we assumed in Theorem 18 that $Z + Z^*$ is positive, we conclude that for any vector $h \neq 0$, (Zh, h) has positive real part.

Let h be an eigenvector for Z of norm $\| h \| = 1$, with z the corresponding eigenvalue, $Zh = zh$. Then $(Zh, h) = z$ has positive real part.   □

In Appendix 14 we give a far-reaching extension of Theorem 20.

Theorem 20 can be used to give another proof of Theorem 4 about symmetrized products: Let A and B be self-adjoint maps, and assume that A and $AB + BA = S$ are positive. We claim that then B is positive.

Second Proof of Theorem 4. Since A is positive, it has according to Theorem 1 a square root $A^{1/2}$ that is invertible. We multiply the relation

$$AB + BA = S$$

by $A^{-1/2}$ from the right and the left:

$$A^{1/2}BA^{-1/2} + A^{-1/2}BA^{1/2} = A^{-1/2}SA^{-1/2}. \tag{55}$$

We introduce the abbreviation

$$A^{1/2}BA^{-1/2} = Z \tag{56}$$

and rewrite (55) as

$$Z + Z^* = A^{-1/2}SA^{-1/2}. \tag{55$'$}$$

Since S is positive, so, according to Theorem 1, is $A^{-1/2}SA^{-1/2}$; it follows from (55)$'$ that $Z + Z^*$ is positive. By Theorem 20 the eigenvalues of Z have positive real part.

Formula (56) shows that Z and B are similar; therefore they have the same eigenvalues. Since B is self-adjoint, it has real eigenvalues; so we conclude that the eigenvalues of B are positive. This, according to Theorem 1, guarantees that B is positive.   □

EXERCISE 12. Prove that if the self-adjoint part of Z is positive, then Z is invertible, and the self-adjoint part of Z^{-1} is positive.

The decomposition of an arbitrary Z as a sum of its self-adjoint and anti-self-adjoint parts is analogous to writing a complex number as the sum of its real and imaginary parts, and the norm is analogous to the absolute value. The next result strengthens this analogy. Let a denote any complex number with positive real part; then

$$z \longrightarrow \frac{1 - az}{a + \bar{a}z} = w$$

maps the right half-plane Re $z > 0$ onto the unit disc $|w| < 1$. Analogously, we claim the following:

Theorem 21. Let a be a complex number with Re $a > 0$. Let Z be a mapping whose self-adjoint part $Z + Z^*$ is positive. Then

$$W = (I - aZ)(I + \bar{a}Z)^{-1} \tag{57}$$

is a mapping of norm less than 1. Conversely, $\| W \| < 1$ implies that $Z + Z^* > 0$.

Proof. According to Theorem 20 the eigenvalues z of Z have positive real part. It follows that the eigenvalues of $I + \bar{a}Z$, $1 + aZ$ are $\neq 0$; therefore $I + \bar{a}Z$ is invertible. For any vector x, denote $(I + \bar{a}Z)^{-1}x = y$; then by (57),

$$(I - aZ)y = Wx,$$

and by definition of y,

$$(I + \bar{a}Z)y = x.$$

The condition $\|\,W\,\| < 1$ means that $\|\,Wx\,\|^2 < \|\,x\,\|^2$ for all $x \neq 0$; in terms of y this can be expressed as

$$\|\,y - aZy\,\|^2 < \|\,y + \bar{a}Zy\,\|^2 \tag{58}$$

Expanding both sides gives

$$\|\,y\,\|^2 + |a|^2\|\,Zy\,\|^2 - a\,(Zy, y) - \bar{a}(y, Zy) < \|\,y\,\|^2 + |a|^2\|\,Zy\,\|^2 \\ + \bar{a}(Zy, y) + a(y, Zy). \tag{59}$$

Cancelling identical terms and rearranging gives

$$0 < (a + \bar{a})[(Zy, y) + (y, Zy)] = 2\,\mathrm{Re}\,a[Z + Z^*]y, y). \tag{60}$$

Since we have assumed that $\mathrm{Re}\ a$ is positive and that $Z + Z^* > 0$, (60) is true. Conversely, if (60) holds for all y, $Z + Z^*$ is positive. □

Complex numbers z have not only additive but multiplicative decompositions: $z = re^{i\theta}$, $r > 0$, $|e^{i\theta}| = 1$. Mappings of Euclidean spaces have similar decompositions.

Theorem 22. Let A be a linear mapping of a complex Euclidean space into itself. Then A can be factored as

$$A = RU, \tag{61}$$

where R is a nonnegative self-adjoint mapping, and U is unitary. When A is invertible, R is positive.

Proof. Take first the case that A is invertible; then so is A^*. For any $x \neq 0$,

$$(AA^*x, x) = (A^*x, A^*x) = \|\,A_x^*\,\|^2 > 0.$$

This proves that AA^* is a positive mapping. According to Theorem 1, AA^* has a unique positive square root R:

$$AA^* = R^2. \tag{62}$$

Define U as $R^{-1}A$; then $U^* = A^*R^{-1}$, and so by (62),

$$UU^* = R^{-1}AA^*R^{-1} = R^{-1}R^2R^{-1} = I.$$

It follows that U is unitary. By definition of U as $R^{-1}A$,

$$A = RU,$$

as asserted in (61).

When A is not invertible, AA^* is a nonnegative self-adjoint map; it has a uniquely determined nonnegative square root R. Therefore

$$\| Rx \|^2 = (Rx, Rx) = (R^2x, x) = (AA^*x, x)$$
$$= (A^*x, A^*x) = \| A^*x \|^2. \tag{63}$$

Suppose $Rx = Ry$; then $\| R(x - y) \| = 0$, and so according to (63), $\| A^*(x - y) \| = 0$, therefore $A^*_x = A^*y$. This shows that for any u in the range of R, $u = Rx$, we can define Vu as A^*x. According to (63), V is an isometry; therefore it can be extended to the whole space as a unitary mapping.

By definition, $A^* = VR$; taking its adjoint gives $A = RV^*$, which is relation (61) with $V^* = U$. □

According to the spectral representation theorem, the self-adjoint map R can be expressed as $R = WDW^*$, where D is diagonal and W is unitary. Setting this into (61) gives $A = WDW^*U$. Denoting W^*U as V, we get

$$A = WDV, \tag{64}$$

where W and V are unitary and D is diagonal, with nonnegative entries. Equation (64) is called the *singular value decomposition* of the mapping A. The diagonal entries of D are called the *singular values* of A; they are the nonnegative square roots of the eigenvalues of AA^*.

Take the adjoint of both sides of (61); we get

$$A^* = U^*R. \tag{61}^*$$

Denote A^* as B, denote U^* as V, and restate $(61)^*$ as

Theorem 22*. Every linear mapping B of a complex Euclidean space can be factored as

$$B = MS,$$

where S is self-adjoint and nonnegative, and M is unitary.

Note. When B maps a *real* Euclidean space into itself, so do S and M.

EXERCISE 13. Let A be any mapping of a Euclidean space into itself. Show that AA^* and A^*A have the same eigenvalues with the same multiplicity.

EXERCISE 14. Let A be a mapping of a Euclidean space into another Euclidean space. Show that AA^* and A^*A have the same *nonzero* eigenvalues with the same multiplicity.

EXERCISE 15. Give an example of a 2×2 matrix Z whose eigenvalues have positive real part but $Z + Z^*$ is not positive.

EXERCISE 16. Verify that the commutator (50) of two self-adjoint matrices is anti-self-adjoint.

CHAPTER 11

Kinematics and Dynamics

In this chapter we shall illustrate how extremely useful the theory of linear algebra in general and matrices in particular are for describing motion in space. There are three sections, on the kinematics of rigid body motions, on the kinematics of fluid flow, and on the dynamics of small vibrations.

1. THE MOTION OF RIGID BODIES

An *isometry* was defined in Chapter 7 as a mapping of a Euclidean space into itself that preserves distances. When the isometry relates the positions of a mechanical system in three-dimensional real space at two different times, it is called a *rigid body motion*. In this section we shall study such motions.

Theorem 10 of Chapter 7 shows that an isometry M that preserves the origin is linear and satisfies

$$M^*M = I. \tag{1}$$

As noted in equation (33) of that chapter, the determinant of such an isometry is plus or minus 1; its value for all rigid body motions is 1.

Theorem 1 (Euler). An isometry M of three-dimensional real Euclidean space with determinant plus 1 that is nontrivial, that is not equal to I, is a rotation; it has a uniquely defined axis of rotation and angle of rotation θ.

Proof. Points f on the axis of rotation remain fixed, so they satisfy

$$Mf = f; \tag{2}$$

Linear Algebra and Its Applications, Second Edition, by Peter D. Lax
Copyright © 2007 John Wiley & Sons, Inc.

that is, they are eigenvectors of M with eigenvalue 1. We claim that a nontrivial isometry, $\det M = 1$, has exactly one eigenvalue equal to 1. To see this, look at the characteristic polynomial of M, $p(s) = \det(sI - M)$. Since M is a real matrix, $p(s)$ has real coefficients. The leading term in $p(s)$ is s^3, so $p(s)$ tends to $+\infty$ as s tends to $+\infty$. On the other hand, $p(0) = \det(-M) = -\det M = -1$. So p has a root on the positive axis; that root is an eigenvalue of M. Since M is an isometry, that eigenvalue can only be plus 1. Furthermore, 1 is a simple eigenvalue; for if a second eigenvalue were equal to 1, then, since the product of all three eigenvalues equals $\det M = 1$, the third eigenvalue of M would also be 1. Since M is a normal matrix, it has a full set of eigenvectors, all with eigenvalue 1; that would make $M = I$, excluded as the trivial case.

To see that M is a rotation around the axis formed by the fixed vectors, we represent M in an orthonormal basis consisting of f satisfying (2), and two other vectors. In this basis the column vector $(1,0,0)$ is an eigenvector of M with eigenvalue 1; so the first column is $(1,0,0)$. Since the columns of an isometry are orthogonal unit vectors and $M = 1$, the matrix M has the form

$$M = \begin{pmatrix} 1 & 0 & 0 \\ 0 & c & -s \\ 0 & s & c \end{pmatrix}, \tag{3}$$

where $c^2 + s^2 = 1$. Thus $c = \cos\theta$, $s = \sin\theta$, θ some angle. Clearly, (3) is rotation around the first axis by angle θ. $\qquad\square$

The rotation angle is easily calculated without introducing a new basis that brings M into form (3). We recall the definition of trace from Chapter 6 and Theorem 2 in that chapter, according to which similar matrices have the same trace. Therefore, M has the same trace in every basis; from (3),

$$\operatorname{tr} M = 1 + 2\cos\theta, \tag{4}$$

hence

$$\cos\theta = \frac{\operatorname{tr} M - 1}{2}. \tag{4$'$}$$

We turn now to rigid motions which keep the origin fixed and which depend on time t, that is, functions $M(t)$ whose values are rotations. We take $M(t)$ to be the rotation that brings the configuration at time 0 into the configuration at time t. Thus

$$M(0) = I. \tag{5}$$

If we change the reference time from 0 to t_1, the function M_1 describing the motion from t_1 to t is

$$M_1(t) = M(t)M(t_1)^{-1}. \tag{6}$$

Equation (1) shows that M^* is left inverse of M; then it is also right inverse:

$$MM^* = I. \tag{7}$$

We assume that $M(t)$ is a differentiable function of t. Differentiating this with respect to t and denoting the derivative by the subscript t gives

$$M_t M^* + MM_t^* = 0. \tag{8}$$

We denote

$$M_t M^* = A. \tag{9}$$

Since differentiation and taking the adjoint commute,

$$A^* = MM_t^*;$$

therefore (8) can be written as

$$A + A^* = 0. \tag{10}$$

This shows that $A(t)$ is antisymmetric. Equation (9) itself can be rewritten by multiplying by M on the right and using (1);

$$M_t = AM. \tag{11}$$

Note that if we differentiate (6) and use (11) we get the same equation

$$M_{1_i} = AM_1. \tag{11}_1$$

This shows the significance of $A(t)$, for the motion is independent of the reference time; $A(t)$ is called the *infinitesimal generator* of the motion.

EXERCISE 1.   Show that if $M(t)$ satisfies a differential equation of form (11), where $A(t)$ is antisymmetric for each t and the initial condition (5), then $M(t)$ is a rotation for every t.

EXERCISE 2.   Suppose that A is independent of t; show that the solution of equation (11) satisfying the initial condition (5) is

$$M(t) = e^{tA}. \tag{12}$$

EXERCISE 3.   Show that when A depends on t, equation (11) is *not* solved by

$$M(t) = e^{\int_0^t A(s)ds},$$

unless $A(t)$ and $A(s)$ commute for all s and t.

We investigate now $M(t)$ near $t = 0$; we assume that $M(t) \neq I$ for $t \neq 0$; then for each $t \neq 0$, $M(t)$ has a unique axis of rotation $f(t)$:

$$M(t)f(t) = f(t).$$

We assume that $f(t)$ depends differentiably on t; differentiating the preceding formula gives

$$M_t f + M f_t = f_t.$$

We assume that both $f(t)$ and $f_t(t)$ have limits as $t \to 0$. Letting $t \to 0$ in this formula gives

$$M_t f(0) + M(0)f_t = f_t. \tag{13}$$

Using (11) and (5), we get

$$A(0)f(0) = 0. \tag{14}$$

We claim that if $A(0) \neq 0$ then this equation has essentially one solution, that is, all are multiples of each other. To see that there is a nontrivial solution, recall that A is antisymmetric; for n odd,

$$\det A = \det A^* = \det(-A) = (-1)^n \det A = -\det A,$$

from which it follows that $\det A = 0$, that is, the determinant of an antisymmetric matrix of *odd* order is zero. This proves that A is not invertible, so that (14) has a nontrivial solution. This fact can also be seen directly for 3×3 matrices by writing out

$$A = \begin{pmatrix} 0 & a & b \\ -a & 0 & c \\ -b & -c & 0 \end{pmatrix}. \tag{15}$$

Inspection shows that

$$f = \begin{pmatrix} -c \\ b \\ -a \end{pmatrix}, \tag{16}$$

lies in the nullspace of A.

EXERCISE 4. Show that if A in (15) is not equal to 0, then all vectors annihilated by A are multiples of (16).

EXERCISE 5. Show that the two other eigenvalues of A are $\pm i\sqrt{a^2 + b^2 + c^2}$.

EXERCISE 6. Show that the motion $M(t)$ described by (12) is rotation around the axis through the vector f given by formula (16). Show that the angle of rotation is $t\sqrt{a^2 + b^2 + c^2}$. (*Hint:* Use formula (4)′.)

The one-dimensional subspace spanned by $f(0)$ satisfying (14), being the limit of the axes of rotation $f(t)$, is called the *instantaneous axis of rotation* of the motion at $t = 0$.

Let $\theta(t)$ denote the angle through which $M(t)$ rotates. Formula (4)′ shows that $\theta(t)$ is a differentiable function of t; since $M(0) = I$, it follows that $\operatorname{tr} M(0) = 3$, and so by (4)′ $\cos \theta(0) = 1$. This shows that $\theta(0) = 0$.

We determine now the derivative of θ at $t = 0$. For this purpose we differentiate (4)′ twice with respect to t. Since trace is a linear function of matrices, the derivative of the trace is the trace of the derivative, and so we get

$$-\theta_{tt} \sin \theta - \theta_t^2 \cos \theta = \frac{1}{2} \operatorname{tr} M_{tt}.$$

Setting $t = 0$ gives

$$\theta_t^2(0) = -\frac{1}{2} \operatorname{tr} M_{tt}(0). \tag{17}$$

To express $M_{tt}(0)$ we differentiate (11):

$$M_{tt} = A_t M + A M_t = A_t M + A^2 M.$$

Setting $t = 0$ gives

$$M_{tt}(0) = A_t(0) + A^2(0).$$

Take the trace of both sides. Since $A(t)$ is antisymmetric for every t, so is A_t; the trace of an antisymmetric matrix being zero, we get $\operatorname{tr} M_{tt}(0) = \operatorname{tr} A^2(0)$. Using formula (15), a brief calculation gives

$$\operatorname{tr} A^2(0) = -2(a^2 + b^2 + c^2).$$

Combining the last two relations and setting it into (17) gives

$$\theta_t^2(0) = a^2 + b^2 + c^2.$$

Compare this with (16); we get

$$|\theta_t| = |f|. \tag{18}$$

The quantity θ_t is called the *instantaneous angular velocity* of the motion; the vector f given by (16) is called the *instantaneous angular velocity vector*.

EXERCISE 7. Show that the commutator

$$[A, B] = AB - BA$$

of two antisymmetric matrices is antisymmetric.

EXERCISE 8. Let A denote the 3×3 matrix (15); we denote the associated null vector (16) by f_A. Obviously, f depends linearly on A.
(a) Let A and B denote two 3×3 antisymmetric matrices. Show that

$$\text{tr } AB = -2(f_A, f_B),$$

where (,) denotes the standard scalar product for vectors in $\mathbb{R}^3$.

EXERCISE 9. Show that the cross product can be expressed as

$$f_{|A,B|} = f_A \times f_B.$$

2. THE KINEMATICS OF FLUID FLOW

The concept of angular velocity vector is also useful for discussing motions that are not rigid, such as the motion of fluids. We describe the motion of a fluid by

$$x = x(y, t); \tag{19}$$

here x denotes the position of a point in the fluid at time t that at time zero was located at y:

$$x(y, 0) = y. \tag{19_0}$$

The partial derivative of x with respect to t, y fixed, is the *velocity* v of the flow:

$$\frac{\partial}{\partial t} x(y, t) = x_t(y, t) = v(y, t). \tag{20}$$

The mapping $y \to x$, t fixed, is described locally by the Jacobian matrix

$$J(y, t) = \frac{\partial x}{\partial y}, \qquad \text{that is, } J_{ij} = \frac{\partial x_i}{\partial y_j}. \tag{21}$$

It follows from $(19)_0$ that

$$J(y, 0) = I. \tag{21_0}$$

We learn in the integral calculus of functions of several variables that the determinant of the Jacobian $J(y, t)$ is the factor by which the volume of the fluid initially at y is *expanded* at time t. We assume that the fluid is never compressed to zero. Since at $t = 0$, $\det J(y, 0) = \det I = 1$ is positive, it follows that $\det J(y, t)$ is positive for all t.

We appeal now to Theorem 22[*] of Chapter 10 to factor the matrix J as

$$J = MS, \tag{22}$$

$M = M(y, t)$ a rotation, $S = S(y, t)$ selfadjoint and positive. Since J is real, so are M and S. Since det J and det S are positive, so is det M. Since $J(t) \to I$ as $t \to 0$, it follows, see the proof of Theorem 22 in Chapter 10, that also S and $M \to I$ as $t \to 0$.

It follows from the spectral theory of self-adjoint matrices that S acts as *compression* or *dilation* along the three axes that are the eigenvectors of S. M is rotation; we shall calculate now the rate of rotation by the action of M. To do this we differentiate (22) with respect to t:

$$J_t = MS_t + M_t S. \tag{22$'$}$$

We multiply (22) by M^* on the left; since $M^* M = I$ we get

$$M^* J = S.$$

We multiply this relation by M_t from the left, make use of the differential equation $M_t = AM$, see (11), and that $MM^* = I$.

$$M_t S = AMM^* J = AJ.$$

Setting this into (22)$'$ gives

$$J_t = MS_t + AJ. \tag{23}$$

Set $t = 0$:

$$J_t(0) = S_t(0) + A(0). \tag{23$_0$}$$

We recall from (10) that $A(0)$ is anti-self-adjoint. S_t on the other hand, being the derivative of self-adjoint matrices, is itself self-adjoint. Thus (23)$_0$ is the decomposition of $J_t(0)$ into its self-adjoint and anti-self-adjoint parts.

Differentiating (21) with respect to t and using (20) gives

$$J_t = \frac{\partial v}{\partial y}, \tag{24}$$

that is,

$$J_{t_{ij}} = \frac{\partial v_i}{\partial y_j}. \tag{24$'$}$$

Thus the self-adjoint and anti-self-adjoint parts of $J_t(0)$ are

$$S_{t_{ij}}(0) = \frac{1}{2}\left(\frac{\partial v_i}{\partial y_j} + \frac{\partial v_j}{\partial y_i}\right), \tag{25}$$

$$A_{ij}(0) = \frac{1}{2}\left(\frac{\partial v_i}{\partial y_j} - \frac{\partial v_j}{\partial y_i}\right), \tag{25}'$$

In (15) we have given the names a, b, c to the entries of A:

$$a = \frac{1}{2}\left(\frac{\partial v_1}{\partial y_2} - \frac{\partial v_2}{\partial y_1}\right), \qquad b = \frac{1}{2}\left(\frac{\partial v_1}{\partial y_3} - \frac{\partial v_3}{\partial y_1}\right),$$
$$c = \frac{1}{2}\left(\frac{\partial v_2}{\partial y_3} - \frac{\partial v_3}{\partial y_2}\right).$$

Set this into formula (16) for the instantaneous angular velocity vector:

$$f = \frac{1}{2}\begin{pmatrix} \dfrac{\partial v_3}{\partial y_2} - \dfrac{\partial v_2}{\partial y_3} \\[2mm] \dfrac{\partial v_1}{\partial y_3} - \dfrac{\partial v_3}{\partial y_1} \\[2mm] \dfrac{\partial v_2}{\partial y_1} - \dfrac{\partial v_1}{\partial y_2} \end{pmatrix} = \frac{1}{2}\,\mathrm{curl}\,v. \tag{26}$$

In words: A fluid that is flowing with velocity v has instantaneous angular velocity equal to $\frac{1}{2}$ curl v, called its *vorticity*. A flow for which curl $v = 0$ is called *irrotational*.

We recall from advanced calculus that a vector field v whose curl is zero can, in any simply connected domain, be written as the gradient of some scalar function ϕ. Thus for an irrotational flow, the velocity is

$$v = \mathrm{grad}\,\phi;$$

ϕ is called the *velocity potential*.

We calculate now the rate at which the fluid is being expanded. We saw earlier that expansion is det J. Therefore the rate at which fluid is expanded is (d/dt) det J. In Chapter 9, Theorem 4, we have given a formula, equation (10), for the logarithmic derivative of the determinant:

$$\frac{d}{dt}\log \det J = \mathrm{tr}(J^{-1}J_t). \tag{27}$$

We set $t = 0$; according to $(21)_0$, $J(0) = I$; therefore we can rewrite equation (27) as

$$\frac{d}{dt}\det J(0) = \mathrm{tr}\,J_t(0).$$

By (24)', $J_{t_{ij}} = \partial v_i/\partial y_i$; therefore

$$\frac{d}{dt}\det J = \sum \frac{\partial v_i}{\partial y_i} = \operatorname{div} v. \tag{27}'$$

In words: A fluid that is flowing with velocity v is being expanded at the rate div v. That is why the velocity field of an *incompressible fluid* is *divergence free*.

3. THE FREQUENCY OF SMALL VIBRATIONS

By small vibrations we mean motions of small amplitude about a point of equilibrium. Since the amplitude is small, the equation of motion can be taken to be linear. Let us start with the one-dimensional case, the vibration of a mass m under the action of a spring. Denote by $x = x(t)$ displacement of the mass from equilibrium $x = 0$. The force of the spring, restoring the mass toward equilibrium, is taken to be $-kx$, k a positive constant. Newton's law of motion, force equals mass times acceleration, says that,

$$m\ddot{x} + kx = 0; \tag{28}$$

here the dot symbol $\cdot$ denotes differentiation with respect to t.
    Multiply (28) by $\dot{x}$:

$$m\ddot{x}\dot{x} + kx\dot{x} = \frac{d}{dt}\left[\frac{1}{2}m\dot{x}^2 + \frac{k}{2}x^2\right] = 0;$$

therefore

$$\frac{1}{2}m\dot{x}^2 + \frac{k}{2}x^2 = E \tag{29}$$

is a constant, independent of t. The first term in (29) is the *kinetic energy* of a mass m moving with velocity $\dot{x}$; the second term is the *potential energy* stored in a spring displaced by the amount x. That their sum, E, is constant expresses the *conservation* of *total energy*.
    The equation of motion (28) can be solved explicitly: All solutions are of the form

$$x(t) = a\sin\left(\sqrt{\frac{k}{m}}t + \theta\right); \tag{30}$$

a is called the amplitude, θ the phase. All solutions (30) are *periodic* in t, with period $p = 2\pi\sqrt{m/k}$. The *frequency*, defined as the reciprocal of the period, is the number of vibrations the system performs per unit time:

$$\text{frequency} = \frac{1}{2\pi}\sqrt{\frac{k}{m}}. \tag{31}$$

We note that part of this result can be deduced by dimensional analysis. From the fact that kx is a force, we deduce that

$$\dim k \cdot \text{length} = \dim \text{force}$$
$$= \text{mass} \cdot \text{acceleration} = \frac{\text{mass} \cdot \text{length}}{\text{time}^2}.$$

So

$$\dim k = \frac{\text{mass}}{\text{time}^2}.$$

The only quantity constructed out of the two parameters m and k whose dimension is time is const $\sqrt{m/k}$. So we conclude that the period p of motion is given by

$$p = \text{const} \sqrt{m/k}.$$

Formula (31) shows that *frequency is an increasing function of k, and a decreasing function of m.* Intuitively this is clear; increasing k makes the spring stiffer and the vibration faster; the smaller the mass, the faster the vibration.

We present now a far-reaching generalization of this result to the motion of a system of n masses on a line, each linked elastically to each other and to the origin. Denote by x_i the position of the ith particle; Newton's second law of motion for the ith particle is

$$m_i \ddot{x}_i - f_i = 0, \tag{32}$$

where f_i is the total force acting on the ith particle and m_i is its mass. We take the origin to be a point of equilibrium for the system, that is, all f_i are zero when all the x_i are zero.

We denote by f_{ij} the force exerted by the jth particle on the ith. According to Newton's third law, the force exerted by the ith particle on the jth is $-f_{ij}$. We take f_{ij} to be proportional to the distance of x_i and x_j:

$$f_{ij} = k_{ij}(x_j - x_i), \qquad i \neq j. \tag{33}$$

To satisfy $f_{ij} = -f_{ji}$ we take $k_{ij} = k_{ji}$. Finally, we take the force exerted from the origin on particle i to be $-k_i x_i$. Altogether we have

$$f_i = \sum_j k_{ij} x_j, \qquad k_{ii} = -k_i - \sum_j k_{ij}. \tag{33'}$$

We now rewrite the system (32) in matrix form as

$$M\ddot{x} + Kx = 0; \tag{32'}$$

here x denotes the vector $(x_1, x_2, \ldots, x_n)'$, M is a *diagonal* matrix with entries m_i, and the elements of K are $-k_{ij}$ from (33)'. The matrix K is real and *symmetric*; then taking the scalar product of (32)' with $\dot{x}$ we obtain

$$(\dot{x}, M\ddot{x}) + (\dot{x}, Kx) = 0.$$

Using the symmetry of K and M we can rewrite this as

$$\frac{d}{dt} \left[\frac{1}{2} (\dot{x}, M\dot{x}) + \frac{1}{2} (x, Kx) \right] = 0,$$

from which we conclude that

$$\frac{1}{2} (\dot{x}, M\dot{x}) + \frac{1}{2} (x, Kx) = E \tag{34}$$

is a constant independent of t. The first term on the left-hand side is the *kinetic energy* of the masses, the second term the *potential energy* stored in the system when the particles have been displaced from the origin to x. That their sum, E, is constant during the motion is an expression of the *conservation of total energy*.

We assume now that all the forces are attractive, that is, that k_{ij} and k_i are positive. We claim that then the matrix K is *positive*. For proof see Theorem 5 at the end of this chapter. According to inequality (5)' of Chapter 10, a positive matrix K satisfies for all x,

$$a \, \| x \|^2 \leq (x, Kx), \qquad a \text{ positive.}$$

Since the diagonal matrix M is positive, combining the above inequality with (34) gives

$$a \, \| x \|^2 \leq E.$$

This shows that the *amplitude* $\| x \|$ is *uniformly bounded* for all time, and furthermore if the total energy E is sufficiently small, the amplitude $\| x \|$ is small.

A second important consequence of the positivity of K is

Theorem 2. Solutions of the differential equation (32)' are uniquely determined by their initial data $x(0)$ and $\dot{x}(0)$. That is, two solutions that have the same initial data are equal for all time.

Proof. Since equation (32)' is linear, the difference of two solutions is again a solution. Therefore it is sufficient to prove that if a solution x has zero initial data, then $x(t)$ is zero for all t. To see this, we observe that if $x(0) = 0$, $\dot{x}(0) = 0$, then energy E at $t = 0$ is zero. Therefore energy is zero for all t. But energy defined by (34) is the sum of two nonnegative terms; therefore each is zero for all t.   □

Since equation (32)′ is linear, its solutions form a linear space. We shall show that the dimension of this space of solutions is $\leq 2n$, where n is the number of masses. To see this, map each solution $x(t)$ into its initial data $x(0)$, $\dot{x}(0)$. Since there are n particles, their initial data belong to a $2n$-dimensional linear space. This mapping is linear; we claim that it is 1-to-1. According to Theorem 2, two solutions with the same initial data are equal; in particular the nullspace of this map is $\{0\}$. Then it follows from Theorem 1 of Chapter 3 that the dimension of the space of solutions is $\leq 2n$.

We turn now to finding all solutions of the equations of motion (32)′. Since the matrices M and K are constant, differentiating equation (32)′ with respect to t gives

$$M\ddot{x} + K\dot{x} = 0.$$

In words: If $x(t)$ is a solution of (32)′, so is $\dot{x}(t)$.

The solutions of (32)′ form a finite-dimensional space. The mapping $x \to \dot{x}$ maps this space into itself. According to the spectral theorem, the eigenfunctions and generealized eigenfunctions of this mapping span the space.

Eigenfunctions of the map $x \to \dot{x}$ satisfy the equation $\dot{x} = ax$; the solutions of this are $x(t) = e^{-at} h$, where a is a complex number, h is a vector with n components, and n is the number of particles. Since we have shown above that each solution of (32)′ is uniformly bounded for all t, it follows that a is pure imaginary: $a = ic$, c real. To determine c and h we set $x = e^{ict}h$ into (32)′. We get, after dividing by e^{ict}, that

$$c^2 Mh = Kh. \tag{35}$$

This is an eigenvalue problem we have already encountered in Chapter 8, equation (48). We can reduce (35) to a standard eigenvalue problem by introducing $M^{1/2}h = k$ as new unknown vector into (35) and then multiplying equation (35) on the left by $M^{-1/2}$. We get

$$c^2 k = M^{-1/2}KM^{-1/2}k. \tag{35'}$$

Since $M^{-1/2}KM^{-1/2}$ is self-adjoint, it has n linearly independent eigenvectors $k_1, \ldots, k_n$, with corresponding eigenvalues $c_1^2, \ldots, c_n^2$. Since, as we shall show, K is a positive matrix, so is $M^{-1/2}KM^{-1/2}$. Therefore the c_j are real numbers; we take them to be positive.

The corresponding n solutions of the differential equation (32)′ are $e^{ic_jt}h_j$, whose real and imaginary parts also are solutions:

$$(\cos c_j t)h_j, \qquad (\sin c_j t)h_j, \tag{36}$$

as are all linear combinations of them:

$$\sum a_j(\cos c_j t)h_j + \sum b_j(\sin c_j t)h_j = x(t); \tag{36'}$$

the a_j and b_j are arbitrary real numbers.

Theorem 3. Every solution of the differential equation (32)' is of form (36).

Proof. Solutions of the form (36)' form a $2n$-dimensional space. We have shown that the set of all solutions is a linear space of dimension $\leq 2n$. It follows that all solutions are of form (36)'.   □

EXERCISE 10. Verify that solutions of the form (36) form a $2n$-dimensional linear space.

The special solutions (36); are called *normal modes*; each is periodic, with period $2\pi/c_j$ and frequency $c_j/2\pi$. These are called the *natural frequencies* of the mechanical system governed by equation (32)'.

Theorem 4. Consider two differential equations of form (32)':

$$M\ddot{x} + Kx = 0, \qquad N\ddot{y} + Ly = 0, \tag{37}$$

M, K, N, L positive, real $n \times n$ matrices. Suppose that

$$M \geq N \quad \text{and} \quad K \leq L. \tag{38}$$

Denote 2π times the natural frequencies of the first system, arranged in increasing order by $c_1 \leq \ldots \leq c_n$, and those of the second system by $d_1 \leq \ldots \leq d_n$. We claim that

$$c_j \leq d_j, \qquad j = 1, \ldots, n. \tag{39}$$

Proof. We introduce an intermediate differential equation

$$M\ddot{z} + Lz = 0,$$

Denote its natural frequencies by $f_j/2\pi$. In analogy with equation (35), the f_j satisfy

$$f^2 Mh = Lh,$$

where h is an eigenvector. In analogy with equation (35)', we can identify the numbers f^2 as the eigenvalues of

$$M^{-1/2}LM^{-1/2}.$$

We recall that the numbers c^2 are eigenvalues of

$$M^{-1/2}KM^{-1/2}.$$

Since K is assumed to be $\leq$ L, it follows from Theorem 1 of Chapter 10 that also

$$M^{-1/2}KM^{-1/2} \leq M^{-1/2}LM^{-1/2}.$$

Then it follows from Theorem 16 of Chapter 10 that

$$c_j^2 \leq f_j^2, \qquad j = 1,\ldots,n. \tag{39}'$$

On the other hand, in analogy with equation (35)″, we can identify the reciprocals $1/f^2$ as the eigenvalues of

$$L^{-1/2}ML^{-1/2},$$

whereas the reciprocals $1/d^2$ are the eigenvalues of

$$L^{-1/2}NL^{-1/2}.$$

Since N is assumed, to be $\leq$ M, it follows as before that

$$L^{-1/2}NL^{-1/2} \leq L^{-1/2}ML^{-1/2},$$

so by Theorem 16 of Chapter 10

$$\frac{1}{d_j^2} \leq \frac{1}{f_j^2}. \tag{39}''$$

We can combine inequalities (39)′ and (39)″ to deduce (39).   □

Note: If either of the inequalities in (38) is strict, then all the inequalities in (39) are strict.

The intuitive meaning of Theorem 4 is that if in a mechanical system we stiffen the forces binding the particles to each other and reduce the mass of all the particles, then *all* natural frequencies of the system increase.

We supply now the proof of the positivity of the matrix K.

Theorem 5. Suppose that the numbers k_i and k_{ij}, $i \neq j$ are positive. Then the symmetric matrix K,

$$K_{ij} = -k_{ij}, \quad i \neq j; \qquad K_{ii} = k_i + \sum_{i \neq j} k_{ij} \tag{40}$$

is positive.

Proof. It suffices to show that every eigenvalue a of K is positive:

$$Ku = au. \tag{41}$$

Normalize the eigenvector u of K so that the largest component, say u_i, equals 1, and all others are ≤ 1. The ith component of (41) is

$$K_{ii} + \sum_{j \neq i} K_{ij} u_j = a.$$

Using the definition (40) of the entries of K, this can be rewritten as

$$k_i + \sum_{j \neq i} k_{ij}(1 - u_j) = a.$$

The left-hand side is positive; therefore, so is the right-hand side a.   □

For a more general result, see Appendix 7.

CHAPTER 12

Convexity

Convexity is a primitive notion, based on nothing but the bare bones of the structure of linear spaces over the reals. Yet some of its basic results are surprisingly deep; furthermore, these results make their appearance in an astonishingly wide variety of topics.

X is a linear space over the reals. For any pair of vectors x, y in X, the *line segment* with endpoints x and y is defined as the set of points in X of form

$$ax + (1 - a)y, \qquad 0 \le a \le 1. \tag{1}$$

Definition. A set K in X is called *convex* if, whenever x and y belong to K, all points of the line segment with endpoints x, y also belong to K.

Examples of Convex Sets

(a) $K =$ the whole space X.
(b) $K = \phi$, the empty set.
(c) $K = \{x\}$, a single point.
(d) $K =$ any line segment.
(e) Let l be a linear function in X; then the sets

$$l(x) = c, \text{called a } hyperplane, \tag{2}$$
$$l(x) < c, \text{called an } open\ half\text{-}space, \tag{3}$$
$$l(x) \le c, \text{called a } closed\ half\text{-}space, \tag{4}$$

are all convex sets.

Concrete Examples of Convex Sets

(f) X the space of all polynomials with real coefficients, K the subset of all polynomials that are positive at every point of the interval $(0, 1)$.

(g) X the space of real, self-adjoint matrices, K the subset of positive matrices.

EXERCISE 1.   Verify that these are convex sets.

Theorem 1.   **(a)** The intersection of any collection of convex sets is convex.

(b) The sum of two convex sets is convex, where the sum of two sets K and H is defined as the set of all sums $x + y, x$ in K, y in H.

EXERCISE 2.   Prove these propositions.

Using Theorem 1, we can build an astonishing variety of convex sets out of a few basic ones. For instance, a triangle in the plane is the intersection of three half-planes.

Definition.   A point x is called an *interior point* of a set S in X if for every y in X, $x + yt$ belongs to S for all sufficiently small positive t.

Definition.   A convex set K in X is called *open* if every point in it is an interior point.

EXERCISE 3.   Show that an open half-space (3) is an open convex set.

EXERCISE 4.   Show that if A is an open convex set and B is convex, then $A + B$ is open and convex.

Definition.   Let K be an open convex set that contains the vector 0. We define its *gauge function* $p_K = p$ as follows: For every x in X,

$$p(x) = \inf r, \qquad r > 0 \quad \text{and} \quad \frac{x}{r} \text{ in } K. \tag{5}$$

EXERCISE 5.   Let X be a Euclidean space, and let K be the open ball of radius a centered at the origin: $\| x \| < a$.

(i) Show that K is a convex set.

(ii) Show that the gauge function of K is $p(x) = \| x \|/a$.

EXERCISE 6.   In the (u, v) plane take K to be the quarter-plane $u < 1, v < 1$. Show that the gauge function of K is

$$p(u, v) = \begin{cases} 0 & \text{if } u \le 0, \quad v \le 0, \\ v & \text{if } 0 < v, \quad u \le 0, \\ u & \text{if } 0 < u, \quad v \le 0, \\ \max(u, v) & \text{if } 0 < u, \quad 0 < v. \end{cases}$$

Theorem 2. **(a)** The gauge function p of an open convex set K that contains the origin is well-defined for every x.

(b) p is positive homogeneous:

$$p(ax) = ap(x) \qquad \text{for } a > 0. \tag{6}$$

(c) p is subadditive:

$$p(x+y) \le p(x) + p(y). \tag{7}$$

(d) $p(x) < 1$ iff x is in K.

Proof. Call the set of $r > 0$ for which x/r is in K *admissible* for x. To prove **(a)** we have to show that for any x the set of admissible r is nonempty. This follows from the assumption that 0 is an interior point of K.

(b) follows from the observation that if r is admissible for x and $a > 0$, then ar is admissible for ax.

(c) Let s and t be positive numbers such that

$$p(x) < s, \qquad p(y) < t. \tag{8}$$

Then by definition of p as inf, it follows that s and t are admissible for x and y; therefore x/s and y/t belong to K. The point

$$\frac{x+y}{s+t} = \frac{s}{s+t}\frac{x}{s} + \frac{t}{s+t}\frac{y}{t} \tag{9}$$

lies on the line segment connecting x/s and y/t. By convexity, $(x+y)/s+t$ belongs to K. This shows that $s+t$ is admissible for $x+y$; so by definition of p,

$$p(x+y) \le s+t. \tag{10}$$

Since s and t can be chosen arbitrarily close to $p(x)$ and $p(y)$, **(c)** follows.

(d) Suppose $p(x) < 1$; by definition there is an admissible $r < 1$. Since r is admissible, x/r belongs to K. The identity $x = rx/r + (1-r)0$ shows that x lies on the line segment with endpoints 0 and x/r, so by convexity belongs to K.

Conversely, suppose x belongs to K; since x is assumed to be an interior point of K the point $x + \epsilon x$ belongs to K for $\epsilon > 0$ but small enough. This shows that $r = 1/(1+\epsilon)$ is admissible, and so by definition

$$p(x) \le \frac{1}{1+\epsilon}.$$

This completes the proof of the theorem. $\qquad\square$

EXERCISE 7.   Let p be a positive homogeneous, subadditive function. Prove that the set K consisting of all x for which $p(x) < 1$ is convex and open.

Theorem 2 gives an analytical description of the open convex sets. There is another, dual description. To derive it we need the following basic, and geometrically intuitive results.

Theorem 3.   Let K be an open convex set, and let y be a point not in K. Then there is an open half-space containing K but not y.

Proof. An open half-space is by definition a set of points satisfying inequality $l(x) < c$; see (3). So we have to construct a linear function l and a number c such that

$$l(x) < c \qquad \text{for all } x \text{ in } K, \tag{11}$$
$$l(y) = c \tag{12}$$

We assume that 0 lies in K; otherwise shift K. Set $x = 0$ in (11); we get $0 < c$. We may set $c = 1$. Let p be the gauge function of K; according to Theorem 2, points of K are characterized by $p(x) < 1$;. It follows that (11) can be stated so:

$$\text{If } p(x) < 1, \quad \text{then } l(x) < 1. \tag{11$'$}$$

This will certainly be the case if

$$l(x) \le p(x) \qquad \text{for all } x. \tag{13}$$

So Theorem 3 is a consequence of the following: there exists a linear function l which satisfies (13) for all x and whose value at y is 1. We show first that the two requirements are compatible. Requiring $l(y) = 1$ implies by linearity that $l(ky) = k$ for all k. We show now that (13) is satisfied for all x of form ky; that is, for all k.

$$k = l(ky) \le p(ky), \tag{14}$$

For k positive, we can by (6) rewrite this as

$$k \le kp(y), \tag{14$'$}$$

true because y does not belong to K and so by part (d) of Theorem 2, $p(y) \ge 1$. On the other hand, inequality (14) holds for k negative; since the left-hand side is less than 0, the right-hand side, by definition (5) of gauge function, is positive.

The remaining task is to extend l from the line through y to all of X so that (13) is satisfied. The next theorem asserts that this can be done.

Theorem 4 (Hahn–Banach).   Let p be a real-valued positive homogeneous subadditive function defined on a linear space X over $\mathbb{R}$. Let U be a subspace of X on which a linear function is defined, satisfying (13):

$$l(u) \le p(u) \qquad \text{for all } u \text{ in } U. \tag{13}_U$$

Then l can be extended to all of X so that (13) is satisfied for all x.

Proof. Proof is by induction; we show that l can be extended to a subspace V spanned by U and any vector z not in U. That is, V consists of all vectors of form

$$v = u + tz, \qquad u \text{ in } U, \ t \text{ any real number.}$$

Since l is linear

$$l(v) = l(u) + tl(z);$$

this shows that the value of $l(z) = a$ determines the value of l on V:

$$l(v) = l(u) + ta.$$

The task is to choose a so that (13) is satisfied: $l(v) \le p(v)$, that is,

$$l(u) + ta \le p(u + tz) \tag{13}_v$$

for all u in U and all real t.

We divide $(13)_v$ by $|t|$. For $t > 0$, using positive homogeneity of p and linearity of l we get

$$l(u^*) + a \le p(u^* + z), \tag{14}_+$$

where u^* denotes u/t. For $t < 0$ we obtain

$$l(u^{**}) - a \le p(u^{**} - z), \tag{14}_-$$

where u^{**} denotes $-u/t$. Clearly, $(13)_v$ holds for all u in U and all real t iff $(14)_+$ and $(14)_-$ hold for all u^* and u^{**}, respectively, in U.

We rewrite $(14)_\pm$ as

$$l(u^{**}) - p(u^{**} - z) \le a \le p(u^* + z) - l(u^*);$$

the number a has to be so chosen that this holds for all u^* and u^{**} in U. Clearly, this is possible iff every number on the left is less than or equal to any number on the right, that is, if

$$l(u^{**}) - p(u^{**} - z) \le p(u^* + z) - l(u^*) \tag{15}$$

for all u^*, u^{**} in U. We can rewrite this inequality as

$$l(u^{**}) + l(u^*) \leq p(u^* + z) + p(u^{**} - z). \tag{15}'$$

By linearity, the left-hand side can be written as $l(u^{**} + u^*)$; since $(13)_U$ holds,

$$l(u^{**} + u^*) \leq p(u^{**} + u^*).$$

Since p is subadditive,

$$p(u^{**} + u^*) = p(u^{**} - z + u^* + z) \leq p(u^{**} - z) + p(u^* + z).$$

This proves $(15)'$, which shows that l can be extended to V. Repeating this n times, we extend l to the whole space X.   □

This completes the proof of Theorem 3.   □

Note. The Hahn–Banach Theorem holds in infinite-dimensional spaces. The proof is the same, with some added logical prestidigitation.

The following result is an easy extension of Theorem 3.

Theorem 5. Let K and H be open convex sets that are disjoint. Then there is a hyperplane that separates them. That is, there is a linear function l and a constant d such that

$$l(x) < d \text{ on } K, \qquad l(y) > d \text{ on } H.$$

Proof. Define the difference $K - H$ to consist of all differences $x - y, x$ in K, y in H. It is easy to verify that this is an open, convex set. Since K and H are disjoint, $K - H$ does not contain the origin. Then by Theorem 3, with $y = 0$, and therefore $c = 0$, there is a linear function l that is negative on $K - H$:

$$l(x - y) < 0 \qquad \text{for } x \text{ in } K, y \text{ in } H.$$

We can rewrite this as

$$l(x) < l(y) \qquad \text{for all } x \text{ in } K, y \text{ in } H.$$

It follows from the completeness of real numbers that there is a number d such that for x in K, y in H,

$$l(x) \leq d \leq l(y).$$

Since both K and H are open, the sign of equality cannot hold; this proves Theorem 5.   □

We show next how to use Theorem 3 to give a dual description of open convex sets.

Definition. Let S be any set in X. We define its *support function* q_S on the dual X' of X as follows:

$$q_S(l) = \sup_{x \text{ in } S} l(x), \tag{16}$$

where l is any linear function.

Remark. $q_S(l)$ may be ∞ for some l.

EXERCISE 8. Prove that the support function q_S of any set is *subadditive*; that is, it satisfies $q_S(m + l) \le q_S(m) + q_S(l)$ for all l, m in X'.

EXERCISE 9. Let S and T be arbitrary sets in X. Prove that $q_{S+T}(l) = q_S(l) + q_T(l)$.

EXERCISE 10. Show that $q_{S \cup T}(l) = \max\{q_S(l), q_T(l)\}$.

Theorem 6. Let K be an open convex set, q_K its support function. Then x belongs to K iff

$$l(x) < q_K(l) \tag{17}$$

for all l in X'.

Proof. It follows from definition (16) that for every x in K $l(x) \le q_K(l)$ for every l; therefore the strict inequality (17) holds for all interior points x in K. To see the converse, suppose that y is not in K. Then by Theorem 3 there is an l such that $l(x) < 1$ for all x in K, but $l(y) = 1$ Thus

$$l(y) = 1 \ge \sup_{x \text{ in } K} l(x) = q_K(l); \tag{18}$$

this shows that y not in K fails to satisfy (17) for some l. This proves Theorem 6. $\square$

Definition. A convex set K in X is called *closed* if every open segment $ax + (1 - a)y$, $0 < a < 1$, that belongs to K has its endpoints x and y in K.

Examples

The whole space X is closed.
The empty set is closed.

A set consisting of a single point is closed.

An interval of form (1) is closed.

EXERCISE 11.   Show that a closed half-space as defined by (4) is a closed convex set.

EXERCISE 12.   Show that the closed unit ball in Euclidean space, consisting of all points $\| x \| \leq 1$, is a closed convex set.

EXERCISE 13.   Show that the intersection of closed convex sets is a closed convex set.

Theorems 2, 3, and 6 have their analogue for closed convex sets.

Theorem 7.   Let K be a closed, convex set, and y a point not in K. Then there is a closed half-space that contains K but not y.

Sketch of Proof. Suppose K contains the origin. If K has no interior points, it lies in a lower-dimensional subspace. If it has an interior point, we choose it to be the origin. Then the gauge function p_K of K can be defined as before. If x belongs to K, we may choose in the definition (5) of p_K the value $r = 1$; this shows that for x in K, $p(x) \leq 1$, Conversely, if $p_K(x) < 1$, then by (5) x/r belongs to K for some $r < 1$. Since 0 belongs to K, by convexity so does x. If $p_K(x) = 1$, then for all $r > 1, x/r$ belongs to K. Since K is closed, so does the endpoint x. This shows that K consists of all points x which satisfy $p_K(x) \leq 1$. We then proceed as in the proof of Theorem 3.   □

Theorem 7 can be rephrased as follows.

Theorem 8.   Let K be a closed, convex set, q_K its support function. Then x belongs to K iff

$$l(x) \leq q_K(l) \tag{19}$$

for all l in X'.

EXERCISE 14.   Complete the proof of Theorems 7 and 8.

Both Theorems 6 and 8 describe convex sets as intersections of half-spaces, open and closed, respectively.

Definition.   Let S be an arbitrary set in X. The *closed convex hull* of S is defined as the intersection of all closed convex sets containing S.

Theorem 9. The closed convex hull of any set S is the set of points x satisfying $l(x) \leq q_S(l)$ for all l in X'.

EXERCISE 15. Prove Theorem 9.

Let $x_1, \ldots, x_m$ denote m points in X, and $p_1, \ldots, p_m$ denote m nonnegative numbers whose sum is 1.

$$p_j \geq 0, \qquad \sum_1^m p_j = 1. \tag{20}$$

Then

$$x = \sum p_j x_j \tag{20}'$$

is called a *convex combination* of $x_1, \ldots, x_m$.

EXERCISE 16. Show that if $x_1, \ldots, x_m$ belong to a convex set, then so does any convex combination of them.

Definition. A point of a convex set K that is not an interior point is called a *boundary point* of K.

Definition. Let K be a closed, convex set. A point e of K is called an *extreme point* of K if it is not the interior point of a line segment in K. That is, x is not an extreme point of K if

$$x = \frac{y+z}{2}, \qquad y \text{ and } z \text{ in } K, \quad y \neq z.$$

EXERCISE 17. Show that an interior point of K cannot be an extreme point.

All extreme points are boundary points of K, but not all boundary points are extreme points. Take for example, K to be a convex polygon. All edges and vertices are boundary points, but only the vertices are extreme points.

In three-dimensional space the set of extreme points need not be a closed set. Take K to be the convex hull of the points $(0, 0, 1), (0, 0, -1)$ and the circle $(1 + \cos\theta, \sin\theta, 0)$. The extreme points of K are all the above points except $(0, 0, 0)$.

Definition. A convex set K is called *bounded* if it does not contain a ray, that is, a set of points of the form $x + ty, 0 \leq t$.

Theorem 10 (Carathéodory). Let K be a nonempty closed bounded convex set in X, $\dim X = n$. Then every point of K can be represented as a convex combination of at most $(n + 1)$ extreme points of K.

Proof. We prove this inductively on the dimension of X. We distinguish two cases:

(i) *K has no interior points.* Suppose K contains the origin, which can always be arranged by shifting K appropriately. We claim that K does not contain n linearly independent vectors; for if it did, the convex combination of these vectors and the origin would also belong K; but these points constitute an n-dimensional simplex, full of interior points. Let m be the largest number of linearly independent vectors in K, and let $x_1, \ldots, x_m$ be m linearly independent vectors. Then $m < n$, and being maximal, every other vector in K is a linear combination of $x_1, \ldots, x_m$. This proves that K is contained in an m-dimensional subspace of X. By the induction hypothesis. Theorem 10 holds for K.

(ii) *K has interior points.* Denote by K_0 the set of all interior points of K. It is easy to show that K_0 is convex and that K_0 is open. We claim that K has boundary points; for, since K is bounded, any ray issuing from any interior point of K intersects K in an interval; since K is closed, the other endpoint is a boundary point y of K.

Let y be a boundary point of K. We apply Theorem 3 to K_0 and y; clearly y does not belong to K_0, so there is a linear functional l such that

$$l(y) = 1, \qquad l(x_0) < 1 \quad \text{for all } x_0 \text{ in } K_0. \tag{21}$$

We claim that $l(x_1) \leq 1$ for all x_1 in K. Pick any interior point x_0 of K; then all points x on the open segment bounded by x_0 and x_1 are interior points of K, and so by (21), $l(x) < 1$. It follows that at the endpoint $x_1, l(x_1) \leq 1$.

Denote by K_1 the set of those points x of K for which $l(x) = 1$. Being the intersection of two closed, convex sets, K_1 is closed and convex; since K is bounded, so is K_1. Equation (21) shows that y belongs to K_1, so K_1 is nonempty.

We claim that every extreme point e of K_1 is also an extreme point of K; for, suppose that

$$e = \frac{z + w}{2}, \qquad z \text{ and } w \text{ in } K.$$

Since e belongs to K_1,

$$l = l(e) = \frac{l(z) + l(w)}{2}. \tag{22}$$

Both z and w are in K; as we have shown before, $l(z)$ and $l(w)$ are both less than or equal to 1. Combining this with (22), we conclude that

$$l(z) = l(w) = 1.$$

This puts both z and w into K_1. But since e is an extreme point of $K_1, z = w$. This proves that extreme points of K_1 are extreme points of K.

Since K_1 lies in a hyperplane of dimension less than n, it follows from the induction assumption that K_1 has a sufficient number of extreme points, that is, every point in K_1 can be written as a convex combination of n extreme points of K_1. Since we have shown that extreme points of K_1 are extreme points of K, this proves Theorem 10 for boundary points of K.

Let x_0 be an interior point of K. We take any extreme point e of K (the previous argument shows that there are such things) and look at the intersection of the line through x_0 and e with K. Being the intersection of two closed convex sets, of which one, K, is bounded, this intersection is a closed interval. Since e is an extreme point of K, e is one of the end points; denote the other end point by y. Clearly, y is a boundary point of K. Since by construction x_0 lies on this interval, it can be written in the form

$$x_0 = py + (1 - p)e, \qquad 0 < p < 1. \tag{23}$$

We have shown above that y can be written as a convex combination of n extreme points of K. Setting this into (23) gives a representation of x_0 as the convex combination of $(n + 1)$ extreme points. The proof of Theorem 10 is complete.   □

We now give an application of Carathéodory's theorem.

Definition.   An $n \times n$ matrix $S = (s_{ij})$ is called *doubly stochastic* if

(i) $$s_{ij} \geq 0 \qquad \text{for all } i, j,$$

(ii) $$\sum_i s_{ij} = 1 \quad \text{for all } j, \tag{24}$$

(iii) $$\sum_j s_{ij} = 1 \quad \text{for all } i.$$

Such matrices arise, as the name indicates, in probability theory.

Clearly, the doubly stochastic matrices form a bounded, closed convex set in the space of all $n \times n$ matrices.

Example.   In Exercise 8 of Chapter 5 we defined the *permutation matrix P* associated with the permutation p of the integers $(1, \ldots, n)$ as follows:

$$P_{ij} = \begin{cases} 1, & \text{if } j = p(i), \\ 0, & \text{otherwise.} \end{cases} \tag{25}$$

EXERCISE 18.   Verify that every permutation matrix is a doubly stochastic matrix.

Theorem 11 (Dénes König, Garrett Birkhoff).   The permutation matrices are the extreme points of the set of doubly stochastic matrices.

Proof. It follows from (i) and (ii) of (24) that no entry of a doubly stochastic matrix can be greater than 1. Thus $0 \leq s_{ij} \leq 1$.

We claim that all permutation matrices P are extreme points; for, suppose

$$P = \frac{A + B}{2},$$

A and B doubly stochastic. It follows that if an entry of P is 1, the corresponding entries of A and B both must be equal to 1, and if an entry of P is zero, so must be the corresponding entries of A and B. This shows that $A = B = P$.

Next we show the converse. We start by proving that if S is doubly stochastic and has an entry which lies between 0 and 1:

$$0 < s_{i_0 j_0} < 1, \qquad\qquad (26)_{00}$$

S is not extreme. To see this we construct a sequence of entries, all of which lie between 0 and 1, and which lie alternatingly on the same row or on the same column.

We choose j_1 so that

$$0 < s_{i_0 j_1} < 1. \qquad\qquad (26)_{01}$$

This is possible because the sum of elements in the i_0th row must be $= 1$, and since $(26)_{00}$ holds. Similarly, since the sum of elements in the j_1st column $= 1$, and since $(26)_{01}$ holds, we can choose a row i_1 so that

$$0 < s_{i_1 j_1} < 1. \qquad\qquad (26)_{11}$$

We continue in this fashion, until the same position is traversed twice. Thus a closed chain has been constructed.

$$S_{i_k j_k} \to S_{i_k j_{k+1}} \to \ldots \to S_{i_m j_m} = S_{i_k j_k}$$

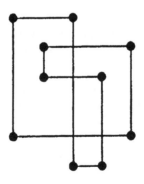

We now define a matrix N as follows:

(a) The entries of N are zero except for those points that lie on the chain.

(b) The entries of N on the points of the chain are $+1$ and -1, in succession.

The matrix N has the following property:

(c) The row sums and column sums of N are zero.

We now define two matrices A, B by

$$A = S + \epsilon N, \qquad B = S - \epsilon N.$$

It follows from (c) that the row sums and columns sums of A and B are both 1. By (a) and the construction the elements of S are positive at all points where N has a nonzero entry. It follows therefore that ϵ can be chosen so small that both A and B have nonnegative entries. This shows that A and B both are doubly stochastic. Since $A \neq B$, and

$$S = \frac{A + B}{2},$$

it follows that S is not an extreme point.

It follows that extreme points of the set of doubly stochastic matrices have entries either 0 or 1. It follows from (24) that each row and each column has exactly one 1. It is easy to check that such a matrix is a permutation matrix. This completes the proof of the converse. □

Applying Theorem 10 in the situation described in Theorem 11, we conclude: Every doubly stochastic matrix can be written as a convex combination of permutation matrices:

$$S = \sum c(P)P, \qquad c(P) \geq 0, \qquad \sum c(P) = 1.$$

EXERCISE 19. Show that, except for two dimensions, the representation of doubly stochastic matrices as convex combinations of permutation matrices is not unique.

Carathéodory's theorem has many applications in analysis. Its infinite-dimensional version is the Krein-Milman Theorem.

The last item in the chapter is a kind of a dual of Carathéodory's theorem.

Theorem 12 (Helly). Let X be a linear space of dimension n over the reals. Let $\{K_1, \ldots, K_N\}$ be a collection of N convex sets in X. Suppose that every subcollection of $n + 1$ sets K has a nonempty intersection. Then all K in the whole collection have a common point.

Proof (Radon). We argue by induction on N, the number of sets, starting with the trivial situation $N = n + 1$. Suppose that $N > n + 1$ and that the assertion is true for $N - 1$ sets. It follows that if we omit any one of the sets, say K_i, the rest have a point x_i in common:

$$x_i \in K_j, \qquad j \neq i. \tag{27}$$

We claim that there are numbers $a_i, \ldots, a_N$, not all zero, such that

$$\sum_1^N a_i x_i = 0 \tag{28}$$

and

$$\sum_1^N a_i = 0. \tag{28$'$}$$

These represent $n + 1$ equations for the N unknowns. According to Corollary A$'$ (concrete version) of Theorem 1 of Chapter 3, a homogeneous system of linear equations has a nontrivial (i.e., not all unknowns are equal to 0) solution if the number of equations is less than the number of unknowns. Since in our case $n + 1$ is less than N, (28) and (28)$'$ have a nontrivial solution.

It follows from (28)$'$ that not all a_i can be of the same sign; there must be some positive ones and some negative ones. Let us renumber them so that $a_i, \ldots, a_p$ are positive, the rest nonpositive.

We define a by

$$a = \sum_1^p a_i. \tag{29}$$

Note that it follows from (28)$'$ that

$$a = -\sum_{p+1}^N a_i. \tag{29$'$}$$

We define y by

$$y = \frac{1}{a} \sum_1^p a_i x_i. \tag{30}$$

Note that it follows from (28) and (30) that

$$y = \frac{-1}{a} \sum_{p+1}^N a_i x_i. \tag{30$'$}$$

Each of the points $x_i, i = 1, \ldots, p$ belongs to each of the sets $K_j, j > p$. It follows from (29) that (30) represents y as a convex combination of $x_1, \ldots, x_p$. Since K_j is convex, it follows that y belongs to K_j for $j > p$.

On the other hand, each $x_i, i = p + 1, \ldots, N$ belongs to each $K_j, j \leq p$. It follows from $(29)'$ that $(30)'$ represents y as a convex combination of $x_{p+1}, \ldots, x_N$. Since K_j is convex, it follows that y belongs to K_j for $j \leq p$. This concludes the proof of Helly's theorem. □

Remark. Helly's theorem is nontrivial even in the one-dimensional case. Here each K_j is an interval, and the hypothesis that every K_j and K_i intersects implies that the lower endpoint a_i of any K_i is less than or equal to the upper endpoint b_j of any other K_j. The point in common to all is then sup a_i or inf b_i, or anything in between.

Remark. In this chapter we have defined the notions of open convex set, closed convex set, and bounded convex set purely in terms of the linear structure of the space containing the convex set. Of course the notions open, closed, bounded have a usual topological meaning in terms of the Euclidean distance. It is easy to see that if a convex set is open, closed, or bounded in the topological sense, then it is open, closed, or bounded in the linear sense used in this chapter.

EXERCISE 20. Show that if a convex set in a finite-dimensional Euclidean space is open, or closed, or bounded in the linear sense defined above, then it is open, or closed, or bounded in the topological sense, and conversely.

The Duality Theorem

Let X be a linear space over the reals, dim $X = n$. Its dual X' consists of all linear functions on X. If X is represented by column vectors x of n components $x_1, \ldots, x_n$, then elements of X' are traditionally represented as row vectors ξ with n components $\xi_1, \ldots, \xi_n$. The value of ξ at x is

$$\xi_1 x_1 + \cdots + \xi_n x_n. \tag{1}$$

If we regard ξ as a $1 \times n$ matrix and regard x as an $n \times 1$ matrix, (1) is their matrix product ξx.

Let Y be a subspace of X; in Chapter 2 we have defined the annihilator $Y^\perp$ of Y as the set of all linear functions ξ that vanish on Y, that is, satisfy

$$\xi y = 0 \qquad \text{for all } y \text{ in } Y. \tag{2}$$

According to Theorem 3 of Chapter 2, the dual of X' is X itself, and according to Theorem 5 there, the annihilator of $Y^\perp$ is Y itself. In words: *if $\xi x = 0$ for all ξ in $Y^\perp$, then x belongs to Y.*

Suppose Y is defined as the linear space spanned by m given vectors $y_1, \ldots, y_m$ in X. That is, Y consists of all vectors y of the form

$$y = \sum_1^m a_j y_j. \tag{3}$$

Clearly, ξ belongs to $Y^\perp$ iff

$$\xi y_j = 0, \qquad j = 1, \ldots, m. \tag{4}$$

So for the space Y defined by (3), the duality criterion stated above can be formulated as follows: *a vector y can be written as a linear combination* (3) *of m given vectors y_j iff every ξ that satisfies* (4) *also satisfies $\xi y = 0$.*

We are asking now for a criterion that a vector y be the linear combination of m given vectors y_j with *nonnegative* coefficients:

$$y = \sum_1^m p_j y_j, \qquad p_j \geq 0. \qquad (5)$$

Theorem 1 (Farkas–Minkowski). A vector y can be written as a linear combination of given vectors y_j with nonnegative coefficients as in (5) iff every ξ that satisfies

$$\xi y_j \geq 0, \qquad j = 1, \ldots, m \qquad (6)$$

also satisfies

$$\xi y \geq 0. \qquad (6)'$$

Proof. The necessity of condition $(6)'$ is evident upon multiplying (5) on the left by ξ. To show the sufficiency we consider the set K of all points y of form (5). Clearly, this is a convex set; we claim it is closed. To see this we first note that any vector y which may be represented in form (5) may be represented so in various ways. Among all these representations there is by local compactness one, or several, for which $\sum p_j$ is as small as possible. We call such a representation of y a *minimal representation.*

Now let $\{z_n\}$ be a sequence of points of K converging to the limit z in the Euclidean norm. Represent each z_n minimally:

$$z_n = \sum p_{n,j} y_j. \qquad (5)'$$

We claim that $\sum p_{n,j} = P_n$ is a bounded sequence. For suppose on the contrary that $P_n \to \infty$. Since the sequence z_n is convergent, it is bounded; therefore z_n / P_n tends to zero:

$$\frac{z_n}{P_n} = \sum \frac{p_{n,j}}{P_n} y_j \to 0. \qquad (5)''$$

The numbers $p_{n,j}/P_n$ are nonnegative and their sum is 1. Therefore by compactness we can select a subsequence for which they converge to limits:

$$\frac{p_{n,j}}{P_n} \to q_j.$$

These limits satisfy $\sum q_j = 1$. It follows from $(5)''$ that

$$\sum q_j y_j = 0.$$

Subtract this from $(5)'$:

$$z_n = \sum (p_{n,j} - q_j) y_j.$$

For each j for which $q_j > 0$, $p_{n,j} \to \infty$; therefore for n large enough, this is a positive representation of z_n, showing that $(5)'$ is not a minimal representation. This contradiction shows that the sequence $P_n = \sum p_{n,j}$ is bounded. But then by local compactness we can select a subsequence for which $p_{n,j} \to p_j$ for all j. Let n tend to ∞ in $(5)'$; we obtain

$$z = \lim z_n = \sum p_j y_j.$$

Thus the limit z can be represented in the form (5); this proves that the set K of all points of form (5) is closed in the Euclidean norm.

We note that the origin belongs to K.

Let y be a vector that does not belong to K. Since K is closed and convex, according to the hyperplane separation Theorem 7 of Chapter 12 there is a closed halfspace

$$\eta x \geq c \tag{7}$$

that contains K but not y:

$$\eta y < c. \tag{8}$$

Since 0 belongs to K, it follows from (7) that $0 \geq c$. Combining this with (8), we get

$$\eta y < 0. \tag{9}$$

Since $k y_j$ belongs to K for any positive constant k, it follows from (7) that

$$k \eta y_j \geq c, \qquad j = 1, \ldots, m$$

for all $k > 0$; this is the case only if

$$\eta y_j \geq 0, \qquad j = 1, \ldots, m. \tag{10}$$

Thus if y is not of form (5), there is an η that according to (10) satisfies (6) but according to (9) violates $(6)'$. This completes the proof of Theorem 1. $\qquad\square$

EXERCISE 1.   Show that K defined by (5) is a convex set.

We reformulate Theorem 1 in matrix language by defining the $n \times m$ matrix Y as

$$Y = (y_1, \ldots, y_m),$$

that is, the matrix whose columns are y_j. We denote the column vector formed by $p_1, \ldots, p_m$ by p:

$$p = \begin{pmatrix} p_1 \\ \vdots \\ p_m \end{pmatrix}.$$

We shall call a vector, column or row, nonnegative, denoted as ≥ 0, if all its components are nonnegative. The inequality $x \geq z$ means $x - z \geq 0$.

EXERCISE 2.   Show that if $x \geq z$ and $\xi \geq 0$, then $\xi x \geq \xi z$.

Theorem 1′.   Given an $n \times m$ matrix Y, a vector y with n components can be written in the form

$$y = Yp, \qquad p \geq 0 \tag{11}$$

iff every row vector ξ that satisfies

$$\xi Y \geq 0 \tag{12}$$

also satisfies

$$\xi y \geq 0. \tag{12′}$$

For the proof, we merely observe that (11) is the same as (5), (12) the same as (6), and (12)′ the same as (6)′.   □

The following is a useful extension.

Theorem 2.   Given an $n \times m$ matrix Y and a column vector y with n components, the inequality

$$y \geq Yp, \qquad p \geq 0 \tag{13}$$

can be satisfied iff every ξ that satisfies

$$\xi Y \geq 0, \quad \xi \geq 0 \tag{14}$$

also satisfies

$$\xi y \geq 0. \tag{15}$$

Proof. To prove necessity, multiply (13) by ξ on the left and use (14) to deduce (15). Conversely by definition of ≥ 0 for vectors, (13) means that there is a column vector z with n components such that

$$y = Yp + z, \qquad z \geq 0, \qquad p \geq 0. \tag{13$'$}$$

We can rewrite (13)$'$ by introducing the $n \times n$ identity matrix I, the augmented matrix (Y, I) and the augmented vector $\begin{pmatrix} p \\ z \end{pmatrix}$. In terms of these (13)$'$ can be written as

$$y = (Y, I) \begin{pmatrix} p \\ z \end{pmatrix}, \qquad \begin{pmatrix} p \\ z \end{pmatrix} \geq 0 \tag{13$''$}$$

and (14) can be written as

$$\xi(Y, I) \geq 0. \tag{14$'$}$$

We now apply Theorem 1$'$ to the augmented matrix and vector to deduce that if (15) is satisfied whenever (14)$'$ is, then (13)$''$ has a solution, as asserted in Theorem 2.   □

Theorem 3 (Duality Theorem). Let Y be a given $n \times m$ matrix, y a given column vector with n components, and γ a given row vector with m components.

We define two quantities, S and s, as follows:

Definition

$$S = \sup_{p} \gamma p \tag{16}$$

for all column vectors p with m components satisfying

$$y \geq Yp, \qquad p \geq 0. \tag{17}$$

We call the set of p satisfying (17) admissible for the sup problem (16).

Definition

$$s = \inf_{\xi} \xi y \tag{18}$$

for all row vectors ξ with n components satisfying the admissibility conditions

$$\gamma \leq \xi Y, \qquad \xi \geq 0. \tag{19}$$

We call the set of ξ satisfying (19) admissible for the inf problem (18).

Assertion. Suppose that there are admissible vectors p and ξ; then S and s are finite, and

$$S = s.$$

Proof. Let p and ξ be admissible vectors. Multiply (17) by ξ on the left, (19) by p on the right. Using Exercise 2 we conclude that

$$\xi y \geq \xi Y p \geq \gamma p.$$

This shows that any γp is bounded from above by every ξy; therefore

$$s \geq S. \tag{20}$$

To show that equality actually holds, it suffices to display a single p admissible for the sup problem (16) for which

$$\gamma p \geq s. \tag{21}$$

To accomplish this, we combine (17) and (21) into a single inequality by augmenting the matrix Y with an extra row $-\gamma$, and the vector y with an extra component $-s$:

$$\begin{pmatrix} y \\ -s \end{pmatrix} \geq \begin{pmatrix} Y \\ -\gamma \end{pmatrix} p, \qquad p \geq 0. \tag{22}$$

If this inequality has no solution, then according to Theorem 2 there is a row vector ξ and a scalar α such that

$$(\xi, \alpha) \begin{pmatrix} Y \\ -\gamma \end{pmatrix} \geq 0, \qquad (\xi, \alpha) \geq 0, \tag{23}$$

but

$$(\xi, \alpha) \begin{pmatrix} y \\ -s \end{pmatrix} < 0. \tag{24}$$

We claim that $\alpha > 0$; for, if $\alpha = 0$, then (23) implies that

$$\xi Y \geq 0, \qquad \xi \geq 0, \tag{23}'$$

and (24) that

$$\xi y < 0. \tag{24}'$$

According to the "only if" part of Theorem 2 this shows that (13), the same as (17), cannot be satisfied; this means that there is no admissible p, contrary to assumption.

Having shown that α is necessarily positive, we may, because of the homogeneity of (23) and (24), take $\alpha = 1$. Writing out these inequalities gives

$$\xi Y \geq \gamma, \qquad \xi \geq 0 \tag{25}$$

and

$$\xi y < s. \tag{26}$$

Inequality (25), the same as (19), shows that ξ is admissible; (26) shows that s is not the infimum (18), a contradiction we got into by denying that we can satisfy (21). Therefore (21) can be satisfied; this implies that equality holds in (20). This proves that $S = s$. $\square$

EXERCISE 3. Show that the sup and inf in Theorem 3 is a maximum and minimum. [*Hint*: The sign of equality holds in (21).]

We give now an application of the duality theorem in economics.

We are keeping track of n different kinds of food (milk, meat, fruit, bread, etc.) and m different kinds of nutrients (protein, fat, carbohydrates, vitamins, etc.). We denote

y_{ij} = number of units of the jth nutrient present in one unit of the ith food item.
γ_j = minimum daily requirement of the jth nutrient.
y_i = price of one unit of the ith food item.

Note that all these quantities are nonnegative.

Suppose our daily food purchase consists of ξ_i units of the ith food item. We insist on satisfying all the daily minimum requirements:

$$\sum_i \xi_i y_{ij} \geq \gamma_j, \qquad j = 1, \ldots, m. \tag{27}$$

This inequality can be satisfied, provided that each nutrient is present in at least one of the foods.

The total cost of the purchase is

$$\sum_i \xi_i y_i. \tag{28}$$

A natural question is, *What is the minimal cost of food that satisfies the daily minimum requirements?* Clearly, this is the minimum of (28) subject to (27) and $\xi \geq 0$, since we cannot purchase negative amounts. If we identify the column vector formed by the y_i with y, the row vector formed by the γ_j with γ, and the matrix y_{ij} with Y, the quantity (28) to be minimized is the same as (18), and (27) is the same as (19). Thus the infimum s in the duality theorem can in this model be identified with minimum cost.

To arrive at an interpretation of the *supremum S* we denote by $\{p_j\}$ a possible set of *values* for the nutrients that is consistent with the prices. That is, we require that

$$y_i \geq \sum_j Y_{ij} p_j, \qquad i = 1, \ldots, n. \tag{29}$$

The value of the minimum daily requirement is

$$\sum \gamma_j p_j. \tag{30}$$

Since clearly p_j are nonnegative, the restriction (29) is the same as (17). The quantity (30) is the same as that maximized in (16). Thus the quantity S in the duality theorem is the *largest possible value of the total daily requirement*, consistent with the prices.

A second application comes from *game theory*. We consider two-person, deterministic, zero-sum games. Such a game can (by definition) always be presented as a matrix game, defined as follows:

An $n \times m$ matrix Y, called the *payoff matrix*, is given. The game consists of player C picking one of the columns and player R picking one of the rows; neither player knows what the other has picked but both are familiar with the payoff matrix. If C chooses column j and R chooses row i, then the outcome of the game is the payment of the amount Y_{ij} by player C to player R. If Y_{ij} is a negative number, then R pays C.

We think of this game as being played repeatedly many times. Furthermore, the players do not employ the same strategy each time, that is, do not pick the same row, respectively, column, each time, but employ a so-called *mixed strategy* which consists of picking rows, respectively columns, *at random* but according to a set of frequencies which each player is free to choose. That is, player C will choose the jth column with frequency x_j, where x is a *probability vector*, that is,

$$x_j \geq 0, \qquad \sum_j x_j = 1. \tag{31}$$

Player R will choose the ith row with frequency η_i,

$$\eta_i \geq 0, \qquad \sum_i \eta_i = 1. \tag{31}'$$

Since the choices are made at random, the choices of C and R are *independent* of each other. It follows that the frequency with which C chooses column j *and* R chooses row i in the same game is the *product* $n_j x_j$.

Since the payoff of C to R is Y_{ij}, the *average payoff* over a long time is

$$\sum_{i,j} \eta_i x_j Y_{ij}.$$

In vector–matrix notation that is

$$\eta Y x. \tag{32}$$

If C has picked his mix x of strategies, then by observing over a long time R can determine the relative frequencies that C is using, and therefore will choose his own mix η of strategies so that he maximizes his gain:

$$\max_{\eta} \eta Y x. \tag{33}$$

Suppose C is a conservative player, that is, C anticipates that R will adjust his mix so as to gain the maximum amount (33). Since R's gain is C's loss, C chooses his mix x to minimize his loss—that is, so that (33) is a minimum:

$$\min_{x} \max_{\eta} \eta Y x, \tag{34}$$

x and η probability vectors.

If, on the other hand, we suppose that R is the conservative player, R will assume that C will guess R's mix η first and therefore C will choose x so that C's loss is minimized:

$$\min_{x} \eta Y x. \tag{33}'$$

R therefore picks his mix η so that the outcome (33)' is as large as possible:

$$\max_{\eta} \min_{x} \eta Y x. \tag{34}'$$

Theorem 4 (Minmax Theorem). The minmax (34) and the maxmin (34)', where η and x are required to be probability vectors, are equal:

$$\min_{x} \max_{\eta} \eta Y x = \max_{\eta} \min_{x} \eta Y x. \tag{35}$$

The quantity (35) is called the ˙*value* of the matrix game Y.

Proof. Denote by E the $n \times m$ matrix of all 1s. For any pair of probability vectors η and x, $\eta E x = 1$. Therefore if we replace Y by $Y + kE$, we merely add k to both (34) and (34)'. For k large enough all entries of $Y + kE$ are positive; so we may consider only matrices Y with all positive entries.

We shall apply the duality theorem with

$$\gamma = (1, \ldots, 1) \quad \text{and} \quad y = \begin{pmatrix} 1 \\ \vdots \\ 1 \end{pmatrix}. \tag{36}$$

Since y is positive, the maximum problem

$$S = \max_p \gamma p, \qquad y \geq Yp, p \geq 0 \tag{37}$$

has positive admissible vectors p. Since the entries of γ are positive, $S > 0$. We denote by p_0 a vector where the maximum is achieved.

Since $Y > 0$, the minimum problem

$$s = \min_\xi \xi y, \qquad \xi Y \geq \gamma, \xi \geq 0, \tag{37'}$$

has admissible vectors ξ. We denote by ξ_0 a vector where the minimum is reached.

According to (36), all components of γ are 1; therefore γp_0 is the sum of the components of p_0. Since $\gamma p_0 = S$,

$$x_0 = \frac{p_0}{S} \tag{38}$$

is a probability vector. Using an analogous argument, we deduce that

$$\eta_0 = \frac{\xi_0}{s} \tag{38'}$$

is a probability vector.

We claim that x_0 and η_0 are solutions of the minmax and maxmin problems (34) and (34)', respectively. To see this, set p_0 into the second part of (37), and divide by S. Using the definition $x_0 = p_0/S$, we get

$$\frac{y}{S} \geq Yx_0. \tag{39}$$

Multiply this on the left with any probability vector η. Since according to (36) all components of y are 1, $\eta y = 1$, and so

$$\frac{1}{S} \geq \eta Yx_0. \tag{40}$$

It follows from this that

$$\frac{1}{S} \geq \max_\eta \eta Yx_0,$$

from which it follows that

$$\frac{1}{S} \geq \min_{x} \max_{\eta} \eta Y x. \tag{41}$$

On the other hand, we deduce from (40) that for all η,

$$\frac{1}{S} \geq \min_{x} \eta Y x,$$

from which it follows that

$$\frac{1}{S} \geq \max_{\eta} \min_{x} \eta Y x. \tag{42}$$

Similarly we set ξ_0 for ξ into the second part of (37)$'$, divide by s, and multiply by any probability vector x. By definition (38)$'$, $\eta_0 = \xi_0/s$; since according to (36) all components of γ are 1, $\gamma x = 1$. So we get

$$\eta_0 Y x \geq \frac{1}{s}. \tag{40$'$}$$

From this we deduce that for any probability vector x,

$$\max_{\eta} \eta Y x \geq \frac{1}{s},$$

from which it follows that

$$\min_{x} \max_{\eta} \eta Y x \geq \frac{1}{s}. \tag{41$'$}$$

On the other hand, it follows from (40)$'$ that

$$\min_{x} \eta_0 Y x \geq \frac{1}{s},$$

from which it follows that

$$\max_{\eta} \min_{x} \eta Y x \geq \frac{1}{s}. \tag{42$'$}$$

Since by the duality theorem $S = s$, (41) and (41)$'$ together show that

$$\min_{x} \max_{\eta} \eta Y x = \frac{1}{s} = \frac{1}{S},$$

while (42) and (42)$'$ show that

$$\max_{\eta} \min_{x} \eta Y x = \frac{1}{s} = \frac{1}{S}.$$

This proves the minmax theorem. □

The minmax theorem is due to von Neumann. It has important implications for economic theory.

CHAPTER 14

Normed Linear Spaces

In Chapter 12, Theorem 2, we saw that every open, convex set K in a linear space X over $\mathbb{R}$ containing the origin can be described as the set of vectors x satisfying $p(x) < 1$, where p, the gauge function of K, is a subadditive, positive homogeneous function, positive except at the origin. Here we consider such functions with one additional property: evenness, that is, $p(-x) = p(x)$. Such a function is called a *norm*, and is denoted by the symbol $|x|$, the same as absolute value. We list now the properties of a norm:

(i) Positivity:	$\|x\| > 0$ for $x \neq 0$, $\|0\| = 0$.	
(ii) Subadditivity:	$\|x+y\| \leq \|x\| + \|y\|$.	(1)
(iii) Homogeneity:	for any real number k, $\|kx\| = \|k\|\|x\|$.	

A linear space with a norm is called a *normed linear space*. Except for Theorem 4, in this chapter X denotes a finite-dimensional normed linear space.

Definition. The set of points x in X satisfying $|x| < 1$ is called the *open unit ball* around the origin; the set $|x| \leq 1$ is called the *closed unit ball*.

EXERCISE 1. **(a)** Show that the open and closed unit balls are convex.
(b) Show that the open and closed unit balls are symmetric with respect to the origin, that is, if x belongs to the unit ball, so does $-x$.

Definition. The distance of two vectors x and y in X is defined as

$$|x - y|.$$

Linear Algebra and Its Applications, Second Edition, by Peter D. Lax
Copyright © 2007 John Wiley & Sons, Inc.

EXERCISE 2. Prove the *triangle inequality*, that is, for all x, y, z in X,

$$|x - z| \leq |x - y| + |y - z|. \tag{2}$$

Definition. Given a point y and a positive number r, the set of x satisfying $|x - y| < r$ is called the open ball of radius r, center y; it is denoted $B(y, r)$.

Examples

$$X = \mathbb{R}^n, \qquad x = (a_1 \ldots a_n).$$

(a) Define

$$|x|_\infty = \max_j |a_j|. \tag{3}$$

Properties (i) and (iii) are obvious; property (ii) is easy to show.
   (b) Define $|x|_2$ as the Euclidean norm:

$$|x|_2 = \left(\sum |a_j|^2 \right)^{1/2}. \tag{4}$$

Properties (i) and (iii) are obvious; property (ii) was shown in Theorem 3 of Chapter 7.
   (c) Define

$$|x|_1 = \sum |a_j|. \tag{5}$$

EXERCISE 3. Prove that $|x|_1$ defined by (5) has all three properties (1) of a norm. The next example includes the first three as special cases:

(d) p any real number, $1 \leq p$; we define

$$|x|_p = \left(\sum |a_j|^p \right)^{1/p}. \tag{6}$$

Theorem 1. $|x|_p$ defined by (6) is a norm, that is, it has properties (1).

Proof. Properties (i) and (iii) are obvious. To prove (ii), we need the following: □

Hölder's Inequality. Let p and q be positive numbers that satisfy

$$\frac{1}{p} + \frac{1}{q} = 1. \tag{7}$$

Let $(a_1, \ldots, a_n) = x$ and $(b_1, \ldots, b_n) = y$ be two vectors; then

$$xy \leq |x|_p \, |y|_q, \tag{8}$$

where the product xy is defined as

$$xy = \sum a_j b_j; \tag{9}$$

$|x|_p, |y|_q$ are defined by (6). Equality in (8) holds iff $|a_j|^p$ and $|b_j|^q$ are proportional and $\operatorname{sgn} a_j = \operatorname{sgn} b_j, j = 1, \ldots, n$.

EXERCISE 4. Prove or look up a proof of Hölder's inequality.

Note. For $p = q = 2$, Hölder's inequality is the Schwarz inequality (see Theorem 1, Chapter 7).

EXERCISE 5. Prove that

$$|x|_\infty = \lim_{p \to \infty} |x|_p,$$

where $|x|_\infty$ is defined by (3).

Corollary. For any vector x

$$|x|_p = \max_{|y|_q = 1} xy \tag{10}$$

Proof. Inequality (8) shows that when $|y|_q = 1$, xy cannot exceed $|x|_p$. Therefore to prove (10) we have to exhibit a single vector y_0, $|y_0|_q = 1$, for which $xy_0 = |x|_p$. Here it is:

$$y_0 = \frac{z}{|x|_p^{p/q}}, \qquad z = (c_1, \ldots, c_n), \qquad c_j = \operatorname{sgn} a_j |a_j|^{p/q}. \tag{11}$$

Clearly

$$|y_0|_q = \frac{|z|_q}{|x|_p^{p/q}}, \tag{12}$$

and

$$|z|_q^q = \sum |c_j|^q = \sum |a_j|^p = |x|_p^p. \tag{12$'$}$$

Combining (12) and (12$'$)

$$|y_0|_q = \frac{|x|_p^{p/q}}{|x|_p^{p/q}} = 1. \tag{13}$$

From (11)

$$xy_0 = \frac{xz}{|x|_p^{p/q}} = \frac{\sum |a_j||a_j|^{p/q}}{|x|_p^{p/q}} = \frac{\sum |a_j|^{1+p/q}}{|x|_p^{p/q}} = |x|_p^{p-p/q} = |x|_p, \qquad (13)'$$

where we have used (7) to set $1 + p/q = p$. Formulas (13) and (13)$'$ complete the proof of the corollary.   □

To prove subadditivity for $|x|_p$ we use the corollary. Let x and z be any two vectors; then by (10),

$$|x + z|_p = \max_{|y|_q=1}(x+z)y \leq \max_{|y|_q=1} xy + \max_{|y|_q=1} zy = |x|_p + |z|_p.$$

This proves that the l^p norm is subadditive.   □

We return now to arbitrary norms.

Definition. Two norms in a finite-dimensional linear space X, $|x|_1$ and $|x|_2$, are called *equivalent* if there is a constant c such that for all x in X,

$$|x|_1 \leq c|x|_2, \qquad |x|_2 \leq c|x|_1. \qquad (14)$$

Theorem 2. In a finite-dimensional linear space, all norms are equivalent; that is, any two satisfy (14) with some c, depending on the pair of norms.

Proof. Any finite-dimensional linear space X over $\mathbb{R}$ is isomorphic to $\mathbb{R}^n$, $n = \dim X$; so we may take X to be $\mathbb{R}^n$. In Chapter 7 we introduced the Euclidean norm:

$$\| x \| = \left(\sum_1^n a_j^2 \right)^{1/2}, \qquad x = (a_1, \ldots, a_n). \qquad (15)$$

Denote by e_j the unit vectors in $\mathbb{R}^n$:

$$e_j = (0, \ldots, 1, 0, \ldots, 0), \qquad j = 1, \ldots, n.$$

Then $x = (a_1, \ldots, a_n)$ can be written as

$$x = \sum a_j e_j. \qquad (16)$$

Let $|x|$ be any other norm in $\mathbb{R}^n$. Using subadditivity and homogeneity repeatedly we get

$$|x| \leq \sum |a_j||e_j|. \qquad (16)'$$

Applying the Schwarz inequality to (16)' (see Theorem 1, Chapter 7), we get, using (15),

$$|x| \leq \left(\sum |e_j|^2 \right)^{1/2} \left(\sum a_j^2 \right)^{1/2} = c \| x \|, \tag{17}$$

where c abbreviates $(\sum |e_j|^2)^{1/2}$. This gives one half of inequalities (14).

To get the other half, we show first that $|x|$ is a continuous function with respect to the Euclidean distance. By subadditivity,

$$|x| \leq |x - y| + |y|, \qquad |y| \leq |x - y| + |x|,$$

from which we deduce that

$$||x| - |y|| \leq |x - y|.$$

Using inequality (17), we get

$$||x| - |y|| \leq c \| x - y \|,$$

which shows that $|x|$ is a continuous function in the Euclidean norm.

It was shown in Chapter 7 that the unit sphere S in a finite-dimensional Euclidean space, $\| x \| = 1$, is a compact set. Therefore the continuous function $|x|$ achieves its minimum on S. Since by (1), $|x|$ is positive at every point of S, it follows that the minimum m is positive. Thus we conclude that

$$0 < m \leq |x| \qquad \text{when } \| x \| = 1. \tag{18}$$

Since both $|x|$ and $\| x \|$ are homogeneous functions, we conclude that

$$m \| x \| \leq |x| \tag{19}$$

for all x in $\mathbb{R}^n$. This proves the second half of the inequalities (14), and proves that any norm in $\mathbb{R}^n$ is equivalent in the sense of (14) with the Euclidean norm.

The notion of equivalence is *transitive*; if $|x|_1$ and $|x|_2$ are both equivalent to the Euclidean norm, then they are equivalent to each other. This completes the proof of Theorem 2.   □

Definition.   A sequence $\{x_n\}$ in a normed linear space is called *convergent* to the limit x, denoted as $\lim x_n = x$ if $\lim |x_n - x| = 0$.

Obviously, the notion of convergence of sequences is the same with respect to two equivalent norms; so by Theorem 2, it is the same for any two norms.

Definition.   A set S in a normed linear space is called *closed* if it contains the limits of all convergent sequences $\{x_n\}, x_n$ in S.

EXERCISE 6.   Prove that every subspace of a finite-dimensional normed linear space is closed.

Definition.   A set S in a normed linear space is called *bounded* if it is contained in some ball, that is, if there is an R such that for all points z in S, $|z| \leq R$. Clearly, if a set is bounded in the sense of one norm, it is bounded in the sense of any equivalent norm, and so by Theorem 2 for all norms.

Definition.   A sequence of vectors $\{x_k\}$ in a normed linear space is called a *Cauchy sequence* if $|x_k - x_j|$ tends to zero as k and j tend to infinity.

Theorem 3.   **(i)** In a finite-dimensional normed linear space X, every Cauchy sequence converges to a limit.

(ii) Every bounded infinite sequence $\{x_n\}$ in a finite-dimensional normed linear space X has a convergent subsequence.

Property (i) of X is called *completeness*, and property (ii) is called *local compactness*.

Proof. **(i)** Introduce a Euclidean structure in X. According to Theorem 2, the Euclidean norm and the norm in X are equivalent. Therefore a Cauchy sequence in the norm of X is also a Cauchy sequence in the Euclidean norm. According to Theorem 16 in Chapter 7, a Cauchy sequence in a finite-dimensional Euclidean space converges. But then the sequence also converges in the norm of X.

(ii) A sequence $\{X_n\}$ that is bounded in the norm of X is also bounded in the Euclidean norm imposed on X. According to Theorem 16 of Chapter 7, it contains a subsequence that converges in the Euclidean norm. But then that subsequence also converges in the norm of X. $\square$

Just as in Euclidean space, see Theorem 17 in Chapter 7, part (ii) of Theorem 3 has a converse:

Theorem 4.   Let X be a normed linear space that is locally compact—that is, in which every bounded sequence has a convergent subsequence. Then X is finite-dimensional.

Proof. We need the following result. $\square$

Lemma 5.   Let Y be a finite-dimensional subspace of a normed linear space X. Let x be a vector in X that does not belong to Y. Then

$$d = \inf_{y \text{ in } Y} |x - y|$$

is positive.

Proof. Suppose not; then there would be a sequence of vectors $\{y_n\}$ in Y such that

$$\lim |x - y_n| = 0.$$

In words, y_n tends to x. It follows that $\{y_n\}$ is a Cauchy sequence; according to part (i) of Theorem 3, y_n converges to a limit in Y. This would show that the limit x of $\{y_n\}$ belongs to Y, contrary to the choice of x. $\square$

Suppose X infinite-dimensional; we shall construct a sequence $\{y_k\}$ in X with the following properties:

$$|y_n| < 2, \quad |y_k - y_l| \geq 1 \qquad \text{for } k \neq l. \tag{20}$$

Clearly, such a sequence is bounded and, equally clearly, contains no convergent subsequence.

We shall construct the sequence recursively. Suppose $y_1, \ldots, y_n$ have been chosen; denote by Y the space spanned by them. Since X is infinite-dimensional, there is an x in X that does not belong to Y. We appeal now to Lemma 5,

$$d = \inf_{y \text{ in } Y} |x - y| > 0.$$

By definition of infimum, there is a vector y_0 in Y which satisfies

$$|x - y_0| < 2d.$$

Define

$$y_{n+1} = \frac{x - y_0}{d} \tag{21}$$

It follows from the inequality above that $|y_{n+1}| < 2$. For any y in Y, $y_0 + dy$ belongs to Y. Therefore by definition of infimum,

$$|x - y_0 - dy| \geq d.$$

Dividing this by d and using the definition of y_{n+1}, we get

$$|y_{n+1} - y| \geq 1.$$

Since every y_l, $l = 1, \ldots, n$, belongs to Y,

$$|y_{n+1} - y_l| \geq 1 \qquad \text{for } l = 1, \ldots, n.$$

This completes the recursive construction of the sequence $\{y_k\}$ with property (20). $\square$

Theorem 4 is due to Frederic Riesz.

EXERCISE 7.   Show that the infimum in Lemma 5 is a minimum.

We have seen in Theorem 5 of Chapter 7 that every linear function l in a Euclidean space can be written in the form of a scalar product $l(x) = (x, y)$. Therefore by the Schwarz inequality, Theorem 1 of Chapter 7,

$$|l(x)| \le \| x \| \| y \|.$$

Combining this with (19), we deduce that

$$|l(x)| \le c|x|, \qquad c = \frac{\| y \|}{m}.$$

We can restate this as Theorem 6.

Theorem 6.   Let X be a finite-dimensional normed linear space, and let l be a linear function defined on X. Then there is a constant c such that

$$|l(x)| \le c|x| \tag{22}$$

for all x in X.

Corollary 6'.   Every linear function on a finite-dimensional normed linear space is continuous.

Proof. Using the linearity of l and inequality (22), we deduce that

$$|l(x) - l(y)| = |l(x - y)| \le c|x - y|. \qquad \square$$

Definition.   Denote by c_0 the infimum of all numbers c for which (22) holds for all x. Clearly, (22) holds for $c = c_0$, and c_0 is the smallest number c for which (22) holds; c_0 is called the *norm* of the linear function l, denoted as $|l|'$.

The norm of l can also be characterized as

$$|l|' = \sup_{x \ne 0} \frac{|l(x)|}{|x|}. \tag{23}$$

It follows from (23) that for all x and all l,

$$|l(x)| \le |l|'|x|. \tag{24}$$

Theorem 7. X is a finite-dimensional normed linear space.

(i) Given a linear function l defined on X, there is an x in X, $x \ne 0$, for which equality holds in (24).

(ii) Given a vector x in X, there is a linear function l defined on X, $l \ne 0$, for which equality holds in (24).

Proof. **(i)** We shall show that the supremum definition (23) of $|l|'$ is a maximum. We note that the ratio $|l(x)|/|x|$ doesn't change if we replace x by any multiple of x. Therefore it suffices to take the supremum (23) over the unit sphere $|x| = 1$.

According to Corollary 6', $l(x)$ is a continuous function; then so is $|l(x)|$. Since the space X is locally compact, the continuous function $|l(x)|$ takes on its maximum value at some point x of the unit sphere. At this point, equality holds in (24).

(ii) If $x = 0$, any l will do.

For $x \neq 0$, we define $l(x) = |x|$; since l is linear, we set for any scalar k

$$l(kx) = k|x|. \tag{25}$$

We appeal now to the Hahn–Banach Theorem, Theorem 4 in Chapter 12. We choose the positive homogeneous, subadditive function $p(x)$ to be $|x|$, and the subspace U on which l is defined consists of all multiples of x. It follows from (25) that for all u in U, $l(u) \leq |u|$. According to Hahn–Banach, l can be extended to all y of X so that $l(y) \leq |y|$ for all y. Setting $-y$ for y, we deduce that $|l(y)| \leq |y|$ as well. So by definition (23) of the norm of l, it follows that $|l|' \leq 1$. Since $l(x) = |x|$, it follows that $|l|' = 1$, so equality holds in (24). □

In Chapter 2 we have defined the *dual* of a finite-dimensional linear space X as the set of all linear functions l defined on X. These functions form a linear space, denoted as X'. We have shown in Chapter 2 that the dual of X' can be identified with X itself: $X'' = X$, as follows. For each x in X we define a linear function f over X' by setting

$$f(l) = l(x). \tag{26}$$

We have shown in Chapter 2 that these are all the linear functions on X.

When X is a finite-dimensional *normed* linear space, there is an *induced norm* $|l|'$ in X', defined by formula (23). This, in turn, induces a norm in the dual X'' of X'.

Theorem 8. The norm induced in X'' by the induced norm in X' is the same as the original norm in X.

Proof. The norm of a linear function of on X' is, according to formula (23),

$$|f|'' = \sup_{l \neq 0} \frac{|f(l)|}{|l|'}. \tag{27}$$

The linear functions f on X' are of the form (26); setting this into (27) gives

$$|f|'' = \sup_{l \neq 0} \frac{|l(x)|}{|l|'}. \tag{28}$$

According to (24), $|l(x)|/|l|' \leq |x|$ for all $l \neq 0$. According to part (ii) of Theorem 7, equality holds for some l. This proves that $|f|'' = |x|$. $\quad\square$

EXERCISE 8.   Show that $|l|'$ defined by (23) satisfies all postulates for a norm listed in (1).

Note. The dual of an *infinite-dimensional* normed linear space X consists of all linear functions on X that are *bounded* in the sense of (22). The induced norm on X' is defined by (24). Theorem 7 holds in infinite-dimensional spaces.

The dual of X' is defined analogously. For each x in X, we can define a linear function f by formula (25); f is bounded and its bound equals $|x|$. So f lies in X''; but for many spaces X that are used in analysis, it is no longer true that all elements f in X'' are of the form (26).

Part (ii) of Theorem 7 can be stated as follows:

$$|x| = \max_{|l|'=1} l(x) \tag{29}$$

for every vector x.

The following is an interesting generalization of (29).

Theorem 9.   Let Z be a subspace of X, y any vector in X. The distance $d(y,Z)$ of y to Z is defined to be

$$d(y,z) = \inf_{z \text{ in } Z} |y - z|. \tag{30}$$

Then

$$d(y,Z) = \max l(y) \tag{31}$$

over all ξ in X' satisfying

$$|l|' \leq 1, \qquad l(z) = 0 \text{ for } z \text{ in } Z. \tag{32}$$

Proof. By definition of distance, for any $\epsilon > 0$ there is a z_0 in Z such that

$$|y - z_0| < d(y,Z) + \epsilon. \tag{33}$$

For any l satisfying (32) we get, using (33) that

$$l(y) = l(y) - l(z_0) = l(y - z_0) \leq |l||y - z_0| < d(y,Z) + \epsilon.$$

Since $\epsilon > 0$ is arbitrary, this shows that for all l satisfying (32).

$$l(y) \leq d(y,z). \tag{34}$$

To show the opposite inequality we shall exhibit a linear function m satisfying (32), such that $m(y) = d(y, z)$. Since for y in Z the result is trivial, we assume that the vector y does not belong to Z. We define the linear subspace U to consist of all vectors u of the form

$$u = z + ky, \qquad z \text{ in } Z, k \text{ any real number.} \tag{35}$$

We define the linear function $m(u)$ in U by

$$m(u) = kd(y, Z). \tag{36}$$

Obviously, m is zero for u in Z; it follows from (35), (36), and the definition (30) of d that

$$m(u) \leq |u| \qquad \text{for } u \text{ in } U. \tag{37}$$

By Hahn–Banach we can extend m to all of X so that (37) holds for all x; then

$$|m|' \leq 1. \tag{37}'$$

Clearly, m satisfies (32); on the other hand, we see by combining (35) and (36) that

$$m(y) = d(y, Z).$$

Since we have seen in (34) that $l(y) \leq d(y, Z)$ for all l satisfying (32), this completes the proof of Theorem 9. $\qquad \square$

In Chapter 1 we have introduced the notion of the *quotient* of a linear space X by one of its subspaces Z. We recall the definition: two vectors x_1 and x_2 in X are congruent mod Z,

$$x_1 \equiv x_2 \bmod Z$$

if $x_1 - x_2$ belongs to Z. We saw that this is an equivalence relation, and therefore we can partition the vectors in X into congruence classes $\{\}$. The set of congruence classes $\{\}$ is denoted as X/Z and can be made into a linear space; all this is described in Chapter 1. We note that the subspace Z is one of the congruence classes, which serves as the zero element of the quotient space.

Suppose X is a normed linear space; we shall show that then there is a natural way of making X/Z into a normed linear space, by defining the following norm for the congruence classes:

$$|\{\}| = \inf |x|, \qquad x \in \{\}. \tag{38}$$

Theorem 10. Definition (38) is a norm, that is, has all three properties (1).

Proof. Every member x of a given congruence class $\{\}$ can be described as $x = x_0 - z$, x_0 some vector in $\{\}$, z any vector in Z. We claim that property (i), positivity, holds: for $\{\} \neq 0$,

$$|\{\}| > 0. \tag{38}'$$

Suppose on the contrary that $|\{\}| = 0$. In view of definition (38) this means that there is a sequence x_j in $\{\}$ such that

$$\lim |x_j| = 0. \tag{39}$$

Since all x_j belong to the same class, they all can be written as

$$x_j = x_0 - z_j, \qquad z_j \text{ in } Z.$$

Setting this into (39) we get

$$\lim |x_0 - z_j| = 0.$$

Since by Theorem 3 every linear subspace Z is closed, it follows that x_0 belongs to Z. But then every point $x_0 - z$ in $\{\}$ belongs to Z, and in fact $\{\} = Z$. But we saw earlier that $\{\} = Z$ is the zero element of X/Z. Since we have stipulated $\{\} \neq 0$, we have a contradiction, that we got into by assuming $|\{\}| = 0$.

Homogeneity is fairly obvious; we turn now to subaddivity: by definition (38) we can, given any $\epsilon > 0$, choose x_0 and $\{x\}$ and y_0 in $\{y\}$ so that

$$|x_0| < |\{x\}| + \epsilon, \qquad |y_0| < |\{y\}| + \epsilon. \tag{40}$$

Addition of classes is defined so that $x_0 + y_0$ belongs to $\{x\} + \{y\}$. Therefore by definition (38), subadditivity of $|\cdot|$ and (39, 40),

$$|\{x\} + \{y\}| \leq |x_0 + y_0| \leq |x_0| + |y_0| < |\{x\}| + |\{y\}| + 2\epsilon.$$

Since ϵ is an arbitrary positive number,

$$|\{x\} + \{y\}| \leq |\{x\}| + |\{y\}|$$

follows. This completes the proof of Theorem 10. $\square$

We conclude this chapter by remarking that a norm in a linear space over the *complex* numbers is defined entirely analogously, by the three properties (1). The theorems proved in the real case extend to the complex. To prove Theorems 7 and 9 in the complex case, we need a complex version of the Hahn–Banach theorem, due to Bohnenblust–Szobcyk and Sukhomlinov. Here it is:

Theorem 11. X is a linear space over $\mathbb{C}$, and p is a real-valued function defined on X with the following properties:

(i) p is absolute homogeneous; that is, it satisfies

$$p(ax) = |a|p(x).$$

for all complex numbers a and all x in X.

(ii) p is subadditive:

$$p(x+y) \leq p(x) + p(y).$$

Let U be a subspace of X, and l is a linear functional defined on U that satisfies

$$|l(u)| \leq p(u) \tag{41}$$

for all u in U.

Then l can be extended as a linear functional to the whole space so that

$$|l(x)| \leq p(x) \tag{41}'$$

for all x in X.

Proof. The complex linear space X can also be regarded as a linear space over $\mathbb{R}$. Any linear function on complex X can be split into its real and imaginary part:

$$l(u) = l_1(u) + il_2(u),$$

where l_1 and l_2 are real-valued, and linear on real U. l_1 and l_2 are related by

$$l_1(iu) = -l_2(u).$$

Conversely, if l_1 is a real-valued linear function over real X,

$$l(x) = l_1(x) - il_1(ix) \tag{42}$$

is linear over complex X.

We turn now to the task of extending l. It follows from (41) that l_1, the real part of l, satisfies on U the inequality

$$l_1(u) \leq p(u). \tag{43}$$

Therefore, by the real Hahn–Banach Theorem, l_1 can be extended to all of X so that the extended l is linear on real X and satisfies inequality (43). Define l by formula (42); clearly, it is linear over complex X and is an extension of l defined on U. We claim that it satisfies (41)$'$ for all x in X. To see this, we factor $l(x)$ as

$$l(x) = ar, \qquad r \text{ real}, \ |a| = 1.$$

Using the fact that if $l(y)$ is real, it is equal to $l_1(y)$, we deduce that

$$|l(x)| = r = a^{-1}l(x) = l(a^{-1}x) = l_1(a^{-1}x) \le p(a^{-1}x) = p(x). \qquad \square$$

We conclude this chapter by a curious characterization of Euclidean norms among all norms. According to equation (53) of Chapter 7, every pair of vectors u, v in a Euclidean space satisfies the following identity:

$$\| u + v \|^2 + \| u - v \|^2 = 2 \| u \|^2 + 2 \| v \|^2 .$$

Theorem 12. This identity characterizes Euclidean space. That is, if in a real normed linear space X

$$|u + v|^2 + |u - v|^2 = 2|u|^2 + 2|v|^2 \tag{44}$$

for all pairs of vectors u, v, then the norm $|\ |$ is Euclidean.

Proof. We define a scalar product in X as follows:

$$4(x, y) = |x + y|^2 - |x - y|^2. \tag{45}$$

The following properties of a scalar product follow immediately from definition (45):

$$(x, x) = |x|^2, \tag{46}$$

Symmetry:

$$(y, x) = (x, y), \tag{47}$$

and

$$(x, -y) = -(x, y) \tag{48}$$

Next we show that (x, y) as defined in (45) is additive:

$$(x + z, y) = (x, y) + (z, y). \tag{49}$$

By definition (45),

$$4(x + z, y) = |x + z + y|^2 - |x + z - y|^2. \tag{50}$$

We apply now identity (44) four times:

(i) $u = x + y, v = z$:

$$|x + y + z|^2 + |x + y - z|^2 = 2|x + y|^2 + 2|z|^2 \tag{50}_\mathrm{i}$$

(ii) $u = y + z, \quad v = x$:

$$|x + y + z|^2 + |y + z - x|^2 = 2|y + z|^2 + 2|x|^2 \tag{51}_\mathrm{ii}$$

(iii) $u = x - y, \quad v = z$:

$$|x - y + z|^2 + |x - y - z|^2 = 2|x - y|^2 + 2|z|^2 \tag{51}_\mathrm{iii}$$

(iv) $u = z - y, \quad v = x$:

$$|z - y + x|^2 + |z - y - x|^2 = 2|z - y|^2 + 2|x|^2 \tag{51}_\mathrm{iv}$$

Add $(51)_\mathrm{i}$ and $(51)_\mathrm{ii}$, and subtract from it $(51)_\mathrm{iii}$ and $(51)_\mathrm{iv}$; we get, after dividing by 2,

$$\begin{aligned} |x + y + z|^2 &- |x - y + z|^2 \\ &= |x + y|^2 - |x - y|^2 + |y + z|^2 - |y - z|^2. \end{aligned} \tag{52}$$

The left-hand side of (52) equals $4(x + z, y)$, and the right-hand side is $4(x, y) + 4(z, y)$. This proves (49).   □

EXERCISE 9.   **(i)** Show that for all rational r,

$$(rx, y) = r(x, y).$$

(ii) Show that for all real k,

$$(kx, y) = k(x, y).$$

CHAPTER 15

Linear Mappings Between Normed Linear Spaces

Let X and Y be a pair of finite-dimensional normed linear spaces over the reals; we shall denote the norm in both spaces by $|\ |$, although they have nothing to do with each other. The first lemma shows that every linear map of one normed linear space into another is bounded.

Lemma 1. For any linear map T: $X \rightarrow Y$, there is a constant c such that for all x in X,

$$|Tx| \leq c|x|. \tag{1}$$

Proof. Express x with respect to a basis $\{x_j\}$:

$$x = \sum a_j x_j; \tag{2}$$

then

$$Tx = \sum a_j Tx_j.$$

By properties of the norm in Y,

$$|Tx| \leq \sum |a_j||Tx_j|.$$

From this we deduce that

$$|Tx| \leq k|x|_\infty, \tag{3}$$

Linear Algebra and Its Applications, Second Edition, by Peter D. Lax
Copyright © 2007 John Wiley & Sons, Inc.

where

$$|x|_\infty = \max_j |a_j|, \qquad k = \sum |Tx_j|.$$

We have noted in Chapter 14 that $|\ |_\infty$ is a norm. Since we have shown in Chapter 14, Theorem 2, that all norms are equivalent, $|x|_\infty \leq$ const $|x|$ and (1) follows from (3).   □

EXERCISE I.   Show that every linear map T: $X \to Y$ is continuous, that is, if lim $x_n = x$, then lim $Tx_n = Tx$.

In Chapter 7 we have defined the norm of a mapping of one Euclidean space into another. Analogously, we have the following definition.

Definition.   The norm of the linear map T: $X \to Y$, denoted as $|T|$, is

$$|T| = \sup_{x \neq 0} \frac{|Tx|}{|x|}. \tag{4}$$

Remark 1.   It follows from (1) that $|T|$ is finite.

Remark 2.   It is easy to see that $|T|$ is the *smallest value* we can choose for c in inequality (1).

Because of the homogeneity of norms, definition (4) can be phrased as follows:

$$|T| = \sup_{|x|=1} |Tx|. \tag{4}'$$

Theorem 2. $|T|$ as defined in (4) and (4)$'$ is a norm in the linear space of all linear mappings of X into Y.

Proof.   Suppose T is nonzero; that means that for some vector $x_0 \neq 0$, $Tx_0 \neq 0$. Then by (4),

$$|T| \geq \frac{|Tx_0|}{|x_0|};$$

since the norms in X and Y are positive, the positivity of $|T|$ follows.

To prove subadditivity we note, using (4)', that when S and T are two mappings of $X \to Y$, then

$$|T + S| = \sup_{|x|=1} |(T + S)x| \leq \sup_{|x|=1} (|Tx| + |Sx|)$$

$$\leq \sup_{|x|=1} |Tx| + \sup_{|x|=1} |Sx| = |T| + |S|.$$

The crux of the argument is that the supremum of a function that is the sum of two others is less than or equal to the sum of the separate suprema of the two summands.

Homogeneity is obvious; this completes the proof of Theorem 2.   □

Given any mapping T from one linear space X into another Y, we explained in Chapter 3 that there is another map, called the *transpose* of T and denoted as T', mapping Y', the dual of Y, into X', the dual of X. The defining relation between the two maps is given in equation (9) of Chapter 3:

$$(T'l, x) = (l, Tx), \tag{5}$$

where x is any vector in X and l is any element of Y'. The scalar product on the right, (l, y), denotes the bilinear pairing of elements y of Y and l of Y'. The scalar product (m, x) on the left is the bilinear pairing of elements x in X and m in X'. Relation (5) defines $T'l$ as an element of X'. We have noted in Chapter 3 that (5) is a symmetric relation between T and T' and that

$$T'' = T, \tag{6}$$

just as X'' is X and Y'' is Y.

We have shown in Chapter 14 that there is a natural way of introducing a dual norm in the dual X' of a normed linear space X, see Theorem 7; for m in X',

$$|m|' = \sup_{|x|=1} (m, x). \tag{7}$$

The dual norm for l in Y' is defined similarly as $\sup(l, y), |y| = 1$; from this definition, [see equation (24) of Chapter 14], it follows that

$$(l, y) \leq |l|' \, |y|. \tag{8}$$

Theorem 3.   Let T be a linear mapping from a normed linear space X into another normed linear space Y, T' its transpose, mapping Y' into X'. Then

$$|T'| = |T|, \tag{9}$$

where X' and Y' are equipped with the dual norms.

Proof. Apply definition (7) to $m = T'l$:

$$|T'l|' = \sup_{|x|=1} (T'l, x).$$

Using definition (5) of the transpose, we can rewrite the right-hand side as

$$|T'l|' = \sup_{|x|=1} (l, Tx).$$

Using the estimate (8) on the right, with $y = Tx$, we get

$$|T'l|' \leq \sup_{|x|=1} |l|' \, |Tx|.$$

Using (4)' to estimate $|Tx|$ we deduce that

$$|T'l| \leq |l|' \, |T|.$$

By definition (4) of the norm of T', this implies

$$|T'| \leq |T|. \tag{10}$$

We replace now T by T' in (10); we obtain

$$|T''| \leq |T'|. \tag{10}'$$

According to (6), $T'' = T$, and according to Theorem 8 of Chapter 14, the norms in X'' and Y'', the spaces between which T'' acts, are the same as the norms in X and Y. This shows that $|T''| = |T|$; now we can combine (10) and (10)' to deduce (9). This completes the proof of Theorem 3. ☐

Let T be a linear map of a linear space X into Y, S another linear map of Y into another linear space Z. Then, as remarked in Chapter 3, we can define the *product* ST as the *composite* mapping of T followed by S.

Theorem 4. Suppose X, Y, and Z above are normed linear spaces; then

$$|ST| \leq |S||T|. \tag{11}$$

Proof. By definition (4),

$$|Sy| \leq |S||y|, \qquad |Tx| \leq |T||x| \tag{12}$$

Hence

$$|STx| \leq |S||Tx| \leq |S||T||x|. \tag{13}$$

Applying definition (4) to ST completes the proof of inequality (11). □

We recall that a mapping T of one linear space X into another is called *invertible* if it maps X *onto Y*, and *is one-to-one*. In this case T has an *inverse*, denoted as T^{-1}.

In Chapter 7, Theorem 15, we have shown that if a mapping B of a Euclidean space into itself doesn't differ too much from another mapping A that is invertible, then B, too, is invertible. We present now a straightforward extension of this result to normed linear spaces.

Theorem 5.   Let X and Y be finite-dimensional normed linear spaces of the same dimension, and let T be a linear mapping of X into Y that is invertible. Let S be another linear map of X into Y that does not differ too much from T in the sense that

$$|S - T| < k, \qquad k = \frac{1}{|T^{-1}|}. \tag{14}$$

Then S is invertible.

Proof. We have to show that S is one-to-one and onto. We show first that S is one-to-one. We argue indirectly; suppose that for $x_0 \neq 0$,

$$Sx_0 = 0. \tag{15}$$

Then

$$Tx_0 = (T - S)x_0.$$

Since T is invertible,

$$x_0 = T^{-1}(T - S)x_0.$$

Using Theorem 4 and (14) and that $|x_0| > 0$, we get

$$|x_0| \leq |T^{-1}||T - S||x_0| < |T^{-1}|k|x_0| = |x_0|,$$

a contradiction; this shows that (15) is untenable and so S is one-to-one.

According to Corollary B of Theorem 1 in Chapter 3, a mapping S of a linear space X into another linear space of the same dimension that is one-to-one is onto. Since we have shown that S is one-to-one, this completes the proof of Theorem 5. □

Theorem 5 holds for normed linear spaces that are not finite dimensional, provided that they are complete. Corollary B of Theorem 1 of Chapter 3 does not hold in spaces of infinite dimension; therefore we need a different, more direct argument to invert S. We now present such an argument. We start by recalling the notion of convergence in a normed linear space applied to the space of linear maps.

Definition. Let X, Y be a pair of finite-dimensional normed linear spaces. A sequence $\{T_n\}$ of linear maps of X into Y is said to converge to the linear map T, denoted as $\lim_{n \to x} T_n = T$, if

$$\lim_{n \to \infty} |T_n - T| = 0. \tag{16}$$

Theorem 6. Let X be a normed finite-dimensional linear space, R a linear map of X into itself whose norm is less than 1:

$$|R| < 1. \tag{17}$$

Then

$$S = I - R \tag{18}$$

is invertible, and

$$S^{-1} = \sum_0^\infty R^k. \tag{18}'$$

Proof. Denote $\sum_0^n R^k$ as T_n, and denote $T_n x$ as y_n. We claim that $\{y_n\}$ is a Cauchy sequence; that is, $|y_n - y_l|$ tends to zero as n and l tend to ∞. To see this, we write

$$y_n - y_l = T_n x - T_l x = \sum_{j+1}^n R^k x.$$

By the triangle inequality

$$|y_n - y_l| \le \sum_{j+1}^n |R^k x|. \tag{19}$$

Using repeatedly the multiplicative property of the norm of operators, we conclude that

$$|R^k| \le |R|^k.$$

It follows that

$$|R^k x| \le |R^k||x| \le |R|^k |x|.$$

Set this estimate into (19); we get

$$|y_n - y_j| \le \left(\sum_{j+1}^n |R|^k \right) |x|. \tag{20}$$

Since $|R|$ is assumed to be less than one, the right-hand side of (20) tends to zero as n and j tend to ∞. This shows that $y_n = T_n x = \sum_0^n R^n x$ is a Cauchy sequence.

According to Theorem 3 of Chapter 14, every Cauchy sequence in a finite-dimensional normed linear space has a limit. We define the mapping T as

$$Tx = \lim_{n \to \infty} T_n x. \tag{21}$$

We claim that T is the inverse of $I - R$. According to Exercise 1, the mapping $I - R$ is continuous; therefore it follows from (21) that

$$(I - R)Tx = \lim_{n \to \infty} (1 - R)T_n x$$

Since

$$T_n = \sum_0^n R^k,$$

$$(I - R)T_n x = (I - R) \sum_0^n R^k x = x - R^{n+1} x.$$

as $n \to \infty$, the left-hand side tends to $(I - R)Tx$ and the right-hand side tends to x; this proves that T is the inverse of $I - R$. □

EXERCISE 2.   Show that if for every x in X, $|T_n x - Tx|$ tends to zero as $n \to \infty$, then $|T_n - T|$ tends to zero.

EXERCISE 3.   Show that $T_n = \sum_0^n R^k$ converges to S^{-1} in the sense of definition (16).

Theorem 6 is a special case of Theorem 5, with $Y = X$ and $T = I$.

EXERCISE 4.   Deduce Theorem 5 from Theorem 6 by factoring $S = T + S - T$ as $T[I - T^{-1}(S - T)]$.

EXERCISE 5.   Show that Theorem 6 remains true if the hypothesis (17) is replaced by the following hypothesis. For some positive integer m,

$$|R^m| < 1. \tag{22}$$

EXERCISE 6.   Take $X = Y = \mathbb{R}^n$, and $T: X \to X$ the matrix (t_{ij}). Take for the norm $|x|$ the maximum norm $|x|_\infty$ defined by formula (3) of Chapter 14. Show that the norm $|T|$ of the matrix (t_{ij}), regarded as a mapping of X into X, is

$$|T| = \max_i \sum_j |t_{ij}|. \tag{23}$$

EXERCISE 7.   Take X to be $\mathbb{R}^n$ normed by the maximum norm $|x|_\infty$, Y to be $\mathbb{R}^n$ normed by the 1−norm $|x|_1$, defined by formulas (3) and (4) in Chapter 14. Show that the norm of the matrix (t_{ij}) regarded as a mapping of X into Y is bounded by

$$|T| \leq \sum_{i,j} |t_{ij}|.$$

EXERCISE 8. X is any finite-dimensional normed linear space over $\mathbb{C}$, and T is a linear mapping of X into X. Denote by t_j the eigenvalues of T, and denote by r (T) its spectral radius:

$$r(\mathrm{T}) = \max |t_j|.$$

(i) Show that $|T| \geq r(\mathrm{T})$.
(ii) Show that $|T^n| \geq r(\mathrm{T})^n$.
(iii) Show, using Theorem 18 of Chapter 7, that

$$\lim_{n \to \infty} |T^n|^{1/n} = r(\mathrm{T}).$$

CHAPTER 16

Positive Matrices

Definition. A real $l \times l$ matrix P is called *entrywise positive* if all its entries p_{ij} are positive real numbers.

Caution: This notion of positivity, used only in this chapter, is not to be confused with self-adjoint matrices that are positive in the sense of Chapter 10.

Theorem 1 (Perron). Every positive matrix P has a *dominant eigenvalue*, denoted by $\lambda(P)$ which has the following properties:

(i) $\lambda(P)$ is positive and the associated eigenvector h has positive entries:

$$Ph = \lambda(P)h, \qquad h > 0. \tag{1}$$

(ii) $\lambda(P)$ is a simple eigenvalue.

(iii) Every other eigenvalue κ of P is less than $\lambda(P)$ in absolute value:

$$|\kappa| < \lambda(P). \tag{2}$$

(iv) P has no other eigenvector f with nonegative entries.

Proof. We recall from Chapter 13 that inequality between vectors in $\mathbb{R}^n$ means that the inequality holds for all corresponding components. We denote by $p(P)$ the set of all nonnegative numbers λ for which there is a nonnegative vector $x \neq 0$ such that

$$Px \geq \lambda x, \qquad x \geq 0. \tag{3}$$

Lemma 2. For P positive,

   (i) $p(P)$ is nonempty, and contains a positive number,

  (ii) $p(P)$ is bounded,

 (iii) $p(P)$ is closed.

Proof. Take any positive vector x; since P is positive, Px is a positive vector. Clearly, (3) will hold for λ small enough positive; this proves (i) of the lemma.

Since both sides of (3) are linear in x, we can normalize x so that

$$\xi x = \sum x_i = 1, \qquad \xi = (1, \ldots, 1). \tag{4}$$

Multiply (3) by ξ on the left:

$$\xi P x \geq \lambda \xi x = \lambda. \tag{5}$$

Denote the largest component of ξP by b; then $b\xi \geq \xi P$. Setting this into (5) gives $b \geq \lambda$; this proves part (ii) of the lemma.

To prove (iii), consider a sequence of λ_n in $p(P)$; by definition there is a corresponding $x_n \neq 0$ such that (3) holds:

$$Px_n \geq \lambda_n, x_n, \qquad x_n \geq 0. \tag{6}$$

We might as well assume that the x_n are normalized by (4):

$$\xi x_n = 1.$$

The set of nonnegative x_n normalized by (4) is a closed bounded set in $\mathbb{R}^n$ and therefore compact. Thus a subsequence of x_n tends to a nonnegative x also normalized by (4), while λ_n tends to λ. Passing to the limit of (6) shows that x, λ satisfy (3); therefore $p(P)$ is closed. This proves part (iii) of the lemma. $\quad\square$

Having shown that $p(P)$ is closed and bounded, it follows that it has a maximum $\lambda_{\max}$; by (i), $\lambda_{\max} > 0$. We shall show now that $\lambda_{\max}$ is the dominant eigenvalue.

The first thing to show is that $\lambda_{\max}$ is an eigenvalue. Since (3) is satisfied by $\lambda_{\max}$, there is a nonnegative vector h for which

$$Ph \geq \lambda_{\max} h, \qquad h \geq 0, h \neq 0; \tag{7}$$

we claim that equality holds in (7); for, suppose not, say in the kth component:

$$\begin{aligned}
\sum p_{ij} h_j &\geq \lambda_{\max} h_i, \qquad i \neq k \\
\sum p_{kj} h_j &> \lambda_{\max} h_k.
\end{aligned} \tag{7$'$}$$

Define the vector $x = h + \epsilon e_k$, where $\epsilon > 0$ end e_k has kth component equal to 1, all other components zero. Since P is positive, replacing h by x in (7) increases each component of the left-hand side: $Px > Ph$. But only the kth component of the right-hand side is increased when h is replaced by x. It follows therefore from (7)' that for ϵ small enough positive,

$$Px > \lambda_{max}x. \tag{8}$$

Since this is a strict inequality, we may replace λ_{max} by $\lambda_{max} + \delta$, δ positive but so small that (8) still holds. This shows that $\lambda_{max} + \delta$ belongs to $p(P)$, contrary to the maximal character of λ_{max}. This proves that λ_{max} is an eigenvalue of P and that there is a corresponding eigenvector h that is nonnegative.

We claim now that the vector h is positive. For certainly, since P is positive and $h \geq 0$, it follows that $Ph > 0$. Since $Ph = \lambda_{max}h, h > 0$ follows. This proves part (i) of Theorem 1.

Next we show that λ_{max} is simple. We observe that all eigenvectors of P with eigenvalue λ_{max} must be proportional to h; for if there were another eigenvector y not a multiple of h, then we could construct $h + cy$, c so chosen that $h + cy \geq 0$ but one of the components of $h + cy$ is zero. This contradicts our argument above that an eigenvector of P is nonnegative is in fact positive.

To complete the proof of (ii) we have to show that P has no generalized eigenvectors for the eigenvalue λ_{max}, that is, a vector y such that

$$Py = \lambda_{max}y + ch. \tag{9}$$

By replacing y by $-y$ if necessary we can make sure that $c > 0$; by replacing y by $y + bh$ if necessary we can make sure that y is positive; it follows then from (9) and $h > 0$ that $Py > \lambda_{max}y$. But then for δ small enough, greater than 0,

$$Py > (\lambda_{max} + \delta)y,$$

contrary to λ_{max} being the largest number in $p(P)$.

To show part (iii) of Theorem 1, let κ be another eigenvalue of P, not equal to λ_{max}, y the corresponding eigenvector, both possibly complex: $Py = \kappa y$; componentwise,

$$\sum_j p_{ij}y_j = \kappa y_i.$$

Using the triangle inequality for complex numbers and their absolute values, we get

$$\sum_j p_{ij}|y_j| \geq \left| \sum_j p_{ij}y_j \right| = |\kappa||y_i|. \tag{10}$$

Comparing this with (3), we see that $|\kappa|$ belongs to $p(P)$. If $|\kappa|$ were $= \lambda_{\max}$, the vector

$$\begin{pmatrix} |y_1| \\ \vdots \\ |y_l| \end{pmatrix}$$

would be an eigenvector of P with eigenvalue $\lambda_{\max}$, and thus proportional to h:

$$|y_i| = ch_i. \tag{11}$$

Furthermore, the sign of equality would hold in (10). It is well known about complex numbers that this is the case only if all the y_i have the same complex argument:

$$y_i = e^{i\theta}|y_i|, \qquad i = 1, \ldots, l$$

Combining this with (11) we see that

$$y_i = ce^{i\theta}h_i, \qquad \text{that is, } y = (ce^{i\theta})h.$$

Thus $\kappa = \lambda_{\max}$, and the proof of part (iii) is complete.

To prove (iv) we recall from Chapter 6, Theorem 17, that the product of eigenvectors of P and its transpose P^T pertaining to different eigenvalues is zero. Since P^T also is positive, the eigenvector ξ pertaining to its dominant eigenvalue, which is the same as that of P, has positive entries. Since a positive vector ξ does not annihilate a nonnegative vector f, part (iv) follows from $\xi f = 0$. This completes the proof of Theorem 1.   □

The above proof is due to Bohnenblust; see R. Bellman, *Introduction to Matrix Analysis.*

EXERCISE 1.   Denote by $t(P)$ the set of nonnegative λ such that

$$Px \le \lambda x, \qquad x \ge 0$$

for some vector $x \ne 0$. Show that the dominant eigenvalue $\lambda(P)$ satisfies

$$\lambda(P) = \min_{\lambda \in t(P)} \lambda. \tag{12}$$

We give now some applications of Perron's theorem.

Definition.   A *stochastic matrix* is an $l \times l$ matrix S whose entries are nonnegative:

$$s_{ij} \ge 0, \tag{13}$$

and whose column sums are equal to 1:

$$\sum_i s_{ij} = 1, \qquad j = 1, \ldots, l. \qquad (14)$$

The interpretation lies in the study of collections of l species, each of which has the possibility of changing into another. The numbers s_{ij} are called *transition probabilities*; they represent the fraction of the population of the jth species that is replaced by the ith species. Condition (13) is natural for this interpretation; condition (14) specifies that the total population is preserved. There are interesting applications where this is not so.

The kind of species that can undergo change describable as in the foregoing are atomic nuclei, mutants sharing a common ecological environment, and many others.

We shall first study positive stochastic matrices, that is, ones for which (13) is a strict inequality. To these Perron's theorem is applicable and yields the following theorem.

Theorem 3. Let S be a positive stochastic matrix.

(i) The dominant eigenvalue $\lambda(S) = 1$.

(ii) Let x be any nonnegative vector; then

$$\lim_{N \to \infty} S^N x = ch, \qquad (15)$$

where h the dominant eigenvector and c is some positive constant.

Proof. As remarked earlier, if S is a positive matrix, so is its transpose S^T. Since, according to Theorem 16, Chapter 6, S and S^T have the same eigenvalues, it follows that S and S^T have the same dominant eigenvalue. Now the dominant eigenvalue of the transpose of a stochastic matrix is easily computed: It follows from (14) that the vector with all entries 1,

$$\xi = (1, \ldots, 1).$$

is a left eigenvector of S, with eigenvalue 1. It follows from part (iv) Theorem 1 that this is the dominant eigenvector and 1 is the dominant eigenvalue. This proves part (i).

To prove (ii), we expand x as a sum of eigenvectors h_j of S:

$$x = \sum c_j h_j. \qquad (16)$$

Assuming that all eigenvectors of S are genuine, not generalized, we get

$$S^N x = \sum c_j \lambda_j^N h_j. \qquad (16)_N$$

Here the first component is taken to be the dominant one; so $\lambda_1 = \lambda = 1, |\lambda_j| < 1$ for $j \neq 1$. From this and $(16)_N$ we conclude that

$$S^N x \to ch, \tag{17}$$

where $c = c_1, h = h_1$, the dominant eigenvector.

To prove that c is positive, form the scalar product of (17) with ξ. Since $\xi = S^T \xi = (S^T)^N \xi$, we get

$$(S^N x, \xi) = (x, (S^T)^N \xi) = (x, \xi) \to c(h, \xi). \tag{17$'$}$$

We have assumed that x is nonnegative and not equal to 0; ξ and h are positive. Therefore it follows from (17)$'$ that c is positive. This proves part (ii) of Theorem 3 when all eigenvectors are genuine. The general case can be handled similarly. $\square$

We turn now to applications of Theorem 3 to systems whose change is governed by transition probabilities. Denote by $x_1, \ldots, x_n$ the population size of the jth species, $j = 1, \ldots, n$; suppose that during a unit of time (a year, a day, a nanosecond) each individual of the collection changes (or gives birth to) a member of the other species according to the probabilities s_{ij}. If the population size is so large that fluctuations are unimportant, the new size of the population of the ith species will be

$$y_i = \sum s_{ij} x_j. \tag{18}$$

Combining the components of the old and new population into single column vectors x and y, relation (18) can be expressed in the language of matrices as

$$y = Sx. \tag{18$'$}$$

After N units of time, the population vector will be $S^N x$. The significance of Theorem 3 in such applications is that it shows that as $N \to \infty$, such populations tend to a steady distribution that does not depend on where the population started from.

Theorem 3 is the basis of Google's search strategy.

Theorem 1—and therefore Theorem 3—depend on the positivity of the matrix P; in many applications we have to deal with matrices that are merely nonnegative. How much of Theorem 1 remains true for such matrices?

The three examples,

$$\begin{pmatrix} 0 & 1 \\ 1 & 1 \end{pmatrix}, \quad \begin{pmatrix} 0 & 1 \\ 1 & 0 \end{pmatrix}, \quad \text{and} \quad \begin{pmatrix} 1 & 1 \\ 0 & 1 \end{pmatrix},$$

show different behavior. The first one has a dominant eigenvalue; the second has plus or minus 1 as eigenvalues, neither dominated by the other; the third has 1 as a double eigenvalue.

EXERCISE 2. Show that if some power P^m of P is positive, then P has a dominant positive eigenvalue.

There are other interesting and useful criteria for nonnegative matrices to have a dominant positive eigenvalue. These are combinatorial in nature; we shall not speak about them. There is also the following result, due to Frobenius.

Theorem 4. Every nonnegative $l \times l$ matrix F, $F \neq 0$, has an eigenvalue $\lambda(F)$ with the following properties:

(i) $\lambda(F)$ is nonnegative, and the associated eigenvector has nonnegative entries:

$$Fh = \lambda(F)h, \qquad h \geq 0. \tag{19}$$

(ii) Every other eigenvalue κ is less than or equal to $\lambda(F)$ in absolute value:

$$|\kappa| \leq \lambda(F). \tag{20}$$

(iii) If $|\kappa| = \lambda(F)$, then κ is of the form

$$\kappa = e^{2\pi i k/m}\lambda(F), \tag{21}$$

where k and m are positive integers, $m \leq l$.

Remark. Theorem 4 can be used to study the asymptotically periodic behavior for large N of $S^N x$, where S is a nonnegative stochastic matrix. This has applications to the study of cycles in population growth.

Proof. Approximate F by a sequence F_n of *positive* matrices. Since the characteristic equations of F_n tend to the characteristic equations of F, it follows that the eigenvalues of F_n tend to the eigenvalues of F. Now define

$$\lambda(F) = \lim_{n \to \infty} \lambda(F_n).$$

Clearly, as $n \to \infty$, inequality (20) follows from inequality (2) for F_n. To prove (i), we use the dominant eigenvector h_n of F_n, normalized as in (4):

$$\xi h_n = 1, \qquad \xi = (1, \ldots, 1).$$

By compactness, a subsequence of h_n converges to a limit vector h. Being the limit of normalized positive vectors, h is nonnegative. Each h_n satisfies an equation

$$F_n h_n = \lambda(F_n)h_n;$$

letting n tend to ∞ we obtain relation (19) in the limit.

Part (iii) is trivial when $\lambda(F) = 0$; so we may assume $\lambda(F) > 0$; at the cost of multiplying F by a constant we may assume that $\lambda(F) = 1$. Let κ be a complex eigenvalue of F, $|\kappa| = \lambda(F) = 1$; then κ can be written as

$$\kappa = e^{i\theta}. \tag{22}$$

Denote by $y + iz$ the corresponding eigenvector:

$$F(y + iz) = e^{i\theta}(y + iz). \tag{23}$$

Separate the real and imaginary parts:

$$\begin{aligned} Fy &= \cos\theta\, y - \sin\theta\, z, \\ Fz &= \sin\theta\, z + \cos\theta\, y. \end{aligned} \tag{23$'$}$$

The geometric interpretation of $(23)'$ is that in the plane spanned by the vectors y and z, F is *rotation* around the origin by θ.

Consider now the plane formed by all points x of the form

$$x = h + ay + bz, \tag{24}$$

a and b arbitrary real numbers, h the eigenvector (19). It follows from (19) and $(23)'$ that in this plane F acts as rotation by θ. Consider now the set Q formed by all *nonnegative* vectors x of form (24); if Q contains an open subset of the plane (24), it is a *polygon*. Since F is a nonnegative matrix, it maps Q into itself; since it is a rotation, it maps Q onto itself. Since Q has l vertices, the lth power of F is the identity; this shows that F rotates Q by an angle $\theta = 2\pi k/l$.

It is essential for this argument that Q be a polygon, that is, that it contain an open set of the plane (24). This will be the case when all components of h are positive or when some components of h are zero, but so are the corresponding components of y and z. For then all points x of form (24) with $|a|$, $|b|$ small enough belong to Q; in this case Q is a polygon.

To complete the proof of Theorem 4(iii), we turn to the case when some components of h are zero but the corresponding components of y or z are not. Arrange the components in such an order that the first j components of h are zero, the rest positive. Then it follows from $Fh = h$ that F has the following block form:

$$F = \begin{pmatrix} F_0 & 0 \\ A & B \end{pmatrix}. \tag{25}$$

Denote by y_0 and z_0 the vectors formed by the first j components of y and z. By assumption, $y_0 + iz_0 \neq 0$. Since by (23), $y + iz$ is an eigenvector of F with eigenvalue $e^{i\theta}$, it follows from (25) that $y_0 + iz_0$ is an eigenvector of F_0:

$$F_0(y_0 + iz_0) = e^{i\theta}(y_0 + iz_0).$$

Since F_0 is a nonnegative $j \times j$ matrix, it follows from part (ii) of Theorem 4 already established that the dominant eigenvalue $\lambda(F_0)$ cannot be less than $|e^{i\theta}| = 1$. We claim that equality holds: $\lambda(F_0) = 1$. For, suppose not; then the corresponding eigenvector h_0 would satisfy

$$F_0 h_0 = (1 + \delta)h_0, \qquad h_0 \geq 0, \delta > 0. \tag{26}$$

Denote by k the l-vector whose first j components are those of h_0, the rest are zero. It follows from (26) that

$$Fk \geq (1 + \delta)k. \tag{26$'$}$$

It is easy to show that the dominant eigenvalue $\lambda(F)$ of a nonnegative matrix can be characterized as the largest λ for which (3) can be satisfied. Inequality (26)$'$ would imply that $\lambda(F) \geq 1 + \delta$, contrary to the normalization $\lambda(F) = 1$. This proves that $\lambda(F_0) = 1$.

We do now an induction with respect to j on part (iii) of Theorem 4. Since $e^{i\theta}$ is an eigenvalue of the $j \times j$ matrix F_0, and $\lambda(F_0) = 1$, and since $j < l$, it follows by the induction hypothesis that θ is a rational multiple of 2π with denominator less than or equal to j. This completes the proof of Theorem 4.   □

CHAPTER 17

How to Solve Systems of Linear Equations

To get numerical answers out of any linear model, one must in the end obtain the solution of a system of linear equations. To carry out this task efficiently has therefore a high priority; it is not surprising that it has engaged the attention of some of the leading mathematicians. Two methods still in current use, Gaussian elimination and the Gauss–Seidel iteration, were devised by the Prince of Mathematicians. The great Jacobi invented an iterative method that bears his name.

The availability of programmable, high-performance computers with large memories—and remember, yesterday's high-performance computer is today's pocket computer—has opened the floodgates; the size and scope of linear equations that could be solved efficiently has been enlarged enormously and the role of linear models correspondingly enhanced. The success of this effort has been due not only to the huge increase in computational speed and in the size of rapid access memory, but in equal measure to new, sophisticated, mathematical methods for solving linear equations. At the time von Neumann was engaged in inventing and building a programmable electronic computer, he devoted much time to analyzing the accumulation and amplification of round-off errors in Gaussian elimination. Other notable early efforts were the very stable methods that Givens and Householder found for reducing matrices to Jacobi form (see Chapter 18).

It is instructive to recall that in the 1940s linear algebra was dead as a subject for research; it was ready to be entombed in textbooks. Yet only a few years later, in response to the opportunities created by the availability of high-speed computers, very fast algorithms were found for the standard matrix operations that astounded those who thought there were no surprises left in this subject.

In this chapter we describe a few representative modern algorithms for solving linear equations. Included among them, in Section 4, is the conjugate gradient method developed by Lanczos, Stiefel, and Hestenes.

The systems of linear equations considered in this chapter are of the class that have exactly one solution. Such a system can be written in the form

$$Ax = b, \tag{1}$$

A an invertible square matrix, b some given vector, x the vector of unknowns to be determined.

An algorithm for solving the system (1) takes as its input the matrix A and the vector b and produces as output some approximation to the solution x. In designing and analyzing an algorithm we must first understand how fast and how accurately an algorithm works when all the arithmetic operations are carried out exactly. Second, we must understand the effect of *rounding*, inevitable in computers that do their arithmetic with a finite number of digits.

With algorithms employing billions of operations, there is a very real danger that round-off errors not only accumulate but are magnified in the course of the calculation. Algorithms for which this does not happen are called *arithmetically stable*.

It is important to point out that the use of finite digit arithmetic places an absolute limitation on the accuracy with which the solution can be determined. To understand this, imagine a change δb being made in the vector b appearing on the right in (1). Denote by δx the corresponding change in x:

$$A(x + \delta x) = b + \delta b. \tag{2}$$

since according to (1), $Ax = b$, we deduce that

$$A\delta x = \delta b. \tag{3}$$

We shall compare the *relative change* in x with the relative change in b, that is, the ratio

$$\frac{|\delta x|}{|x|} \Big/ \frac{|\delta b|}{|b|}, \tag{4}$$

where the norm is convenient for the problem. The choice of relative change is natural when the components of vectors are floating point numbers.

We rewrite (4) as

$$\frac{|b|}{|x|} \frac{|\delta x|}{|\delta b|} = \frac{|Ax|}{|x|} \frac{|A^{-1}\delta b|}{|\delta b|}. \tag{4'}$$

The sensitivity of problem (1) to changes in b is estimated by maximum of (4)' over all possible x and δb. The maximum of the first factor on the right in (4)' is $|A|$, the norm of A; the maximum of the second factor is $|A^{-1}|$, the norm of A^{-1}. Thus we conclude that the ratio (4) of the relative error in the solution x to the relative error in b can not be larger than

$$\kappa(A) = |A||A^{-1}|. \tag{5}$$

The quantity $\kappa(A)$ is called the *condition number* of the matrix A.

EXERCISE I.   Show that $\kappa(A)$ is ≥ 1.

Since in k-digit floating point arithmetic the relative error in b can be as large as 10^{-k}, it follows that if equation (1) is solved using k-digit floating point arithmetic, the relative error in x can be as large as $10^{-k}\kappa(A)$.

It is not surprising that the larger the condition number $\kappa(A)$, the harder it is to solve equation (1), for $\kappa(A) = \infty$ when the matrix A is not invertible. As we shall show later in this chapter, the rate of convergence of iterative methods to the exact solution of (1) is slow when $\kappa(A)$ is large.

Denote by β the largest absolute value of the eigenvalues of A. Clearly,

$$\beta \leq |A|. \tag{6}$$

Denote by α the smallest absolute value of the eigenvalues of A. Then applying inequality (6) to the matrix A^{-1} we get

$$\frac{1}{\alpha} \leq |A^{-1}|. \tag{6$'$}$$

Combining (6) and (6)$'$ with (5) we obtain this lower bound for the condition number of A:

$$\frac{|\beta|}{|\alpha|} \leq \kappa(A). \tag{7}$$

An algorithm that, when all arithmetic operations are carried out exactly, furnishes in a finite number of steps the exact solution of (1) is called a *direct* method. Gaussian elimination discussed in Chapter 4 is such a method. An algorithm that generates a sequence of approximations that tend, if all arithmetic operations were carried out exactly, to the exact solution is called an *iterative method*. In this chapter we shall investigate the convergence and rate of convergence of several iterative methods.

Let us denote by $\{x_n\}$ the sequence of approximations generated by an algorithm. The deviation of x_n from x is called the *error* at the nth stage, and is denoted by e_n:

$$e_n = x_n - x. \tag{8}$$

The amount by which the nth approximation fails to satisfy equation (1) is called the nth *residual*, and is denoted by r_n:

$$r_n = Ax_n - b. \tag{9}$$

Residual and error are related to each other by

$$r_n = Ae_n. \tag{10}$$

Note that, since we do not know x, we cannot calculate the errors e_n; but once we have calculated x_n we can by formula (9) calculate r_n.

In what follows, we shall restrict our analysis to the case when the matrix A is *real*, *self-adjoint*, and *positive*; see Chapter 8 and Chapter 10 for the definition of these concepts. We shall use the Euclidean norm, denoted as $\|\,\|$, to measure the size of vectors.

We denote by α and β the smallest and largest eigenvalues of A. Positive definiteness of A implies that α is positive, see Theorem 1 of Chapter 10. We recall from Chapter 8, Theorem 12, that the norm of a positive matrix with respect to the Euclidean norm is its largest eigenvalue;

$$\| A \| = \beta. \tag{11}$$

Since A^{-1} also is positive, we conclude that

$$\| A^{-1} \| = \alpha^{-1}. \tag{11}'$$

Recalling the definitions (5) of the condition number A we conclude that for A self-adjoint and positive,

$$\kappa(A) = \frac{\beta}{\alpha}. \tag{12}$$

1. THE METHOD OF STEEPEST DESCENT

The first iterative method we investigate is based on the variational characterization of the solution of equation (1) in the case when A is positive definite.

Theorem 1.   The solution x of (1) minimizes the functional

$$E(y) = \tfrac{1}{2}(y, Ay) - (y, b); \tag{13}$$

here $(\,,\,)$ denotes the Euclidean scalar product of vectors.

Proof. We add to $E(y)$ a constant, that is, a term independent of y:

$$F(y) = E(y) + \tfrac{1}{2}(x, b). \tag{14}$$

Set (13) into (14); using $Ax = b$ and the self-adjointness of A we can express $F(y)$ as

$$F(y) = \tfrac{1}{2}(y - x, A(y - x)). \tag{14}'$$

Clearly,

$$F(x) = 0.$$

A being positive means that $(v, Av) > 0$ for $v \neq 0$. Thus (14)' shows that $F(y) > 0$ for $y \neq x$. This proves that $F(y)$, and therefore $E(y)$, takes on its minimum at $y = x$.   □

Theorem 1 shows that the task of solving (1) can be accomplished by minimizing E. To find the point where E assumes its minimum we shall use the method of *steepest descent*; that is, given an approximate minimizer y, we find a better approximation by moving from y to a new point along the direction of the negative gradient of E. The gradient of E is easily computed from formula (13):

$$\operatorname{grad} E(y) = Ay - b.$$

So if our nth approximation is x_n, then the $(n + 1)$st, x_{n+1}, is

$$x_{n+1} = x_n - s(Ax_n - b), \tag{15}$$

where s is step length in the direction $-\operatorname{grad} E$. Using the concept (9) of residual, we can rewrite (15) as

$$x_{x+1} = x_n - sr_n. \tag{15}'$$

We determine s so that $E(x_{n+1})$ is as small as possible. This quadratic minimum problem is easily solved; using (13) and (9), we have

$$\begin{aligned}E(x_{n+1}) &= \tfrac{1}{2}(x_n - sr_n, A(x_n - sr_n)) - (x_n - sr_n, b) \\ &= E(x_n) - s(r_n, r_n) + \tfrac{1}{2}s^2(r_n, Ar_n).\end{aligned} \tag{15}''$$

Its minimum is reached for

$$s_n = \frac{(r_n, r_n)}{(r_n, Ar_n)}. \tag{16}$$

Theorem 2.   The sequence of approximations defined by (15), with s given by (16), converges to the solution x of (1).

Proof. We need a couple of inequalities. We recall from Chapter 8 that for any vector r the Rayleigh quotient

$$\frac{(r, Ar)}{(r, r)}$$

of a self-adjoint matrix A lies between the smallest and largest eigenvalues of A. In our case these were denoted by α and β; so we deduce from (16) that

$$\frac{1}{\beta} \le s_n \le \frac{1}{\alpha}. \tag{17}$$

We conclude similarly that for all vectors r,

$$\frac{1}{\beta} \le \frac{(r, A^{-1}r)}{(r, r)} \le \frac{1}{\alpha}. \tag{17$'$}$$

We show now that $F(x_n)$ tends to zero as n tends to ∞. Since we saw in Theorem 1 that $F(y)$, defined in (14), is positive everywhere except at $y = x$, it would follow that x_n tends to x.

We recall the concept (8) of error $e_n = x_n - x$, and its relation (10) to the residual, $Ae_n = r_n$. We can, using (14)$'$ to express F, write

$$F(x_n) = \tfrac{1}{2}(e_n, Ae_n) = \tfrac{1}{2}(e_n, r_n) = \tfrac{1}{2}(r_n, A^{-1}r_n). \tag{18}$$

Since E and F differ only by a constant, we deduce from (15)$''$ that

$$F(x_{n+1}) = F(x_n) - s(r_n, r_n) + \tfrac{1}{2}s^2(r_n, Ar_n).$$

Using the value (16) for s, we obtain

$$F(x_{n+1}) = F(x_n) - \frac{s_n}{2}(r_n, r_n). \tag{18$'$}$$

Using (18), we can restate (18)$'$ as

$$F(x_{n+1}) = F(x_n)\left[1 - s_n \frac{(r_n, r_n)}{(r_n, A^{-1}r_n)}\right]. \tag{19}$$

Using inequalities (17) and (17)$'$, we deduce from (19) that

$$F(x_{n+1}) \le \left(1 - \frac{\alpha}{\beta}\right)F(x_n).$$

Applying this inequality recursively, we get, using (12), that

$$F(x_n) \le \left(1 - \frac{1}{\kappa}\right)^n F(x_0).\tag{20}$$

Using the boundedness of the Rayleigh quotient from below by the smallest eigenvalue, we conclude from (18) that

$$\frac{\alpha}{2}\| e_n \|^2 \le F(x_n).$$

Combining this with (20) we conclude that

$$\| e_n \|^2 \le \frac{2}{\alpha}\left(1 - \frac{1}{\kappa}\right)^n F(x_0).\tag{21}$$

This shows that the error e_n tends to zero, as asserted in Theorem 2.   □

2. AN ITERATIVE METHOD USING CHEBYSHEV POLYNOMIALS

Estimate (21) suggests that when the condition number κ of A is large, x_n converges to x very slowly. This in fact is the case; therefore there is need to devise iterative methods that converge faster; this will be carried out in the present and the following sections.

For the method described in this section we need a priori a positive lower bound for the smallest eigenvalue of A and an upper bound for its largest eigenvalue: $m < \alpha, \beta < M$. It follows that all eigenvalues of A lie in the interval $[m, M]$. According to (12), $\kappa = \frac{\beta}{\alpha}$; therefore $\kappa < \frac{M}{m}$. If m and M are sharp bounds, then κ is $\simeq \frac{M}{m}$.

We generate the sequence of approximations $\{x_n\}$ by the same recursion formula (15) as before.

$$x_{n+1} = (I - s_n A)x_n + s_n b,\tag{22}$$

but we shall choose the step lengths s_n to be optimal after N steps, not after each step; here N is some appropriately chosen number.

Since the solution x of (1) satisfies $x = (1 - s_n A)x + s_n b$, we obtain after subtracting this from (22) that

$$e_{n+1} = (I - s_n A)e_n.\tag{23}$$

From this we deduce recursively that

$$e_N = P_N(A)e_0,\tag{24}$$

where P_N is the polynomial

$$P_N(a) = \prod_1^N (1 - s_n a). \tag{24'}$$

From (24) we can estimate the size of e_N:

$$\| e_N \| \leq \| P_N(A) \| \| e_0 \|. \tag{25}$$

Since the matrix A is self-adjoint, so is $P_N(A)$. It was shown in Chapter 8 that the norm of a self-adjoint matrix is max $|p|, p$ any eigenvalue of $P_N(A)$. According to Theorem 4 of Chapter 6, the spectral mapping theorem, the eigenvalues p of $P_N(A)$ are of the form $p = P_N(a)$, where a is an eigenvalue of A. Since the eigenvalues of A lie in the interval $[m, M]$, we conclude that

$$\| P_N(A) \| \leq \max_{m \leq a \leq M} |P_N(a)|. \tag{26}$$

Clearly, to get the best estimate for $\| e_n \|$ out of inequalities (25) and (26), we have to choose the $s_n, n = 1, \ldots, N$ so that the polynomial P_N has as small a maximum on $[m, M]$ as possible. Polynomials of form (24)' satisfy the normalizing condition

$$P_N(0) = 1. \tag{27}$$

Among all polynomials of degree N that satisfy (27), the one that has smallest maximum on $[m, M]$ is the *rescaled Chebyshev polynomial*. We recall that the Nth Chebyshev polynomial T_N is defined for $-1 \leq u \leq 1$ by

$$T_N(u) = \cos N\theta, \qquad u = \cos \theta. \tag{28}$$

The rescaling takes $[-1, 1]$ into $[m, M]$ and enforces (27):

$$P_N(a) = T_N \left(\frac{M + m - 2a}{M - m} \right) \bigg/ T_N \left(\frac{M + m}{M - m} \right). \tag{29}$$

It follows from definition (28) that $|T_n(u)| \leq 1$ for $|u| \leq 1$. From this and (29) we deduce using $\frac{M}{m} \simeq \kappa$ that

$$\max_{m \leq a \leq M} |P_N(a)| \simeq 1 \bigg/ T_N \left(\frac{\kappa + 1}{\kappa - 1} \right) \tag{29'}$$

Setting this into (26) and using (25), we get

$$\| e_N \| \leq \| e_0 \| \bigg/ T_N \left(\frac{\kappa + 1}{\kappa - 1} \right). \tag{30}$$

Since outside the interval $[-1, 1]$ the Chebyshev polynomials tend to infinity, this proves that e_N tends to zero as N tends to ∞.

How fast e_N tends to zero depends on how large κ is. This calls for estimating $T_N(1 + \epsilon)$, ϵ small; we take θ in (28) imaginary:

$$\theta = i\phi, \qquad u = \cos i\phi = \frac{e^\phi + e^{-\phi}}{2} = 1 + \epsilon.$$

This is a quadratic equation for e^ϕ, whose solution is

$$e^\phi = 1 + \epsilon + \sqrt{2\epsilon + \epsilon^2} = 1 + \sqrt{2\epsilon} + O(\epsilon).$$

So

$$T_N(1 + \epsilon) = \cos iN\phi = \frac{e^{N\phi} + e^{-N\phi}}{2} \simeq \tfrac{1}{2}(1 + \sqrt{2\epsilon})^N.$$

Now set $(\kappa + 1)/(\kappa - 1) = 1 + \epsilon$; then $\epsilon \simeq 2/\kappa$, and

$$T_N\left(\frac{\kappa + 1}{\kappa - 1}\right) \simeq \frac{1}{2}\left(1 + \frac{2}{\sqrt{\kappa}}\right)^N. \tag{31}$$

Substituting this evaluation into (30) gives

$$\| e_N \| \leq 2\left(1 + \frac{2}{\sqrt{\kappa}}\right)^{-N} \| e_0 \| \simeq 2\left(1 - \frac{2}{\sqrt{\kappa}}\right)^N \| e_0 \|. \tag{32}$$

Clearly, e_N tends to zero as N tends to infinity.

When κ is large, $\sqrt{\kappa}$ is very much smaller than κ; therefore for κ large, the upper bound (32) for $\| e_N \|$ is very much smaller than the upper bound (21), $n = N$. This shows that the iterative method described in this section converges faster than the method described in Section 1. Put in another way, to achieve the same accuracy, we need to take far fewer steps when we use the method of this section than the method described in Section 1.

EXERCISE 2. Suppose $\kappa = 100$, $\| e_0 \| = 1$, and $(1/\alpha)F(x_0) = 1$; how large do we have to take N in order to make $\| e_N \| < 10^{-3}$, (a) using the method in Section 1, (b) using the method in Section 2?

To implement the method described in this section we have to pick a value of N. Once this is done, the values of $s_n, n = 1, \ldots, N$ are according to (24)′ determined as the reciprocals of the roots of the modified Chebyshev polynomials (29):

$$s_k^{-1} = \frac{1}{2}\left(M + m - (M - m)\cos\frac{(k + 1/2)\pi}{N}\right),$$

k any integer between 0 and $N - 1$. Theoretically, that is, imagining all arithmetic operations to be carried out exactly, it does not matter in what order we arrange the numbers s_k. Practically, that is, operating with finite floating-point numbers, it matters a great deal. Half the roots of P_N lie in the left half of the interval $[m, M]$; for these roots, $s > 2/(M + m)$, and so the matrix $(I - sA)$ has eigenvalues greater than 1 in absolute value. Repeated application of such matrices could fatally magnify round-off errors and render the algorithm arithmetically unstable.

There is a way of mitigating this instability; the other half of the roots of P_N lie in the other half of the interval $[m, M]$, and for these s all eigenvalues of the matrix $(I - sA)$ are less than 1. The trick is to alternate an unstable s_k with a stable s_k.

3. A THREE-TERM ITERATION USING CHEBYSHEV POLYNOMIALS

We describe now an entirely different way of generating the approximations described in Section 2 based on a recursion relation linking three consecutive Chebyshev polynomials. These are based on the addition formula of cosine:

$$\cos(n \pm 1)\theta = \cos\theta\cos n\theta \mp \sin\theta\sin n\theta.$$

Adding these yields

$$\cos(n + 1)\theta + \cos(n - 1)\theta = 2\cos\theta\cos n\theta.$$

Using the definition (28) of Chebyshev polynomials we get

$$T_{n+1}(u) + T_{n-1}(u) = 2uT_n(u).$$

The polynomials P_n, defined in (29), are rescaled Chebyshev polynomials; therefore they satisfy an analogous recursion relation:

$$P_{n+1}(a) = (u_n a + v_n)P_n(a) + w_n P_{n-1}(a). \tag{33}$$

We will not bother to write down the exact values of u_n, v_n, w_n, except to note that, by construction, $P_n(0) = 1$ for all n; it follows from this and (33) that

$$v_n + w_n = 1. \tag{33'}$$

We define now a sequence x_n recursively; we pick x_0, set $x_1 = (u_0 A + 1)x_0 - u_0 b$, and for $n > 1$

$$x_{n+1} = (u_n A + v_n)x_n + w_n x_{n-1} - u_n b. \tag{34}$$

Note that this is a three-term recursion formula, that is, x_{n+1} is determined in terms of x_n and x_{n-1}. Formulas (15) and (22) used in the last sections are two-term recursion formulas.

Subtract x from both sides of (34); using (33)$'$ and $Ax = b$ we get a recursion formula for the errors:

$$e_{n+1} = (u_n A + v_n)e_n + w_n e_{n-1}. \tag{34$'$}$$

Solving (34)$'$ recursively, it follows that each e_n can be expressed in the form $e_n = Q_n(A)e_0$, where the Q_n are polynomials of degree n, with $Q_0 \equiv 1$. Setting this form of e_n into (34)$'$, we conclude that the polynomials Q_n satisfy the same recursion relation as the P_n; since $Q_0 = P_0 \equiv 1$, it follows that $Q_n = P_n$ for all n. Therefore

$$e_n = P_n(A)e_0 \tag{35}$$

for all n, and not just a single preassigned value N as in equation (24) of Section 2.

4. OPTIMAL THREE-TERM RECURSION RELATION

In this section we shall use a three-term recursion relation of the form

$$x_{n+1} = (s_n A + p_n I)x_n + q_n x_{n-1} - s_n b \tag{36}$$

to generate a sequence of approximations that converges extremely rapidly to x. Unlike (34), the coefficients s_n, p_n, and q_n are not fixed in advance but will be evaluated in terms of r_{n-1} and r_n, the residuals corresponding to the approximations x_{n-1} and x_n. Furthermore, we need no a priori estimates m, M for the eigenvalues of A.

The first approximation x_0 is an arbitrary—or educated—guess. We shall use the corresponding residual, $r_0 = Ax_0 - b$, to completely determine the sequence of coefficients in (36), in a somewhat roundabout fashion. We pose the following minimum problem:

Among all polynomials of degree n that satisfy the normalizing condition

$$Q(0) = 1, \tag{37}$$

determine the one that makes

$$\| Q(A)r_0 \| \tag{38}$$

as small as possible.

We shall show that among all polynomials of degree less than or equal to n satisfying condition (37) there is one that minimizes (38); denote such a polynomial by Q_n.

We formulate now the variational condition characterizing this minimum. Let $R(a)$ be any polynomial of degree less than n; then $aR(a)$ is of degree less than or equal to n. Let ϵ be any real number; $Q_n(a) + \epsilon aR(a)$ is then a polynomial of degree less than or equal to n that satisfies condition (37). Since Q_n minimizes (38), $\| (Q_n(A) + \epsilon AR(A))r_0 \|^2$ takes on its minimum at $\epsilon = 0$. Therefore its derivative with respect to ϵ is zero there:

$$(Q_n(A)r_0, AR(A)r_0) = 0. \tag{39}$$

We define now a *scalar product* for polynomials Q and R as follows:

$$\{Q, R\} = (Q(A)r_0, AR(A)r_0). \tag{40}$$

To analyze this scalar product we introduce the eigenvectors of the matrix A:

$$Af_j = a_j f_j. \tag{41}$$

Since the matrix A is real and self-adjoint, the f_j can be taken to be real and orthonormal; since A is positive, its eigenvalues a_j are positive.

We expand r_0 in terms of the f_j,

$$r_0 = \sum w_j f_j. \tag{42}$$

Since f_j are eigenvectors of A, they are also eigenvectors of $Q(A)$ and $R(A)$, and by the spectral mapping theorem their eigenvalues are $Q(a_j)$, and $R(a_j)$, respectively. So

$$Q(A)r_0 = \sum w_j Q(a_j) f_j, \qquad R(A)r_0 = \sum w_j R(a_j) f_j. \tag{43}$$

Since the f_j are orthonormal, we can express the scalar product (40) for polynomials Q and R as follows:

$$\{Q, R\} = \sum w_j^2 a_j Q(a_j) R(a_j). \tag{44}$$

Theorem 3. Suppose that in the expansion (42) of r_0 none of the coefficients w_j are 0; suppose further that the eigenvalues a_j of A are distinct. Then (44) furnishes a Euclidean structure to the space of all polynomials of degree less than the order K of the matrix A.

Proof. According to Chapter 7, a scalar product needs three properties. The first two—bilinearity and symmetry—are obvious from either (40) or (44). To show positivity, we note that since each $a_j > 0$,

$$\{Q, Q\} = \sum w_j^2 a_j Q^2(a_j) \tag{45}$$

is obviously nonnegative. Since the w_j are assumed nonzero, (45) is zero iff $Q(a_j) = 0$ for all $a_j, j = 1, \ldots, K$. Since the degree of Q is less than K, it can vanish at K points only if $Q \equiv 0$. $\square$

We can express the minimizing condition (39) concisely in the language of the scalar product (40): For $n < K, Q_n$ is *orthogonal to all polynomials of degree less than n.* It follows in particular that Q_n is of degree n.

According to condition (37), $Q_0 \equiv 1$. Using the familiar Gram–Schmidt process we can using the orthogonality and condition (37), determine a unique sequence of polynomials Q_n. We show now that this sequence satisfies a three-term recursion relation. To see this we express $aQ_n(a)$ as linear combination of $Q_j, j = 0, \ldots, n+1$:

$$
aQ_n = \sum_0^{n+1} c_{n,j} Q_j. \tag{46}
$$

Since the Q_j are orthogonal, we can express the $c_{n,j}$ as

$$
c_{n,j} = \frac{\{aQ_n, Q_j\}}{\{Q_j, Q_j\}}. \tag{47}
$$

Since A is self-adjoint, the numerator in (47) can be rewritten as

$$
\{Q_n, aQ_j\}, \tag{47'}
$$

Since for $j < n - 1, aQ_j$ is a polynomial of degree less than n, it is orthogonal to Q_n, and so (47)′ is zero; therefore $c_{n,j} = 0$ for $j < n - 1$. This shows that the right-hand side of (46) has *only three nonzero terms* and can be written in the form

$$
aQ_n = b_n Q_{n+1} + c_n Q_n + d_n Q_{n-1}. \tag{48}
$$

Since Q_n is of degree $n, b_n \neq 0$. For $n = 1, d_1 = 0$.

According to condition (37), $Q_k(0) = 1$ for all k. Setting $a = 0$ in (48) we deduce that

$$
b_n + c_n + d_n = 0. \tag{49}
$$

From (47), with $j = n, n - 1$ we have

$$
c_n = \frac{\{aQ_n, Q_n\}}{\{Q_n, Q_n\}}, \qquad d_n = \frac{\{aQ_n, Q_{n-1}\}}{\{Q_{n-1}, Q_{n-1}\}}. \tag{50}
$$

Since $b_n \neq 0$, we can express Q_{n+1} from (48) as follows:

$$
Q_{n+1} = (s_n a + p_n)Q_n + q_n Q_{n-1}, \tag{51}
$$

where

$$
s_n = \frac{1}{b_n}, \qquad p_n = -\frac{c_n}{b_n}, \qquad q_n = -\frac{d_n}{b_n}. \tag{52}
$$

Note that it follows from (49) and (52) that

$$p_n + q_n = 1. \tag{53}$$

Theoretically, the formulas (50) completely determine the quantities c_n and d_n. Practically, these formulas are quite useless, since in order to evaluate the curly brackets we need to know the polynomials Q_k and evaluate $Q_k(A)$. Fortunately c_n and d_n can be evaluated more easily, as we show next.

We start the algorithm by choosing an x_0; then the rest of the x_n are determined by the recursion (36), with s_n, p_n, and q_n from formulas (52), (50), and (49). We have defined e_n to be $x_n - x$, the nth error; subtracting x from (36), making use of (53), that $b = Ax$, we obtain,

$$e_{n+1} = (s_n A + p_n I)e_n + q_n e_{n-1}. \tag{54}$$

We claim that

$$e_n = Q_n(A)e_0. \tag{55}$$

To see this we replace the scalar argument a in (51) by the matrix argument A:

$$Q_{n+1}(A) = (s_n A + p_n)Q_n(A) + q_n Q_{n-1}(A). \tag{56}$$

Let both sides of (56) act on e_0; we get a recurrence relation that is the same as (54), except that e_k is replaced by $Q_k(A)e_0$. Since $Q_0(A) = I$, the two sequences have the same starting point, and therefore they are the same, as asserted in (55).

We recall now that the residual $r_n = Ax_n - b$ is related to $e_n = x_n - x$ by $r_n = Ae_n$. Applying A to (55) we obtain

$$r_n = Q_n(A)r_0. \tag{57}$$

Applying the mapping A to (54) gives a recursion relation for the residuals:

$$r_{n+1} = (s_n A + p_n I)r_n + q_n r_{n-1}. \tag{58}$$

We now set $Q = Q_n, R = Q_n$ into (40), and use relation (57) to write

$$\{Q_n, Q_n\} = (r_n, Ar_n) \tag{59}$$

Subsequently we set $Q = aQ_n, R = Q_n$ into (40), and use relation (57) to write

$$\{aQ_n, Q_n\} = (Ar_n, Ar_n). \tag{59}'$$

Finally we set $Q = aQ_n$ and $R = Q_{n-1}$ into (40), and we use relation (57) to write

$$\{aQ_n, Q_{n-1}\} = (Ar_n, Ar_{n-1}). \tag{59}''$$

We set these identities into (50):

$$c_n = \frac{(Ar_n, Ar_n)}{(r_n, Ar_n)}, \qquad d_n = \frac{(Ar_n, Ar_{n-1})}{(r_{n-1}, Ar_{n-1})}. \qquad (60)$$

From (49) we determine $b_n = -(c_n + d_n)$. Set these expressions into (52) and we obtain expressions for s_n, p_n, and q_n that are simple to evaluate once r_{n-1} and r_n are known; these residuals can be calculated as soon as we know x_{n-1} and x_n or from recursion (58). This completes the recursive definition of the sequence x_n.

Theorem 4. Let K be the order of the matrix A, and let x_K be the Kth term of the sequence (36), the coefficients being defined by (52) and (60). We claim that x_K satisfies equation (1), $Ax_K = b$.

Proof. Q_K is defined as that polynomial of degree K which satisfies (37) and minimizes (38). We claim that this polynomial is $p_A/p_A(0), p_A$ the characteristic polynomial of A; note that $p_A(0) \neq 0$, since 0 is not an eigenvalue of A. According to the Cayley–Hamilton theorem, Theorem 5 of Chapter 6, $p_A(A) = 0$; clearly, $Q_K(A) = 0$ minimizes $\| Q(A)r_0 \|$. According to (57), $r_K = Q_K(A)r_0$; since according to the above discussion, $Q_K(A) = 0$, this proves that the Kth residual r_K is zero, and therefore x_K exactly solves (1).   □

One should not be misled by Theorem 4; the virtue of the sequence x_n is not that it furnishes the exact answer in K steps, but that, for a large class of matrices of practical interest, it furnishes an excellent approximation to the exact answer in far fewer steps than K. Suppose for instance that A is the discretization of an operator of the form identity plus a compact operator. Then most of the eigenvalues of A would be clustered around 1; say all but the first k eigenvalues a_j of A are located in the interval $(1 - \delta, 1 + \delta)$.

Since Q_n was defined as the minimizer of (38) subject to the condition $Q(0) = 1$, and since according to (57), $Q_n(A)r_0 = r_n$, we conclude that

$$\| r_n \| \leq \| Q(A)r_0 \|$$

for any polynomial Q of degree n that satisfies $Q(0) = 1$. Using formula (45) we write this inequality as

$$\| r_n \|^2 \leq \sum w_j^2 a_j Q^2(a_j), \qquad (61)$$

where the w_j are the coefficients in the expansion of r_0.

We set now $n = k + l$, and we choose Q as follows:

$$Q(a) = \prod_1^k \left(1 - \frac{a}{a_j}\right) T_l\left(\frac{a-1}{\delta}\right) \bigg/ T_l(-1/\delta); \qquad (62)$$

here, as before, T_l denotes the lth Chebyshev polynomial. Clearly, Q satisfies condition (37), $Q(0) = 1$. For a large, $T_l(a)$ is dominated by its leading term, which is $2^{l-1}a^l$. Therefore,

$$\left| T_l\left(\frac{-1}{\delta}\right) \right| \simeq \frac{1}{2}\left(\frac{2}{\delta}\right)^l. \tag{63}$$

By construction, Q vanishes at $a_1, \ldots, a_k$. We have assumed that all the other a_j lie in $(1-\delta, 1+\delta)$; since the Chebyshev polynomials do not exceed 1 in absolute value in $(-1, 1)$, it follows from (62) and (63) that for $j > k$,

$$|Q(a_j)| \leq \text{const}\left(\frac{\delta}{2}\right)^l, \tag{64}$$

where

$$\text{const} \simeq 2\prod\left(1 - \frac{1}{a_j}\right). \tag{65}$$

Setting all this information about $Q(a_j)$ into (61) we obtain

$$\| r_{k+l} \|^2 \leq \text{const}^2\left(\frac{\delta}{2}\right)^{2l}\sum_{k<j}w_j^2 \leq \text{const}^2\left(\frac{\delta}{2}\right)^{2l}\| r_0 \|^2. \tag{66}$$

For example if $|a_j - 1| < 0.2$ for $j > 10$, and if the constant (65) is less than 10, then choosing $l = 20$ in (66) makes $\| r_{30} \|$ less than $10^{-19}\| r_0 \|$.

EXERCISE 3.   Write a computer program to evaluate the quantities s_n, p_n, and q_n.

EXERCISE 4.   Use the computer program to solve a system of equations of your choice.

CHAPTER 18

How to Calculate the Eigenvalues of Self-Adjoint Matrices

1. The basis of one of the most effective methods for calculating approximately the eigenvalues of a self-adjoint matrix is based on the QR decomposition.

Theorem 1. Every real invertible square matrix A can be factored as

$$A = QR, \tag{1}$$

where Q is an orthogonal matrix and R is an upper triangular matrix whose diagonal entries are positive.

Proof. The columns of Q are constructed out of the columns of A by Gram–Schmidt orthonormalization. So the jth column q_j of Q is a linear combination of the first j columns $a_1, \ldots, a_j$ of A:

$$q_1 = c_{11}a_1,$$
$$q_2 = c_{12}a_1 + c_{22}a_2,$$

etc. We can invert the relation between the $q - s$ and the $a - s$:

$$a_1 = r_{11}q_1,$$
$$a_2 = r_{12}q_1 + r_{22}q_2,$$
$$\vdots \tag{2}$$
$$a_n = r_{1n}q_1 + \ldots + r_{nn}q_n.$$

Linear Algebra and Its Applications, Second Edition, by Peter D. Lax
Copyright © 2007 John Wiley & Sons, Inc.

Since A is invertible, its columns are linearly independent. It follows that all coefficients $r_{11}, \ldots, r_{nn}$ in (2) are nonzero.

We may multiply any of the vectors q_j by -1, without affecting their orthonormality. In this way we can make all the coefficients $r_{11}, \ldots, r_{nn}$ in (2) positive. Here A is an $n \times n$ matrix,

Denote the matrix whose columns are $q_1, \ldots, q_n$ by Q, and denote by R the matrix

$$R_{ij} = \begin{cases} r_{ij} & \text{for } i \leq j, \\ 0 & \text{for } i > j. \end{cases} \tag{3}$$

Relation (2) can be written as a matrix product

$$A = QR.$$

Since the columns of Q are orthonormal, Q is an orthogonal matrix.

It follows from the definition (3) of R that R is upper triangular. So $A = QR$ is the sought-after factorization (1). □

The factorization (1) can be used to solve the system of equations

$$Ax = u.$$

Replace A by its factored form,

$$QRx = u$$

and multiply by Q^T on the left. Since Q is an orthogonal matrix, $Q^T Q = I$, and we get

$$Rx = Q^T u. \tag{4}$$

Since R is upper triangular and its diagonal entries are nonzero, the system of equations can be solved recursively, starting with the nth equation to determine x_n, then the $(n-1)$st equation to determine x_{n-1}, and so all the way down to x_1.

In this chapter we shall show how to use the QR factorization of a real symmetric matrix A to find its eigenvalue. The QR algorithm was invented by J.G.F. Francis in 1961; it goes as follows:

Let A be a real symmetric matrix; we may assume that A is invertible, for we may add a constant multiple of the identity to A. Find the QR factorization of A:

$$A = QR.$$

Define A_1 by *switching* the factors Q and R

$$A_1 = RQ. \tag{5}$$

We claim that

   (i) A_1 is real and symmetric, and

   (ii) A_1 has the same eigenvalues as A.

To see these we express R in terms of A and Q by multiplying equation (1) by Q^T on the left. Since $Q^T Q = I$, we get

$$Q^T A = R.$$

Setting this into (5) gives

$$A_1 = Q^T A Q; \tag{6}$$

from which (i) and (ii) follow.

We continue this process, getting a sequence of matrices $\{A_k\}$, each linked to the next one by the relations

$$A_{k-1} = Q_k R_k, \tag{7$_k$}$$
$$A_k = R_k Q_k. \tag{8$_k$}$$

From these we deduce, as before, that

$$A_k = Q_k^T A_{k-1} Q_k. \tag{9$_k$}$$

It follows that all the matrices A_k are symmetric, and they all have the same eigenvalues.

Combining the relations $(9)_k, (9)_{k-1}, \ldots, (9)$, we get

$$A_k = Q^{(k)T} A Q^{(k)}, \tag{10$_k$}$$

where

$$Q^{(k)} = Q_1 Q_2 \ldots Q_k. \tag{11}$$

Define similarly

$$R^{(k)} = R_k R_{k-1} \ldots R_1 \tag{12}$$

We claim that

$$A^k = Q^{(k)} R^{(k)}. \tag{13$_k$}$$

For $k = 1$ this is relation (1). We argue inductively; suppose $(13)_{k-1}$ is true:

$$A^{k-1} = Q^{(k-1)} R^{(k-1)}.$$

Multiply this by A on the left:

$$A^k = AQ^{(k-1)}R^{(k-1)}. \tag{14}$$

Multiply equation $(10)_{k-1}$ by $Q^{(k-1)}$ on the left. Since $Q^{(k-1)}$ is a product of orthogonal matrices, it is itself orthogonal, and so $Q^{(k-1)} Q^{(k-1)T} = I$. So we get that

$$Q^{(k-1)}A_{k-1} = AQ^{(k-1)}.$$

Combining this with (14) gives

$$A^k = Q^{(k-1)}A_{k-1}R^{(k-1)}.$$

Now use $(7)_k$ to express A_{k-1}, and we get relation $(13)_k$.

This completes the inductive proof of $(13)_k$. □

Formula (12) defines $R^{(k)}$ as the product of upper triangular matrices. Therefore $R^{(k)}$ itself is upper triangular, and so $(13)_k$ *is the* QR *factorization of* A^k.

Denote the normalized eigenvectors of A by $u_1, \ldots, u_m$, its corresponding eigenvalues by $d_1, \ldots, d_m$.

Denote by U the matrix whose columns as the eigenvectors,

$$U = (u_1, \ldots, u_m),$$

and by D the diagonal matrix whose entries are $d_1, \ldots, d_m$. The spectral representation of A is

$$A = UDU^T. \tag{15}$$

Therefore the spectral representation of A^k is

$$A^k = UD^kU^T. \tag{15}_k$$

It follows from formula $(15)_k$ that the columns of A^k are linear combinations of the eigenvectors of A of the following form:

$$b_1 d_1^k u_1 + \cdots + b_m d_m^k u_m, \tag{15}'$$

where $b_1, \ldots, b_m$ do not depend on k. We assume now that the eigenvalues of A are *distinct* and positive; arrange them in decreasing order:

$$d_1 > d_2 > \ldots > d_m > 0.$$

It follows then from (15)' that, provided $b_1 \neq 0$, for k large enough the first column of A^k is very close to a constant multiple of u_1. Therefore $q_1^{(k)}$, the first column of

$Q^{(k)}$, is very close to u_1. Similarly, $q_2^{(k)}$, the second column of $Q^{(k)}$, would be very close to u_2, and so on, up to $q_n^{(k)} \simeq u_n$.

We turn now to formula $(10)_k$; it follows that the ith diagonal element of A_k is

$$(A_k)_{ii} = q_i^{(k)T} A q_i^{(k)} = (q_i^k, A q_i^k).$$

The quantity on the right is the Rayleigh quotient of A, evaluated at q_i^k. It was explained in Chapter 8 that if the vector q_i^k differs by ϵ from the ith eigenvector of A, then the Rayleigh quotient differs by less than $O(\epsilon^2)$ from the ith eigenvalue d_i of A. This shows that if the QR algorithm is carried out far enough, the diagonal entries of A_k are very good approximations to the eigenvalue of A, arranged in decreasing order.

EXERCISE 1. Show that the off-diagonal entries of A_k tend to zero as k tends to ∞.

Numerical calculations bear out these contentions.

2. Next we describe another algorithm, due to Alston Householder, for accomplishing the QR factorization of a matrix A. In this algorithm, Q is constructed as a product of particularly simple orthogonal transformations, known as reflections.

A Householder reflection is simply a reflection across a hyperplane, that is, a subspace of form $v^T x = 0$. A reflection H maps all points of the hyperplane into themselves, and it projects points x off the hyperplane into their reflection across the hyperplane. The analytical expression of H is

$$Hx = x - 2 \frac{v^T x}{\| v \|^2} v. \tag{16}$$

Note that if we replace v by a multiple of v, the mapping H is unchanged.

EXERCISE 2. Show that the mapping (16) is norm-preserving.

We shall show now how reflections can be used to accomplish the QR factorization of a matrix A. Q will be constructed as the product of n reflections:

$$Q = H_n H_{n-1} \dots H_1.$$

H_1 is chosen so that the first column of $H_1 A$ is a multiple of $e_1 = (1, 0, \dots, 0)$. That requires that $H_1 a_1$ be a multiple of e_1; since H_1 is norm-preserving, that multiple has to have absolute value $\| a_1 \|$. This leaves two choices:

$$H_1 a_1 = \| a_1 \| e_1 \quad \text{or} \quad H_1 a_1 = - \| a_1 \| e_1.$$

Setting $x = a_1$ into (16), we get for H_1 the relation

$$a_1 - v = \| a_1 \| e_1 \quad \text{or} \quad a_1 - v = - \| a_1 \| e_1,$$

which gives two choices for v:

$$v_+ = a_1 - \| a_1 \| e_1 \quad \text{or} \quad v_- = a_1 + \| a_1 \| e_1. \tag{17}$$

We recall that the arithmetical operations in a computer carry a finite number of digits. Therefore when two nearly equal numbers are subtracted, the relative error in the difference is quite large. To prevent such loss, we choose in (17) the *larger* of the two vectors v_+ or v_- for v.

Having chosen H_1, denote $H_1 A$ as A_1; it is of form

$$A_1 = \begin{pmatrix} \times & \times \ldots \times \\ 0 & \\ 0 & A^{(1)} \end{pmatrix},$$

where $A^{(1)}$ is an $(n-1) \times (n-1)$ matrix.

Choose H_2 to be of the form

$$H_2 = \begin{pmatrix} 1 & 0 \ldots 0 \\ 0 & \\ \vdots & \\ 0 & H^{(2)} \end{pmatrix},$$

where $H^{(2)}$ is chosen as before so that the first column of

$$H^{(2)} A^{(1)}$$

is of the form $(\times, 0, \ldots, 0)^{\mathrm{T}}$. Then the first column of the product $H_2 A_1$ is the same as the first column of A_1, while the second column is of the form $(\times, \times, 0, \ldots, 0)^{\mathrm{T}}$. We continue in this fashion for n steps; clearly, $A_n = H_n \ldots H_1 A$ is upper triangular. Then we set $R = A_n$ and $Q = H_1^{\mathrm{T}} \ldots H_n^{\mathrm{T}}$ and obtain the QR factorization (1) of A. $\quad\square$

Next we show how reflections can be used to bring any symmetric matrix A into tridiagonal form L by an orthogonal similarity transformation:

$$0 A 0^{\mathrm{T}} = L. \tag{18}$$

0 is a product of reflections:

$$0 = H_{n-1} \ldots H_1. \tag{18}'$$

H_1 is of the form

$$H_1 = \begin{pmatrix} 1 & 0\ldots 0 \\ 0 & H^{(1)} \\ \vdots & \\ 0 & \end{pmatrix}. \tag{19}$$

Denote the first column of A as

$$a_1 = \begin{pmatrix} \times \\ a^{(1)} \end{pmatrix},$$

where $a^{(1)}$ is a column vector with $n - 1$ components. Then the action of H_1 of form (19) is as follows:

$H_1 A$ has the same first row as A, and the last $n - 1$ entries of the first column of $H_1 A$ is $H^{(1)} a^{(1)}$.

We choose $H^{(1)}$ as a reflexion in $\mathbb{R}^{n-1}$ that maps $a^{(1)}$ into a vector whose last $n - 2$ components are zero. Thus the first column of $H_1 A$ has zeros in the last $n - 2$ places.

Multiplying an $n \times n$ matrix by a matrix of the form (19) on the right leaves the first column unaltered. Therefore the first column of

$$A_1 = H_1 A H_1^T$$

has zeros in the last $n - 2$ rows.

In the next step we choose H_2 of the form

$$H_2 = \begin{pmatrix} 1 & 0 & \cdots & 0 \\ 0 & 1 & \cdots & 0 \\ \vdots & 0 & & \\ 0 & 0 & H^{(2)} \end{pmatrix}, \tag{20}$$

where $H^{(2)}$ is an $(n - 2) \times (n - 2)$ reflection. Since the first column of A_1 has zeros in the last $n - 2$ rows, the first column of $H_2 A_1$ is the same as the first column of A_1. We choose the reflection $H^{(2)}$ so that the second column of $H_2 A_1$ has zeros in the last $n - 3$ rows.

For H_2 of form (20), multiplication on the right by H_2 leaves the first two columns unchanged. Therefore

$$A_2 = H_2 A_1 H_2^T$$

has $n - 2$ and $n - 3$ zeros, respectively, in the first and second columns. Continuing in this fashion, we construct the reflections $H_3, \ldots, H_{n-1}$. Their product $0 = H_{n-1} \ldots H_1$ has the property that $0 A 0^T$ has all ijth entries zero when $i > j + 1$; But since $0 A 0^T$ is symmetric, so are all entries for $j > i + 1$. This shows that $0 A 0^T$ is tridiagonal.   □

We note that Jacobi proposed an algorithm for tridiagonaliring symmetric matrices. This was implemented by Wallce Givens.

Theorem 2. When the QR algorithm $(7)_k$, $(8)_k$ is applied to a real, symmetric, tridiagonal matrix L, all the matrices L_k produced by the algorithm are real, symmetric, and tridiagonal, and have the same eigenvalues as L.

Proof. We have already shown, see $(9)_k$, that L_k is symmetric and has the same eigenvalues as L. To show that L_k is tridiagonal, we start with $L = L_0$ tridiagonal and then argue by induction on k. Suppose L_{k-1} is tridiagonal and is factored as $L_k = Q_k R_k$. We recall that the jth column q_j of Q_k is a linear combination of the first j columns of L_k; since L_k is tridiagonal, the last $n - j - 1$ entries of q_k are zero. The jth column of $R_k Q_k$ is $R_k q_j$; since R_k is upper triangular, it follows that the last $n - j - 1$ entries of $R_k q_j$ are zero. This shows that the ijth entry of $L_k = R_k Q$ is zero for $i > j + 1$. Since L_k is symmetric, this proves that L_k is tridiagonal, completing the induction. $\square$

Having L, and thereby all subsequent L_k, in tridiagonal form greatly reduces the number of arithmetic operations needed to carry out the QR algorithm.

So the strategy for the tridiagonal case of the QR algorithm is to carry out the QR iteration until the off diagonal entries of L_k are less than a small number. The diagonal elements of L_k are good approximations to the eigenvalues of L.

3. Deift, Nanda, and Tomei observed that the Toda flow is a continuous analogue of the QR iteration. Flaschka has shown that the differential equations for the Toda flow can be put into *commutor form*, that is, in the form

$$\frac{d}{dt}L = BL - LB, \tag{21}$$

where L is a symmetric tridiagonal matrix

$$L = \begin{pmatrix} a_1 & b_1 & & 0 \\ b_1 & a_2 & \ddots & \\ & \ddots & \ddots & b_{n-1} \\ 0 & & b_{n-1} & a_n \end{pmatrix} \tag{22}$$

and B is the antisymmetric tridiagonal matrix

$$B = \begin{pmatrix} 0 & b_1 & & 0 \\ -b_1 & 0 & \ddots & \\ & \ddots & \ddots & b_{n-1} \\ 0 & -b_{n-1} & 0 & 0 \end{pmatrix}. \tag{23}$$

EXERCISE 3.   **(i)** Show that $BL - LB$ is a tridiagonal matrix.
(ii) Show that if L satisfies the differential equation (21), its entries satisfy

$$\frac{d}{dt}a_k = 2(b_k^2 - b_{k-1}^2),$$

$$\frac{d}{dt}b_k = b_k(a_{k+1} - a_k),$$
(24)

where $k = 1, \ldots, n$ and $b_0 = b_n = 0$.

Theorem 3. Solutions $L(t)$ of equations in commutator form (21), where B is antisymmetric, are isospectral.

Proof. Let the matrix $V(t)$ be the solution of the differential equation.

$$\frac{d}{dt}V = BV, \qquad V(0) = I.$$
(25)

Since $B(t)$ is antisymmetric, the transpose of (25) is

$$\frac{d}{dt}V^T = -V^T B, \qquad V^T(0) = I.$$
$(25)^T$

Using the product rule for differentiation and equations (25) and $(25)^T$, we get

$$\frac{d}{dt}V^T V = \left(\frac{d}{dt}V^T\right)V + V^T\frac{d}{dt}V$$
$$= -V^T BV + V^T BV = 0.$$

Since $V^T V = I$ at $t = 0$, it follows that $V^T(t)V(t) = I$ for all t. This proves that for all t, $V(t)$ is an orthogonal matrix.
   We claim that if $L(t)$ is a solution of (21) and $V(t)$ a solution of (25), then

$$V^T(t)L(t)V(t)$$
(26)

is independent of t. Differentiate (26) with respect to t; using the product rule, we get

$$\left(\frac{d}{dt}V^T\right)LV + V^T\left(\frac{d}{dt}L\right)V + V^T L\frac{d}{dt}V.$$
(27)

Using equations (21), (25), and $(25)^\perp$, we can rewrite (27) as

$$-V^T BLV + V^T(BL - LV) + V^T LBV,$$

which is zero. This shows that the derivative of (26) is zero, and therefore (26) is independent of t. At $t = 0$, (26) equals $L(0)$, since $V(0)$ is the identity; so

$$V^T(t)L(t)V(t) = L(0). \tag{28}$$

Since $V(t)$ is an orthogonal matrix, (28) shows that $L(t)$ is related to $L(0)$ by an orthogonal similarity. This completes the proof of Theorem 3.   □

Formula (28) shows that if $L(0)$ is real symmetric—which we assume—then $L(t)$ is symmetric for all t.

The spectral representation of a symmetric matrix L is

$$L = UDU^T, \tag{29}$$

where D is a diagonal matrix whose entries are the eigenvalues of L, and the columns of U are the normalized eigenvectors of L; (29) shows that a set of symmetric matrices whose eigenvalues are uniformly bounded is itself uniformly bounded. So we conclude from Theorem 3 that the set of matrices $L(t)$ are uniformly bounded. It follows from this that the system of quadratic equations (24) have a solution for all values of t.

Lemma 4.   An off-diagonal entry $b_k(t)$ of $L(t)$ is either nonzero for all t, or zero for all t.

Proof. Let $[t_0, t_1]$ be an interval on which $b_k(t)$ is nonzero. Divide the differential equation (24) for b_k by b_k and integrate it from t_0 to t_1:

$$\log b_k(t_1) - \log b_k(t_0) = \int_{t_0}^{t_1} (a_{k+1} - a_k)dt.$$

Since, as we have shown, the functions a_k are uniformly bounded for all t, the integral on the right can tend to ∞ only if t_0 or t_1 tends to ∞. This shows that $\log b_k(t)$ is bounded away from $-\infty$, and therefore $b_k(t)$ is bounded away from zero on any interval of t. This proves that if $b_k(t)$ is nonzero for a single value of t, it is nonzero for all t.   □

If one of the off-diagonal entries b_k of $L(0)$ were zero, the matrix $L(0)$ would fall apart into two matrices. We assume that this is not the case; then it follows from Lemma 4 that the $b_k(t)$ are nonzero for all t all k.

Lemma 5.   Suppose none of the off diagonal terms b_k in L is zero.

(i) The first component u_{1k} of every eigenvector u_k of L is nonzero.
(ii) Each eigenvalue of L is simple.

Proof. (i) The first component of the eigenvalue equation

$$Lu_k = d_k u_k \tag{30}$$

is

$$a_1 u_{1k} + b_1 u_{2k} = d_k u_{1k}. \tag{31}$$

If u_{1k} were zero, it would follow from (31), since $b_1 \neq 0$, that $u_{2,k} = 0$. We can then use the second component of (30) to deduce similarly that $u_{3,k} = 0$; continuing in this fashion, we deduce that all components of u_k are zero, a contradiction.

(ii) Suppose on the contrary that d_k is a multiple eigenvalue; then its eigenspace has dimension greater than 1. In a space of dimension greater than one, we can always find a vector whose first component is zero; but this contradicts part (i) of Lemma 5.   □

Lemma 6. The eigenvalues $d_1, \ldots, d_n$ and the first components $u_{1,k}$, $k = 1, \ldots, n$, of the normalized eigenvectors of L uniquely determine all entries $a_1, \ldots, a_n$ and $b_1, \ldots, b_{k-1}$ of L.

Proof. From the spectral representation (29), we can express the entry $L_{11} = a_1$ of L as follows:

$$a_1 = \sum d_k u_{1k}^2. \tag{32}_1$$

From equation (31) we get

$$b_1 u_{2k} = (d_k - a_1) u_{1k}. \tag{33}_1$$

Squaring both sides and summing with respect to k gives

$$b_1^2 = \sum (d_k - a_1)^2 u_1^2 k; \tag{34}_1$$

here we have used the fact that the matrix U is orthogonal, and therefore

$$\sum u_{2k}^2 = 1.$$

We have shown in Lemma 4 that $b_k(t)$ doesn't change sign; therefore b_k is determined by $(34)_1$. We now set this determination of b_1 into $(33)_1$ to obtain the values of u_{2k}.

Next we use the spectral representation (29) again to express $a_2 = L_{22}$ as

$$a_2 = \sum d_k u_{2k}^2. \tag{32}_2$$

We proceed as before to the second equation in 30, which we write as

$$b_2 u_{3k} = -b_1 u_{1k} + (d_k - a_2) u_{2k}. \tag{33}_2$$

Squaring and summing over k gives

$$b_2^2 = \sum (-b_1 u_k + (d_k - a_2) u_{2k})^2, \tag{34}_2$$

and so on.

Jüngen Moser has determined the asymptotic behavior of $L(t)$ as t tends to ∞. $\square$

Theorem 7. (Moser). $L(t)$ is a solution of equation (21). Denote the eigenvalues of L by $d_1, \ldots, d_n$, arranged in decreasing order, and denote by D the diagonal matrix with diagonal entries $d_1, \ldots, d_n$. Then

$$\lim_{t \to \infty} L(t) = D. \tag{34}$$

Similarly,

$$\lim_{t \to -\infty} L(t) = D_-, \tag{34}_-$$

where D_- is the diagonal matrix whose diagonal entries are $d_n, \ldots, d_1$.

Proof. We start with the following lemma. $\square$

Lemma 8. Denote by $u(t)$ the row vector consisting of the first components of the normalized eigenvectors of $L(t)$:

$$u = (u_{11}, \ldots, u_{1n}). \tag{35}$$

Claim:

$$u(t) = \frac{u(0) e^{Dt}}{\| u(0) e^{Dt} \|}. \tag{36}$$

Proof. We have shown that when $L(t)$ satisfies (21), $L(t)$ and $L(0)$ are related by (28). Multiplying this relation by $V(t)$ on the left gives

$$L(t)V(t) = V(t)L(0). \tag{28}'$$

Denote as before the normalized eigenvectors of $L(t)$ by $u_k(t)$. Let $(28)'$ act on $u_k(0)$. Since

$$L(0)u_k(0) = d_k u_k(0),$$

we get

$$L(t)V(t)u_k(0) = d_k V(t)u_k(0).$$

This shows that $V(t)u_k(0) = u_k(t)$ are the normalized eigenvectors of $L(t)$.

$V(t)$ satisfies the differential equation (25), $\frac{d}{dt}V = BV$. Therefore $u_k(t) = V(t)u_k(0)$ satisfies

$$\frac{d}{dt}u_k = Bu_k. \tag{37}$$

Since B is of form (23), the first component of (37) is

$$\frac{d}{dt}u_{1k} = b_1 u_{2k}. \tag{37}'$$

We now use equation $(33)_1$ to rewrite the right-hand side:

$$\frac{d}{dt}u_{1k} = (d_k - a_1)u_{1k}. \tag{37}''$$

Define $f(t)$ by

$$f(t) = \int_0^t a_1(s)ds.$$

Equation $(37)''$ can be rewritten as

$$\frac{d}{dt}e^{f(t)-d_kt}u_{1k}(t) = 0,$$

from which we deduce that

$$e^{f(t)-d_kt}u_{1k} = c_k,$$

where c_k is a constant. So

$$u_{1k}(t) = c_k e^{d_kt}F(t),$$

where $F(t) = \exp [f(t)]$. Since $f(0) = 0$, $F(0) = 1$, and $c_k = u_{1k}(0)$; we obtain

$$u_{1k}(t) = u_{1k}(0)e^{d_k t}F(t). \tag{38}$$

In vector notation, (38) becomes

$$u(t) = u(0)e^{Dt}F(t).$$

Since $u(t)$ is the first row of an orthogonal matrix, it has norm 1. This shows that $F(t)$ is the normalizing factor, and it proves formula (36). $\square$

(i) Since the eigenvalues of $L(t)$ are distinct (see Lemma 5), formula (36) shows that as $t \to \infty$, the first component $u_{11}(t)$ of $u(t)$ is exponentially larger than the other components. Since the vector $u(t)$ has norm 1, it follows that as $t \to \infty$, $u_{11}(t)$ tends to 1, and $u_{1k}(t), k > 1$, tend to zero at an exponential rate.

(ii) Next we take equation $(32)_1$:

$$a_1(t) = \sum d_k u_{1k}^2(t).$$

From what we have shown about $u_{1k}(t)$, it follows that $a_1(t)$ tends to d_1 at an exponential rate as $t \to \infty$.

(iii) To estimate b_1 we take the representation $(34)_1$:

$$b_1^2(t) = \sum (d_k - a_1(t))^2 u_{1k}^2(t).$$

From $a_1(t) \to d_1$ and the fact that $u(t)$ is a unit vector, we deduce that $b_1(t)$ tends to zero at an exponential rate as $t \to \infty$.

(iv) The first two rows of u are orthogonal:

$$\sum u_{1k}(t)u_{2k}(t) = 0. \tag{39}$$

According to (i), $u_{11}(t) \to 1$ and $u_{1k}(t) \to 0$ exponentially as $t \to \infty$. It follows therefore from (39) that $u_{21}(t) \to 0$ exponentially as $t \to \infty$.

(v) From (31) we deduce that

$$\frac{u_{2k}}{u_{22}} = \frac{d_k - a_1 u_{1k}}{d_2 - a_1 u_{12}}. \tag{40}$$

By the explicit formula (38) we can write this as

$$\frac{u_{2k}(t)}{u_{22}(t)} = \frac{d_k - a_1(t) u_{2k}(0)}{d_2 - a_1(t) u_{22}(0)} e^{(d_k - d_2)t}. \tag{41}$$

Take $k > 2$; then the right-hand side of (41) tends to zero as $t \to \infty$, and therefore $u_{2k}(t) \to 0$ as $t \to \infty$ for $k > 2$. We have shown in (iv) that $u_{21}(t) \to 0$ as $t \to \infty$. Since $(u_{21}, \ldots, u_{2n})$ is a unit vector, it follows that $u_{22}(t) \to 1$ exponentially.

(**vi**) According to formula $(32)_2$,

$$a_2(t) = \sum d_k u_{2k}^2(t).$$

Since we have shown in (v) that $u_{2k}(t) \to 0$ for $k \neq 2$ and that $u_{22} \to 1$, it follows that $a_2(t) \to d_2$.

(**vii**) Formula $(34)_2$ represents $b_2(t)$ as a sum. We have shown above that all terms of this sum tend to zero as $t \to \infty$. It follows that $b_2(t) \to 0$ as $t \to \infty$, at the usual exponential rate.

The limiting behavior of the rest of the entries can be argued similarly; Deift et al. supply all the details.

Identical arguments show that L(t) tends to D$_-$ as $t \to -\infty$.

Moser's proof of Theorem 7 runs along different lines.

We conclude this chapter with four observations.

Note 1. It may surprise the reader that in Lemma 8 we present an *explicit solution.* The explanation is that the Toda lattice, of which (21) is a form, is *completely integrable.* According to Liouville's Theorem, such systems have explicit solutions.

Note 2. Moser's Theorem is a continuous analogue of the convergence of the QR algorithm to D when applied to a tridiagonal matrix.

Note 3. Deift et al. point out that (21) is only one of a whole class of flows of tridiagonal symmetric matrices that tend to D as $t \to \infty$. These flows are in commutator form (21), where the matrix B is taken as

$$B = p(L)_+ - p(L)_-,$$

where p is a polynomial, M$_+$ denotes the upper triangular part of M, and M$_-$ denotes its lower triangular part. The choice (23) for B corresponds to the choice $p(L) = L$.

Note 4. Deift et al. point out that solving numerically the matrix differential equation (21) until such time when $b_1, \ldots, b_n$ become less than a preassigned small number is a valid numerical method for finding approximately the eigenvalues of L. In Section 4 of their paper they present numerical examples comparing the speed of this method with the speed of the QR algorithm.

BIBLIOGRAPHY

Deift, P., Nanda, T., and Tomei, C. Ordinary differential equations and the symmetric eigenvalue problem, *SIAM J. Numer. Anal.* **20** (1983), 1–22.

Flaschka, H. The Joda lattice, I, *Phys. Rev. B* **9** (1974), 1924–1925.

Francis, J. The QR transformation, a unitary analogue of the LR transformation, I, *Comput. J.* **4** (1961), 265–271.

Golub, G. H., and Van Loan, C. F. *Matrix Computations*, Johns Hopkins University Press, Baltimore, 1989.

Lax, P. D. Integrals of nonlinear equations of evolution and solitary waves, *CPAM* **21** (1968), 467–490.

Moser, J. Finitely many mass points on the line under the influence of an exponential potential—an integrable system. In *Dynamical Systems, Theory and Applications*, Springer-Verlag, New York, 1975, pp. 467–497.

Parlett, B. N. *The Symmetric Eigenvalue Problem*, Prentice Hall, Englewood Cliffs, NJ, 1980.

Trefathen, L. N., and Bau, D. *Numerical Linear Algebra*, SIAM, Philadelphia, PA, 1997.

Solutions of Selected Exercises

CHAPTER 1

Ex 1. Suppose z is another zero:

$$x + z = x \text{ for all } x.$$

Set $x = 0$: $0 + z = 0$. But also $z + 0 = z$, so $z = 0$.

Ex 3. Every polynomial p of degree $< n$ can be written as

$$p = a_1 x^{n-1} + a_2 x^{n-2} + \cdots a_n.$$

Then $p \leftrightarrow (a_1, \ldots, a_n)$ is an isomorphism.

Ex 7. If x_1, x_2 belong to X and to Y, then $x_1 + x_2$ belongs to X and Y.

Ex 10. If x_i were 0, then

$$1.x_i + \sum_{j \neq i} 0.x_j = 0.$$

Ex 13. (iii) If $x_1 - x_2$ is in Y, and $x_2 - x_3$ is in Y, then so is their sum $x_1 - x_2 + x_2 - x_3 = x_1 - x_3$.

Ex 14. Suppose $\{x_1\}$ and $\{x_2\}$ have a vector x_3 in common. Then $x_3 \equiv x_1$ and $x_3 \equiv x_2$; but then $x_1 \equiv x_2$, so $\{x_1\} = \{x_2\}$.

Linear Algebra and Its Applications, Second Edition, by Peter D. Lax
Copyright © 2007 John Wiley & Sons, Inc.

Ex 16. (i) Polynomials of degree $< n$ that are zero at $t_1, \ldots, t_j$ can be written in the form

$$q(t) \prod (t - t_i),$$

where q is a polynomial of degree $< n - j$. These clearly form a linear space, whose dimension is $n - j$.

By Theorem 6,

$$\dim X/Y = \dim X - \dim Y = n - (n - j) = j.$$

The quotient space X/Y can be identified with the space of vector

$$(p(t_1), \ldots, p(t_j)).$$

Ex 19. Use Theorem 6 and Exercise 18.

Ex 20. (b) and (d) are subspaces

Ex 21. The statement is false; here is an example to the contrary:

$$X = \mathbb{R}^2 = (x, y)\text{space}$$
$$U = \{y = 0\}, V = \{x = 0\}, W = \{x = y\}.$$
$$U + V + W = \mathbb{R}^2, U \cap V = \{0\}, U \cap W = \{0\}$$
$$V \cap W = \{0\}, U \cap V \cap W = 0.$$

So

$$2 \neq 1 + 1 + 1 - 0 - 0 - 0 - 0.$$

CHAPTER 2

Ex 4. We choose $m_1 = m_3$; then (9) is satisfied for $p(t) = t$. For $p(t) = 1$ and $p(t) = t^2$, (9) says that

$$2 = 2m_1 + m_2, \quad \frac{2}{3} = 2m_1 a^2.$$

So

$$m_1 = \frac{1}{3a^2}, \quad \text{and } m_2 = 2 - \frac{2}{3a^2},$$

from which (ii) follows. (iii) (9) holds for all odd polynomials like t^3 and t^5. For $p(t) = t^4$, (9) says that

$$\frac{2}{5} = 2m_1 a^4 = \frac{2}{3}a^2,$$

which holds for $a = \sqrt{3/5}$.

Ex 5.   Take $m_1 = m_4$, $m_2 = m_3$, in order to satisfy (9) for all odd polynomials. For $p(t) = 1$ and $p(t) = t^2$ we get two equations easily solved.

Ex 6.   (a) Suppose there is a linear relation

$$al_1(p) + bl_2(p) + cl_3(p) = 0,$$

Set $p = p(x) = (x - \xi_2)(x - \xi_3)$. Then $p(\xi_2) = p(\xi_3) = 0, p(\xi_1) \neq 0$; so we get from the above relation that $a = 0$. Similarly $b = 0$, $c = 0$.
   (b) Since dim $P_2 = 3$, also dim $P'_2 = 3$. Since l_1, l_2, l_3 are linearly independent, they span P'_2.
   (c 2) Set

$$p_1(x) = (x - \xi_2)(x - \xi_3)/(\xi_1 - \xi_2)(\xi_1 - \xi_3),$$

and define p_2, p_3 analogously. Clearly

$$l_i(p_j) = \begin{array}{l} 1 \text{ if } i = j \\ 0 \text{ if } i \neq j. \end{array}$$

Ex 7. $\ell(x)$ has to be zero for $x = (1, 0, -1, 2)$ and $x = (2, 3, 1, 1)$. These yield two equations for $c_1, \ldots, c_4$:

$$c_1 - c_3 + 2c_4 = 0, \quad 2c_1 + 3c_2 + c_3 + c_4 = 0.$$

We express c_1 and c_2 in terms of c_3 and c_4. From the first equation, $c_1 = c_3 - 2c_4$. Setting this into the second equation gives $c_2 = -c_3 + c_4$.

CHAPTER 3

Ex 1.   If $Ty_1 = u_1$, $Ty_2 = u_2$, then $T(y_1 + y_2) = u_1 + u_2$, and conversely.

Ex 2.   Suppose we drop the ith equation; if the remaining equations do not determine x uniquely, there is an x that is mapped into a vector whose components except the ith are zero. If this were true for all $i = 1, \ldots, m$, the range of the mapping $x \to u$ would be m-dimensional; but according to Theorem 2, the dimension of the range is $\leq n < m$. Therefore one of the equations may be dropped without using uniqueness; by induction $m - n$ of the equations may be omitted.

Ex 4.   Rotation maps the parallelogram $0, x, y, x + y$ into another parallelogram $0, x', y', z'$; therefore $z' = x' + y'$.
   ST maps $(1, 0, 0)$ into $(0, 1, 0)$; TS maps $(1, 0, 0)$ into $(0, 0, 1)$.

Ex 5. Set $Tx = u$; then $(T^{-1}T)x = T^{-1}u = x$, and $(TT^{-1})u = Tx = u$.

Ex 6. Part (ii) is true for all mappings, linear or nonlinear. Part (iii) was illustrated by Eisenstein as follows: the inverse of putting on your shirt and then your jacket is taking off your jacket and then your shirt.

Ex 7.
$$((ST)l, x) = (l, (ST)'x);$$

also

$$((STl, x) = (Tl, S'x) = (l, T'S'x),$$

from which $(ST)' = T'S'$ follows.

Ex 8. $(Tl, x) = (l, T'x) = (T''l, x)$ for all x; therefore $Tl = T''l$.

Ex 10. If $M = SKS^{-1}$, then $S^{-1}MS = K$, and by Theorem 4,

$$S^{-1}M^{-1}S = K^{-1}.$$

Ex 11. $AB = ABAA^{-1} = A(BA)A^{-1}$, by repeated use of the associative law.

Ex 13. The even part of an even function is the function itself.

CHAPTER 4

Ex 1. $(DA)_{ij} = \sum D_{ik}A_{kj} = d_i A_{ij}, (AD)_{ij} = \sum A_{ik}D_{kj} = A_{ij}d_j.$

Ex 2. In most texts the proof is obscure.

Ex 4. Choose B so that its range is the nullspace of A, but the range of A is *not* the nullspace of B.

CHAPTER 5

Ex 1. $P(p_1 \circ p_2(x)) = \sigma(p_1 \circ p_2)P(x)$. Since $p_1 \circ p_2(x) = p_1(p_2(x))$,

$$P(p_1 \circ p_2(x)) = P(p_1(p_2(x))) = \sigma(p_1)P(p_2(x));$$

also

$$P(p_2(x)) = \sigma(p_2)P(x).$$

Combining these identities yields $\sigma(p_1 \cdot p_2) = \sigma(p_1)\sigma(p_2)$.

Ex 2.   (c) The signature of the transposition of two adjacent variables x_k and x_{k+1} is -1. The transposition of any two variables can be obtained by composing an *odd* number of interchanges of adjacent variables. The result follows from Exercise 1.

(d) To factor $p = \frac{1 \ldots n}{p_1 \ldots p_n}$ as a product $p = t_k \ldots t_1$ of transpositions, set

$$t_1 = \frac{1\ 2 \cdots p_1 \cdots n}{p_1\ 2 \cdots 1 \cdots n},$$

$$t_2 = \frac{1\ 2 \cdots p_2 \cdots n}{1\ p_2 \cdots 2 \cdots n},$$

and so on.

Ex 3.   Follows from $(7)_b$.

Ex 4.   (iii) When $a_1 = e_1, \ldots, a_n = e_n$, the only nonzero term on the right side in (16) is $p = $ identity.

(iv) When a_i and a_j are interchanged, the right side of (16) can be written as

$$\sum \sigma(t \circ p) a_{p_1}, \ldots a_{p_n} n,$$

where t is the transposition of i and j. The result follows from $\sigma(t \circ p) = \sigma(t)\sigma(p) = -\sigma(p)$.

Ex 5.   Suppose two columns of A are equal. Then, by (iv),

$$D(a, a) = -D(a, a),$$

so
$$2D(a, a) = 0.$$

CHAPTER 6

Ex 2.   (a) All terms in $(14)'$ tend to zero.

(b) Each component of $A^N h$ is a sum of exponential functions of N, with distinct positive exponents.

Ex 5.   (25) is a special case of (26), with $q(a) = a^N$. The general case follows by combining relations (25) for various values of N.

Ex 7.   For x in N_d

$$(A - aI)^d A x = A(A - dI)^d x = 0.$$

Ex 8. Let $p(s)$ be a polynomial of degree less than $\sum d_i$. Then some a_j is a root of p of order less than d_j. But then $p(A)$ does not map all of N_{d_j} into 0.

Ex 12.
$$l_1 = (1,-1), \qquad l_2 = (1,2),$$
$$(l_1, h_1) = 3, \qquad (l_1, h_2) = 0,$$
$$(l_2, h_1) = 0, \qquad (l_2, h_2) = 3.$$

CHAPTER 7

Ex 1. According to the Schwarz inequality, $(x,y) \leq \| x \|$ for all unit vectors y. For $y = x/\| x \|$ equality holds.

Ex 2. Let Y denote any subspace of X, x and z any pair of vectors in X. Decompose them as

$$x = y + y^{\perp}, \qquad z = u + u^{\perp},$$

where y and u are in Y, $u^{\perp}$ and $u^{\perp}$ orthogonal to Y; then

$$Px = y, \qquad Pz = u,$$

P orthogonal projection into Y.

$$(Px, z) = (y, u + u^{\perp}) = (y, u);$$
$$(x, Pz) = (y + y', u) = (y, u).$$

This proves that P is its own adjoint.

Ex 3. Reflection across the plane $x_3 = 0$ maps (x_1, x_2, x_3) into (x_1, x_2, x_3). The matrix representing this mapping is

$$\begin{pmatrix} 1 & 0 & 0 \\ 0 & 1 & 0 \\ 0 & 0 & -1 \end{pmatrix},$$

whose determinant is -1.

Ex 5. If the rows of M are pairwise orthogonal unit vectors, then according to the rules of matrix multiplication, $MM^* = I$. Since a right inverse is a left inverse as well, $M^*M = I$; from this it follows from the rules of matrix multiplication that the columns of M are pairwise orthogonal unit vectors.

Ex 6. $a_{ij} = (Ae_j, e_i)$. By the Schwarz inequality

$$|a_{ij}| \leq \| Ae_j \| \| e_i \|.$$

From the definition of norm

$$\| Ae_j \| \leq \| A \| \| e_j \|.$$

Since $\| e_i \| = \| e_j \| = 1$, we deduce that

$$|a_{ij}| \leq \| A \|.$$

Ex 7. Let $x_1, \ldots, x_n$ be an orthonormal basis for x. Then any x in X can be expressed as

$$x = \sum a_j x_j,$$

and

$$\| x \|^2 = \sum |a_j|^2.$$

We can write Ax as

$$Ax = \sum a_j Ax_j,$$

so

$$\| Ax \| \leq \sum |a_j| \| Ax_j \|.$$

Using the classical Schwarz inequality yields

$$\| Ax \|^2 \leq \sum |a_j|^2 \sum \| Ax_j \|^2,$$

from which

$$\| A \|^2 \leq \sum \| Ax_j \|^2$$

follows.

Apply this inequality to $A_n - A$ in place of A to deduce that if $(A_n - A)x_j$ converges to zero for all x_j, so does $\| A_n - A \|$.

Ex 8. According to identity (44),

$$\| x + y \|^2 = \| x \|^2 + 2\text{Re}(x, y) + \| y \|^2,$$

Replace y by ty, where $t = -\text{Re}(x, y)/\| y \|^2$. Since the left side is nonnegative, we get

$$|\text{Re}(x, y)| \leq \| x \| \| y \|.$$

Replace x by kx, $|k| = 1$, and choose k so that the left side is maximized; we obtain

$$|(x, y)| \leq \| x \| \| y \|.$$

Ex 14. For any mapping A,

$$\det A^* = \overline{\det A}.$$

For M unitary, $M^*M = I$; by the multiplicative property of determinants,

$$\det M^* \det M = \det I = 1.$$

Using $\det M^* = \overline{\det M}$ we deduce

$$|\det M|^2 = 1.$$

Ex 17. $\quad (AA^*)_{ii} = \sum_k a_{ik} a_{ki}^* = \sum_k a_{ik} \bar{a}_{ik} = \sum_k |a_{ik}|^2;$

so

$$\text{tr} AA^* = \sum_i (AA^*)_{ii} = \sum_{i,k} |a_{ik}|^2.$$

Ex 19. For $A = \left(\begin{smallmatrix} 1 & 2 \\ 0 & 3 \end{smallmatrix}\right)$, $\text{tr} A = 4$, $\det A = 3$, so the characteristic equation of A is

$$a^2 - 4a + 3 = 0,$$

Its roots are the eigenvalues of A; the larger root is $a = 3$.
   On the other hand, $\sum |a_{ik}|^2 = 1 + 4 + 9 = 14$; $\sqrt{14} \simeq 3.74$, so by (46) and (51)

$$3 < \| A \| < 3.74.$$

For the value of $\| A \| \simeq 3.65$, see Ex. 2 in Chapter 8.

Ex 20. (i) Since $\det (x, y, z)$ is a multilinear function of x and y when the other variables are held fixed, $w(x, y)$ is a bilinear function of x and y,
   (ii) follows from $\det (y, x, z) = -\det(x, y, z)$,
   (iii) is true because $\det (x, y, x) = 0$ and $\det (x, y, y) = 0$.
   (iv) Multiply the matrix (x, y, z) by R: $R(x, y, z) = (Rx, Ry, Rz)$.

By the multiplicative property of determinants, and since det R $= 1$,

$$\det(x, y, z) = \det(Rx, Ry, Rz);$$

therefore

$$(w(x, y), z) = (w(Rx, Ry), Rz) = (R^*w(Rx, Ry, z)),$$

from which

$$w(x, y) = R^*w(Rx, Ry)$$

follows. Multiply both sides by R.
  (v) Take $x_0 = a(1, 0, 0)'$, $y_0 = b(\cos\theta, \sin\theta, 0)'$.

$$(x_0 \times y_0, z) = \det\begin{pmatrix} a & b\cos\theta & z_1 \\ 0 & b\sin\theta & z_2 \\ 0 & 0 & z_3 \end{pmatrix}$$
$$= (ab\sin\theta)z_3$$

Therefore

$$x_0 \times y_0 = ab\sin\theta(0, 0, 1)'.$$

Since $a = \| x_0 \|$, $b = \| y_0 \|$,

$$\| x_0 \times y_0 \| = \| x_0 \|\| y_0 \| \sin\theta.$$

Any pair of vectors x, y that make an angle θ can be rotated into x_0, y_0; using (iv) we deduce

$$\| x \times y \| = \| x \|\| y \| |\sin\theta|.$$

CHAPTER 8

Ex 1.
$$(x, Mx) = M^*(x, x) = (\overline{x, M^*x});$$

$$\mathrm{Re}(x, Mx) = \frac{1}{2}(x, Mx) + \frac{1}{2}(\overline{x, Mx}) = \left(x, \frac{M + M^*}{2}x\right).$$

Ex 4. Multiply $(24)'$ by M on the right; since $M^*M^* = I$,

$$HM = DM.$$

The jth column of the left side is H_{mj}, where m_j is the jth column of M. The jth column of the right side is $d_j m_j$; therefore

$$Hm_j = d_j m_j.$$

Ex 8. Let a be an eigenvalue of $M^{-1}H$, u an eigenvector:

$$M^{-1}Hu = au.$$

Multiply both sides on the left by M, and take the inner product with u:

$$(Hu, u) = a(Mu, u),$$

Since M is positive,

$$\frac{(Hu, u)}{(Mu, u)} = a.$$

This proves that a is real.

Ex 10. A normal matrix N has a full set of orthonormal eigenvectors $f_1, \ldots, f_n$:

$$Nf_j = n_j f_j.$$

Any vector x can be expressed a

$$x = \sum a_j f_j, \quad \| x \|^2 = \sum |a_j|^2,$$

while

$$Nx = \sum a_j n_j f_j, \quad \| Nx \|^2 = \sum |a_j|^2 |n_j|^2;$$

so

$$\| Nx \| \leq \max |n_j| \, \| x \|,$$

with equality holding for $x = f_m$, $|n_m| = \max |n_j|$. This proves that

$$\| N \| = \max |n_j|.$$

Ex 11.   (b) An eigenvector of S, with eigenvalue v, satisfies

$$f_{j-1} = vf_j, \ j = 2, \ldots, n, \ f_n = vf_1.$$

so

$$f_1 = v^{n-1}f_n = v^n f_1.$$

Therefore v is an nth root of unity:

$$v_k = \exp\frac{2\pi i}{n}k, \qquad k = 1, \ldots, n,$$

and

$$f_k = (1, v_k^{-1}, \ldots, v_k^{1-n}).$$

Their scalar product,

$$(f_k, f_l) = \sum \exp\left(\frac{-2\pi}{n}kj\right)\exp\left(\frac{2\pi}{n}ej\right)$$
$$= \sum \exp\left(\frac{2\pi}{n}(l-k)j\right) = 0$$

for $k \neq l$.

Ex 12.   (i) $A^*A = \begin{pmatrix} 1 & 0 \\ 2 & 3 \end{pmatrix}\begin{pmatrix} 1 & 2 \\ 0 & 3 \end{pmatrix} = \begin{pmatrix} 1 & 2 \\ 2 & 13 \end{pmatrix}.$

The characteristic equation of A^*A is

$$\lambda^2 - 14\lambda + 9 = 0.$$

The larger root is

$$\lambda_{\max} = 7 + \sqrt{40} \simeq 13.224.$$

By Theorem 13,

$$\| A \| = \sqrt{\lambda_{\max}} \simeq 3.65.$$

(ii) This is consistent with the estimate obtained in Ex. 19 of Chapter 7:

$$3 < \| A \| < 3.74.$$

Ex 13. $\begin{pmatrix} 1 & 0 & -1 \\ 2 & 3 & 0 \end{pmatrix}\begin{pmatrix} 1 & 2 \\ 0 & 3 \\ -1 & 0 \end{pmatrix} = \begin{pmatrix} 2 & 2 \\ 2 & 13 \end{pmatrix}.$

The characteristic polynomial of the matrix on the right is

$$\lambda^2 - 15\lambda + 22 = 0,$$

whose larger root is

$$\lambda_{max} = \frac{15 + \sqrt{137}}{2} \simeq 13.35.$$

By Theorem 13,

$$\left\| \begin{pmatrix} 1 & 0 & -1 \\ 2 & 3 & 0 \end{pmatrix} \right\| \sqrt{\lambda_{max}} \simeq 3.65.$$

CHAPTER 9

Ex 2. Differentiate $A^{-1}A = I$ using the product rule:

$$\left(\frac{d}{dt} A^{-1} \right) A + A^{-1} \frac{d}{dt} A = 0.$$

Solve for $\frac{d}{dt} A^{-1}$; (3) results.

Ex 3. Denote $\begin{pmatrix} 0 & 1 \\ 1 & 0 \end{pmatrix}$ as C; $C^2 = I$, so $C^n = C$ for n odd, $= I$ for n even.

So

$$\exp C = C \left(1 + \frac{1}{3!} + \cdots \right) + I \left(1 + \frac{1}{2} + \cdots \right)$$
$$= C \frac{e - e^{-1}}{2} + I \frac{e + e^{-1}}{2}$$
$$= \begin{pmatrix} 1.54 & 1.17 \\ 1.17 & 1.54 \end{pmatrix}$$

Ex 6. For $Y(t) = \exp At$,

$$\frac{d}{dt} Y(t) = (\exp At)A, \quad Y^{-1} \frac{dY}{dt} = A.$$

By formula (10)

$$\frac{d}{dt} \log \det \exp At = \text{tr } A,$$

so

$$\log \det \exp At = t \text{ tr } A.$$

Thus

$$\det \exp At = \exp(t \operatorname{tr} A).$$

Ex 7. According to Theorem 4 in Chapter 6, for any polynomial p, the eigenvalues of $p(A)$ are of the form $p(a)$, a an eigenvalue of A. To extend this from polynomials to the exponential function we note that e^s is defined as the limit of polynomials, $e_m(s)$ that are defined by formula (12). To complete the proof we apply Theorem 6.

In Ex. 6 we have shown that $\det \exp A = \exp(\operatorname{tr} A)$; this indicates that the multiplicity of e^a as an eigenvalue of e^A is the same as the multiplicity of a as an eigenvalue of A.

CHAPTER 10

Ex 1. In formula (6) for $\sqrt{H}$ we may take $\sqrt{a_j}$ to be either the positive or negative square root. This shows that if H has n distinct, nonzero eigenvalues, H has 2^n square roots. If one of the nonzero eigenvalues of H has multiplicity greater then one, H has infinitely many square roots.

Ex 3. $A = \begin{pmatrix} 1 & 2 \\ 2 & 5 \end{pmatrix}$ is positive; it maps $(1, 0)$ into $(1, 2)$. $B = \begin{pmatrix} 1 & -2 \\ -2 & 5 \end{pmatrix}$ is positive; it maps $(1, 0)$ into $(1, -2)$. The vectors $(1, 2)$ and $(1, -2)$ make an angle $> \pi/2$, so $AB + BA$ is not positive. Indeed,

$$AB + BA = \begin{pmatrix} -6 & 0 \\ 0 & 42 \end{pmatrix}$$

has one negative eigenvalue.

Ex 4. (a) Apply Theorem 5 twice.
      (b) Apply Theorem 5 k times,

where $2^k = m$.
      (c) The limit

$$m[M^{1/m} - I] < m[N^{1/m} - I]$$

gives

$$\log M \leq \log N.$$

Note. (b) remains true for all positive exponents $m > 1$.

Ex 5. Choose A and B as in Exercise 3, that is positive matrices whose symmetrized product is not positive. Set

$$M = A, \ N = A + tB,$$

t a sufficiently small positive number. Clearly, $M < N$.

$$N^2 = A^2 + t(AB + BA) + t^2B^2;$$

for t small the term $t^2 B$ is negligible compared with the linear term. Therefore for t small N^2 is *not* greater than M^2.

Ex 6. We claim that the functions $f(s) = -(s + t)^{-1}$, t positive, are monotone matrix functions. For if $0 < M < N$, then $0 < M + tI < N + tI$, and so by Theorem 2,

$$(M + tI)^{-1} > (N + tI)^{-1}.$$

The function $f(s)$ defined by (19) is the limit of linear combinations with positive coefficents of functions of the form s and $-(s + t)^{-1}, t > 0$. The linear combinations of monotone matrix functions is monotone, and so are their limits.

   Note 1. Loewner also proved the converse of the theorem stated: Every monotone matrix function is of form (19).

   Note 2. Every function $f(s)$ of form (19) can be extended to an analytic function into the complex upper half plane Im $s > 0$, so that the imaginary part of $f(s)$ is positive there, and zero on the positive real axis $s > 0$. According to a theorem of Herglotz, all such functions $f(s)$ can be represented in the form (19).

   It is easy to verify that the functions $s^m, 0 < m < 1$, and the function log s have positive imaginary parts in the upper half plane.

Ex 7. The matrix

$$G_{ij} = \frac{1}{r_i + r_j + 1}, r_i > 0,$$

is a Gram matrix:

$$G_{ij} = \int_0^1 f_i(t)f_j(t)dt, \ f_j(t) = t^{r_j}.$$

Ex 10. By the Schwarz inequality, and the definition of the norm of $M - N$,

$$(u, (M - N)u) \leq \| u \| \| (M - N)u \|$$
$$\leq \| u \|^2 \| M - N \| = d \| u \|^2.$$

Therefore

$$(u, Mu) \leq (u, Nu) + d \parallel u \parallel^2 = (u, (N + dI)u).$$

This proves that $M \leq N + dI$; the other inequality follows by inter changing the role of M and N.

Ex 11.   Arrange the m_i in increasing order:

$$m_1 \leq \cdots \leq m_k.$$

Suppose the n_i are *not* in increasing order, that is that for a pair of indices $i < j$, $n_i > n_j$. We claim that interchanging n_i and n_j increases the sum (51):

$$n_i m_i + n_j m_j \leq n_j m_i + n_i m_j.$$

For rewrite this inequality as

$$(n_i - n_j)m_i + (n_j - n_i)m_j$$
$$= (n_i - n_j)(m_i - m_j) \leq 0,$$

which is manifestly true. A finite number of interchanges shows that (51) is maximized when the n_i and m_i are arranged in the same order.

Ex 12.   If Z were not invertible, it would have zero as an eigenvalue, contradicting Theorem 20.
     Let h be any vector; denote Z^{-1} by k. Then

$$(Z^{-1}h, h) = (k, Zk);$$

Since the self-adjoint part of Z is positive, the right side above is positive. But then so is the left side, which proves that the self-adjoint part of Z^{-1} is positive.

Ex 13.   When A is invertible, AA^* and A^*A are similar:

$$A^*A = A^{-1}AA^*A,$$

and therefore have the same eigenvalues. Noninvertible A can be obtained as the limit of a sequence of invertible matrices.

Ex 14.   Let u be an eigenvector of A^*A, with nonzero eigenvalue:

$$A^*Au = ru, \quad r \neq 0.$$

Denote Au as v; the vector v is nonzero, for if $Au = 0$, it follows from the above relation that $u = 0$.

Let A act on the above relation:

$$AA^*Au = rAu$$

which can be rewritten as

$$AA^*v = rv;$$

which shows that v is an eigenvector of AA^*, with eigenvalue r.

A maps the eigenspace of A^*A with eigenvalue r into the eigenspace of AA^*; this mapping is 1-to-1. Similarly A^* maps the eigenspace of AA^* into the eigenspace of A^*A in a 1-to-1 fashion. This proves that these eigenspaces have the same dimension.

Ex 15. Take $Z = \left(\begin{smallmatrix} 1 & a \\ 0 & 2 \end{smallmatrix}\right)$, a some real number; its eigenvalues are 1 and 2. But

$$Z + Z^* = \begin{pmatrix} 2 & a \\ a & 4 \end{pmatrix}$$

is not positive when $a > \sqrt{8}$.

CHAPTER 11

Ex 1. If $M_t = AM$, $M_t^* = M^*A^* = -MA$.

Then

$$(M^*M)_t = M_t^*M + M^*M_t = -M^*AM + M^*AM = 0.$$

Since $M^*M = I$ at $t = 0$, $M^*M = I$ for all t. At $t = 0$, $\det M = 1$, therefore $\det M = 1$ for all t. This proves that $M(t)$ is a rotation.

Ex 5. The nonzero eigenvalues of a real antisymmetric matrix A are pure imaginary and come in conjugate pairs ik and $-ik$. The eigenvalues of A^2 are $0, -k^2, -k^2$, so tr $A^2 = -2k^2$. The diagonal entries of A^2 are $-(a^2 + b^2), -(a^2 + c^2)$ and $-(b^2 + c^2)$, so tr $A^2 = -2(a^2 + b^2 + c^2)$. Therefore $k = \sqrt{a^2 + b^2 + c^2}$.

Ex 6. The eigenvalues of e^{At} are e^{at}, where a are the eigenvalues of A. Since the eigenvalues of A are $0, \pm ik$, the eigenvalues of e^{At} are 1 and $e^{\pm ikt}$. From $Af = 0$ we deduce that $e^{At}f = f$; thus f is the axis of the rotation e^{At}. The trace of e^{At} is

$1 + e^{ikt} + e^{-ikt} = 2\cos kt + 1$. According to formula $(4)'$, the angle of rotation θ of $M = e^{At}$ satisfies $2\cos\theta + 1 = \operatorname{tr} e^{At}$. This shows that $\theta = kt = \sqrt{a^2 + b^2 + c^2}\, t$.

Ex 8. $\quad A = \begin{pmatrix} 0 & a & b \\ -a & 0 & c \\ -b & -c & 0 \end{pmatrix}, \quad B = \begin{pmatrix} 0 & d & e \\ -d & 0 & g \\ -e & -g & 0 \end{pmatrix};$

their null vectors are

$$f_A = \begin{pmatrix} -c \\ b \\ -a \end{pmatrix}, \quad f_B = \begin{pmatrix} -g \\ e \\ -d \end{pmatrix}.$$

$$AB = -\begin{pmatrix} ad + be & bg & -ag \\ ce & ad + cg & ae \\ -cd & bd & be + cg \end{pmatrix}.$$

Therefore $\operatorname{tr} AB = -2(ad + be + cg)$, whereas the scalar product of f_A and f_B is $cg + be + ad$.

Ex 9. BA can be calculated like AB, given above. Subtracting we get

$$AB - BA = \begin{pmatrix} 0 & ec - bg & -dc + ag \\ bg - ec & 0 & db - ae \\ dc - ag & ae - db & 0 \end{pmatrix}.$$

Therefore

$$f_{[A,B]} = \begin{pmatrix} ae - db \\ ag - dc \\ bg - ec \end{pmatrix}.$$

We can verify that $f_A \times f_B = f_{[A,B]}$ by using the formula for the cross product in Chapter 7.

CHAPTER 12

Ex 2. (a) Let $\{K_i\}$ be a collection of convex sets, denote their intersection by K. If x and y belong to K, they belong to every K_i. Since K_i is convex, it contains the line segment with endpoints x and y. Since this line segment belongs to all K_i, it belongs to K. This shows that K is convex.

   (b) Let x and y be two points in $H + K$; that means that they are of the form

$$x = u + z, \quad y = v + w, \quad u \text{ and } v \text{ in } H, \ z \text{ and } w \text{ in } K.$$

Since H is convex, $au + (1-a)v$ belongs to H, and since K is convex $a\,Z + (1-a)w$ belongs to H for $0 \le a \le 1$. But then their sum

$$au + (1-a)v + az + (1-a)w = a(u+z) + (1-a)(v+w) = ax + (1-a)y$$

belongs to $H + K$. This proves that $H + K$ is convex.

Ex 6. Denote (u, v) as x. If both u and v are ≤ 0, then x/r belongs to K for any positive r, no matter how small. So $p(x) = 0$.

If $0 < v$ and $u \le v$, then $x/r = \left(\frac{u}{r}, \frac{v}{r}\right)$ belongs to K for $r \ge v$, but no smaller r. Therefore $p(x) = v$. We can argue similarly in the remaining case.

Ex 7. If $p(x) < 1$, $p(y) < 1$, and $0 \le a \le 1$, then by sub-additivity and homogeneity of p,

$$p(ax + (1-a)y) \le p(ax) + p((1-a)y) = ap(x) + (1-a)p(y) < 1.$$

This shows that the set of $x : p(x) < 1$ is convex.

To show that the set $p(x) < 1$ is open we argue as follows. By subadditivity and positive homogeneity

$$p(x + ty) \le p(x) + p(ty) = p(x) + tp(y).$$

Since $p(x) < 1, p(x) + tp(y) < 1$ for all t positive but small enough.

Ex 8.

$$q_S(m + l) = \sup_{x \in S}(m + l)(x)$$
$$= \sup_{x \in S}(m(x) + l(x)) \le \sup_{x \in S} m(x) + \sup_{x \in S} l(x) = q_S(m) + q_S(l).$$

Note. This is a special case of the result that the supremum of linear functions is subadditive.

Ex 10.

$$q_{S \cup T}^{(l)} = \sup_{x \in S \cup T} l(x)$$
$$= \max\{\sup_{x \in S} \ell(x), \sup_{x \in T} \ell(x)\} = \max\{q_S(l), q_T(l)\}.$$

Ex 16. Suppose all p_j are positive. Define

$$y_k = \sum_1^k p_j x_j \Big/ \sum_1^k p_j.$$

Then

$$y_{k+1} = \frac{\sum_1^k p_j}{\sum_1^{k+1} p_j} y_k + \frac{p_{k+1}}{\sum_1^k p_j} x_{k+1}.$$

We claim that all points y_k belong to the convex set containing $x_i, \ldots, x_m$. This follows inductively, since $y_1 = x_1$, and y_{k+1} lie on the line segment whose endpoints are y_k and x_{k+1}. Finally,

$$y_m = \sum_1^m p_j x_j.$$

Ex 19.　Denote by P_1, P_2, P_3 the following 3×3 permutation matrices

$$P_1 = \begin{pmatrix} 1 & 0 & 0 \\ 0 & 1 & 0 \\ 0 & 0 & 1 \end{pmatrix}, \quad P_2 = \begin{pmatrix} 0 & 1 & 0 \\ 0 & 0 & 1 \\ 1 & 0 & 0 \end{pmatrix}, \quad P_3 = \begin{pmatrix} 0 & 0 & 1 \\ 1 & 0 & 0 \\ 0 & 1 & 0 \end{pmatrix}.$$

Then

$$\frac{1}{3}P_1 + \frac{1}{3}P_2 + \frac{1}{3}P_3 = \frac{1}{3}\begin{pmatrix} 1 & 1 & 1 \\ 1 & 1 & 1 \\ 1 & 1 & 1 \end{pmatrix} = M.$$

Similarly define

$$P_4 = \begin{pmatrix} 1 & 0 & 0 \\ 0 & 0 & 1 \\ 0 & 1 & 0 \end{pmatrix}, \quad P_5 = \begin{pmatrix} 0 & 1 & 0 \\ 1 & 0 & 0 \\ 0 & 0 & 1 \end{pmatrix}, \quad P_6 = \begin{pmatrix} 0 & 0 & 1 \\ 0 & 1 & 0 \\ 1 & 0 & 0 \end{pmatrix}.$$

Then

$$\frac{1}{3}P_4 + \frac{1}{3}P_5 + \frac{1}{3}P_6 = M.$$

Ex 20.　A set S in Euclidean space is open if for every point x in S there is a ball $\| y - x \| < \epsilon$ centered at x that belongs to S. Suppose S is convex and open; that means that there exist positive numbers ϵ_i such that $x + te_i$ belongs to S for $|t| < \epsilon_i$; here e_i denotes the unit vectors. Denote min ϵ_i by ϵ; it follows that the points $x \pm \epsilon e_i, i = 1, \ldots, n$ belong to S. Since S is convex, the convex hull of these points belongs to S; this convex hull contains a ball of radius $\epsilon/\sqrt{2}$ centered at x.
　　The converse is obvious.

CHAPTER 13

Ex 3. We claim that the sign of equality holds in (21). For if not, S would be larger than s, contrary to (20); this shows that the supremum in (16) is a maximum.

Replacing Y, y, and j by $-Y$, $-y$, $-j$ turns the sup problem into inf, and vice versa. This shows that the $\ln f$ in (18) is a minimum.

CHAPTER 14

Ex 2. $x - z = (x - y) + (y - z)$; apply the subadditive rule $(1)_{ii}$.

Ex 5. From the definition of the $|x|_p$ and $|x|_\infty$ norms we see that

$$|x|_\infty^p \leq |x|_p^p \leq n|x|_\infty^p.$$

Take the pth root:

$$|x|_\infty \leq |x|_p \leq n^{1/p}|x|_\infty.$$

Since $n^{1/p}$ tends to 1 as p tends to ∞, $|x|_\infty = \lim_{p \to \infty} |x|_p$ follows.

Ex 6. Introduce a basis and represent the points by arrays of real numbers. Since all norms are equivalent, it suffices to prove completeness in the $|x|_\infty$ norm.

Let $\{x_n\}$ be a convergent sequence in the $|x|_\infty$ norm. Denote the components of x_n by $x_{n,j}$. It follows from $|x_n - x_m|_{\max} \to 0$ that

$$|x_{n,j} - x_{m,j}| \to 0$$

for every j. Since the real numbers are complete, it follows that

$$\lim_{n \to \infty} x_{n,j} = x_j.$$

Denote by x the vector with components x_j; it follows that

$$\lim |x_n - x|_\infty = 0.$$

For another proof see Theorem 3.

CHAPTER 15

Ex 1.   According to (i), $|Tx| \leq c|x|$. Apply this to $|T(x_n - x)| \leq c|x_n - x|$.

Ex 3.   We have shown above that

$$(I - R)T^n = I - R^{n+1}$$

Multiply both sides on the left by S^{-1}:

$$T^n = S^{-1} - S^{-1}R^{n+1}$$

Therefore

$$|T^n - S^{-1}| \leq |S^{-1}R^{n+1}| \leq |S^{-1}||R^{n+1}|.$$

Since $|R| < 1, |R^{n+1}| \leq |R|^{n+1}$ tends to zero as $n \to \infty$,

Ex 5.   Decompose n modulo m:

$$n = km + r, \qquad 0 \leq r < m.$$

Then

$$R^n = R^{km+r} = (R^m)^k R^r,$$

therefore

$$|R^n| \leq |R^m|^k |R^r|$$

as n tends to ∞, so does k; therefore if $|R^m| < 1$, $|R^n|$ tends to zero as n tends to ∞. That is all that was used in the proof in Ex. 3, that T^n tends in norm to S^{-1}.

Ex 6.   The components of $y = Tx$ are

$$y_i = \sum_j t_{ij}x_j.$$

Since $|x_j| \leq |x|_\infty$,

$$|y_i| \leq \sum_j |t_{ij}||x|_\infty.$$

So $|y|_\infty \leq \max \sum_j |t_{ij}||X|_\infty$; (23) follows

CHAPTER 16

Ex 1. What has to be shown is that if $Px \leq \lambda x$, then $\lambda \geq \lambda(P)$. To see this consider P^T, the transpose of P; it, too, has positive entries, so by Theorem 1 it has a dominant eigenvalue $\lambda(P^T)$ and a corresponding nonnegative eigenvector k:

$$P^T k = \lambda(P^T)k.$$

Take the scalar product of $Px \leq \lambda x$ with k; since k is a nonnegative vector

$$(Px, k) \leq \lambda(x, k)$$

The left side equals $(x, P^T k) = \lambda(P^T)(x, k)$. Since x and k are nonnegative vectors, (x, k) is positive, and we conclude that $\lambda(P^T) \leq \lambda$. But P and P^T have the same eigenvalues, and therefore the largest eigenvalue $\lambda(P)$ of P equals the largest eigenvalue $\lambda(P^T)$ of P^T,

Ex 2. Denote by μ the dominant eigenvalue of P^m, and by k the associated eigenvector:

$$P^m k = \mu k.$$

Let P act on this relation:

$$P^{m+1}k = P^m Pk = \mu Pk,$$

which shows that Pk, too, is an eigenvector of P^m with eigenvalue μ. Since the dominant eigenvalue has multiplicity one, $Pk = ck$. Repeated application of P shows that $P^m k = c^m k$. Therefore $c^m = \mu$. From $Pk = ck$ we deduce that c is real and positive; therefore it is the real root $\mu^{1/m}$. Since the entries of k are positive, $c = \mu^{1/m}$ is the dominant eigenvalue of P.

Bibliography

Axler, Sheldon. *Linear Algebra Done Right*, Undergraduate Texts in Mathematics, Springer-Verlag, New York, 1996.

Bellman, Richard. *Introduction to Matrix Analysis*, McGraw-Hill, New York, 1960.

Courant, Richard, and Hilbert, David. *Methods of Mathematical Physics*, Vol. I, Wiley Interscience, New York, 1953.

Edwards, Harold M. *Linear Algebra*, Birkhäuser, Boston, 1995.

Golub, Gene, and Van Loan, Charles. *Matrix Computations*, 2nd ed., The John Hopkins University Press, Baltimore, 1989.

Greub, Werner. *Linear Algebra*, 4th ed., Graduate Texts in Mathematics, Springer-Verlag, New York, 1975.

Halmos, Paul R. *Finite Dimensional Vector Spaces*, Undergraduate Texts in Mathematics, Springer-Verlag, New York, 1974.

Halmos, Paul R. *Linear Algebra Problem Book*, Dolciani Mathematical Expositions #16, Mathematical Association of America, Providence, RI, 1995.

Hoffman, Kenneth, and Kunze, Ray. *Linear Algebra*, Prentice Hall, Englewood Cliffs, NJ, 1971.

Horn, R. A., and Johnson, C. R. *Matrix Analysis*, Cambridge University Press, 1985.

Horn, R. A., and Johnson, C. R. *Topics in Matrix Analysis*, Cambridge University Press, 1991.

Lancaster, P., and Tismenetsky, M. *The Theory of Matrices*, Academic Press, New York, 1985.

Lang, Serge. *Linear Algebra*, 3rd ed., Undergraduate Texts in Mathematics, Springer-Verlag, New York, 1987.

Lay, David C. *Linear Algebra and Its Applications*, Addision-Wesley, Reading, MA, 1994.

Roman, Steven. *Advanced Linear Algebra*, Graduate Texts in Mathematics, Springer-Verlag, New York, 1992.

Serre, Denis. *Matrices*: *Theory and Applications*, Graduate Texts in Mathematics, Springer-Verlag, New York, 2002.

Strang, Gilbert. *Linear Algebra and Its Applications*, 3rd ed., Harcourt, Brace, Jovanovich, San Diego, 1988.

Trefethenr, Lloyd N., and Bay, David. *Numerical Linear algebra*, SIAM, 1997.

Valenza, Robert J. *Linear Algebra, An Introduction to Abstract Mathematics*, Undergraduate Texts in Mathematics, Springer-Verlag, New York, 1993.

Special Determinants

There are some classes of matrices whose determinants can be expressed by compact algebraic formulas. We give some interesting examples.

Definition. A *Vandermonde matrix* is a square matrix whose columns form a geometric progression. That is, let $a_1, \ldots, a_n$ be n scalars; then $V(a_1, \ldots, a_n)$ is the matrix

$$V(a_1, \ldots, a_n) = \begin{pmatrix} 1 & \cdots & 1 \\ a_1 & & a_n \\ \vdots & & \vdots \\ a_1^{n-1} & \cdots & a_n^{n-1} \end{pmatrix}. \tag{1}$$

Theorem 1

$$\det V(a_1, \ldots, a_n) = \prod_{j>i} (a_j - a_i). \tag{2}$$

Proof. Using formula (16) of Chapter 5 for the determinant, we conclude that $\det V$ is a polynomial in the a_i of degree less than or equal to $n(n-1)/2$. Whenever two of the scalars a_i and $a_j, i \neq j$, are equal, V has two equal columns and so its determinant is zero; therefore, according to the factor theorem of algebra, $\det V$ is divisible by $a_j - a_i$. It follows that $\det V$ is divisible by the product

$$\prod_{j>i} (a_j - a_i).$$

Linear Algebra and Its Applications, Second Edition, by Peter D. Lax
Copyright © 2007 John Wiley & Sons, Inc.

This product has degree $n(n-1)/2$, the same as the degree of det V. Therefore

$$\det V = c_n \prod_{j>i}(a_j - a_i), \tag{2}'$$

c_n a constant. We claim that $c_n = 1$; to see this we use the Laplace expansion (26) of Chapter 5 for det V with respect to the last column, that is, $j = n$. We get in this way an expansion of det V in powers of a_n; the coefficient of a_n^{n-1} is det $V(a_1, \ldots, a_{n-1})$. On the other hand, the coefficient of a_n^{n-1} on the right of $(2)'$ is $c_n \Pi_{n>j>i}(a_j - a_i)$. Using expression $(2)'$ for $V(a_1, \ldots, a_{n-1})$, we deduce that $c_n = c_{n-1}$. An explicit calculation shows that $c_2 = 1$; hence by induction $c_n = 1$ for all n, and (2) follows. $\square$

Definition. Let $a_1, \ldots, a_n$ and $b_1, \ldots, b_n$ be $2n$ scalars. The *Cauchy matrix* $C(a_1, \ldots, a_n; b_1, \ldots, b_n)$ is the $n \times n$ matrix whose ijth element is $1/(a_i + a_j)$:

$$C(a, b) = \left(\frac{1}{a_i + b_j} \right).$$

Theorem 2.

$$\det C(a, b) = \frac{\prod_{j>i}(a_j - a_i)(b_j - b_i)}{\prod_{i,j}(a_i + b_j)}. \tag{3}$$

Proof. Using formula (16) of Chapter 5 for the determinant of $C(a, b)$, and using the common denominator for all terms we can write

$$\det C(a, b) = \frac{P(a, b)}{\prod_{i,j}(a_i + b_j)}, \tag{4}$$

where $P(a, b)$ is a polynomial whose degree is less than or equal to $n^2 - n$. Whenever two of the scalars a_i and a_j are equal, the ith and jth row of $C(a, b)$ are equal; likewise, when $b_i = b_j$, the ith and jth column of $C(a, b)$ are equal. In either case, $\det C(a, b) = 0$; therefore, by the factor theorem of algebra, the polynomial $P(a, b)$ is divisible by $(a_j - a_i)$ and by $(b_j - b_i)$, and therefore by the product

$$\prod_{j>i}(a_j - a_i)(b_j - b_i).$$

The degree of this product is $n^2 - n$, the same as the degree of P; therefore,

$$P(a, b) = c_n \prod_{j>i}(a_j - a_i)(b_j - b_i), \tag{4}'$$

c_n a constant. We claim that $c_n = 1$; to see this we use the Laplace expansion for $C(a,b)$ with respect to the last column, $j = n$; the term corresponding to the element $1/(a_n + b_n)$ is

$$\det C(a_1, \ldots, a_{n-1}; b_1, \ldots, b_{n-1}) \frac{1}{a_n + b_n}.$$

Now set $a_n = b_n = d$; we get from (4) and (4)' that

$$C(a_1, \ldots, d; b_1, \ldots, d)$$
$$= \frac{c_n \prod_{n>i} (d - a_i)(d - b_i)}{2d \prod_{n>i}(d + a_i)(d + b_i)} \frac{\prod_{n>j>i} (a_j - a_i)(b_j - b_i)}{\prod_{i,j<n}(a_i + b_j)}.$$

From the Laplace expansion we get

$$C(a_1, \ldots, d; b_1, \ldots, d)$$
$$= \frac{1}{2d} C(a_1, \ldots, a_{n-1}; b_1, \ldots, b_{n-1}) + \text{other terms}.$$

Multiply both expressions by $2d$ and set $d = 0$; using (4)' to express $C(a_1, \ldots, a_{n-1}; b_1, \ldots, b_n)$, we deduce that $c_n = c_{n-1}$. An explicit calculation shows that $c_1 = 1$, so we conclude by induction that $c_n = 1$ for all n; (3) now follows from (4) and (4)'. $\quad\square$

Note: Every minor of a Cauchy matrix is a Cauchy matrix.

EXERCISE 1.   Let

$$p(s) = x_1 + x_2 s + \cdots + x_n s^{n-1}$$

be a polynomial of degree less than n. Let $a_1, \ldots, a_n$ be n distinct numbers, and let $p_1, \ldots, p_n$ be n arbitrary complex numbers; we wish to choose the coefficients $x_1, \ldots, x_n$ so that

$$p(a_i) = p_i, \qquad i = 1, \ldots, n.$$

This is a system of n linear equations for the n coefficients x_i. Find the matrix of this system of equations, and show that its determinant is $\neq 0$.

EXERCISE 2.   Find an algebraic formula for the determinant of the matrix whose ijth element is

$$\frac{1}{1 + a_i a_j};$$

here $a_1, \ldots, a_n$ are arbitrary scalars.

The Pfaffian

Let A be an $n \times n$ antisymmetric matrix:

$$A^T = -A.$$

We have seen in Chapter 5 that a matrix and its transpose have the same determinant. We have also seen that the determinant of $-A$ is $(-1)^n \det A$ so

$$\det A = \det A^T = \det(-A) = (-1)^n \det A.$$

When n is odd, it follows that $\det A = 0$; what can we say about the even case?

Suppose the entries of A are real; then the eigenvalues come in complex conjugate pairs. On the other hand, according to the spectral theory of anti-self-adjoint matrices, the eigenvalues of A are purely imaginary. It follows that the eigenvalues of A are $(-i\lambda_1, \ldots, -i\lambda_{n/2}, i\lambda_1, \ldots, i\lambda_{n/2})$. Their product is $(\Pi\lambda_i)^2$, a positive number; since the determinant of a matrix is the product of its eigenvalues, we conclude that the determinant of an antisymmetric matrix of even order with real entries is nonnegative.

Far more is true:

Theorem of Cayley. The determinant of an antisymmetric matrix A of even order is the square of a homogeneous polynomial of degree $n/2$ in the entries of A:

$$\det A = P^2.$$

P is called the Pfaffian.

EXERCISE I.   Verify by a calculation Cayley's theorem for $n = 4$.

Proof. The proof is based on the following lemma.   ☐

Lemma 1.   There is a matrix C whose entries are polynomials in the entries of the antisymmetric matrix A such that

$$B = CAC^T \tag{1}$$

is antisymmetric and tridiagonal, that is, $b_{ij} = 0$ for $|i - j| > 1$. Furthermore, $\det C \neq 0$.

Proof. We construct C as a product

$$C = C_{n-2} \cdots C_2 C_1.$$

C_1 is required to have the following properties:

(i) $B_1 = C_1 A C_1^T$ has zeros for the last $(n - 2)$ entries in its first column.
(ii) The first row of C_1 is $e_1 = (1, 0, \ldots, 0)$, its first column is e_1^T.

It follows from (ii) that C_1 maps e_1^T into e_1^T; therefore the first column of $B_1, B_1 e_1^T$, is $C_1 A C_1^T e_1^T = C_1 A e_1^T = C_1 a$, where a denotes the first column of A. To satisfy (i) we have to choose the rest of C_1 so that the last $(n - 2)$ entries of $C_1 a$ are zero. This requires the last $n - 2$ rows of C_1 to be orthogonal to a. This is easily accomplished: set the second row of C_1 equal to $e_2 = (0, 1, 0, \ldots, 0)$ the third row $(0, a_3, -a_2, 0, \ldots, 0)$, the fourth row $(0, 0, a_4, -a_3, 0, \ldots, 0)$, and so on, where $a_1, \ldots, a_n$ are the entries of the vector a. Clearly

$$\det C_1 = a_2 a_3 \cdots a_{n-1}$$

is a nonzero polynomial.

We proceed recursively; we construct C_2 so its first two rows are e_1 and e_2, its first two columns e_1^T and e_2^T. Then the first column of $B_2 = C_2 B_1 C_2^T$ has zero for its last $n - 2$ entires. As before, we fill in the rest of C_2 so that the second column of B_2 has zeros for its last $n - 3$ entries. Clearly, $\det C_2$ is a nonzero polynomial.

After $(n - 2)$ steps we end with $C = C_{n-2} \cdots C_1$, having the property that $B = CAC^T$ has zero entries below the first subdiagonal, that is, $b_{ij} = 0$ for $i > j + 1$. $B^T = CA^TC^T = -B$, that is, B is antisymmetric. It follows that its only nonzero entries lie on the sub and super diagonals $j = i + 1$. Since $B^T = -B, b_{i,i-1} = -b_{i,i+1}$. Furthermore, by construction,

$$\det C = \prod \det C_i \neq 0. \qquad\qquad ☐ \tag{2}$$

What is the determinant of an antisymmetric, tridiagonal matrix of even order? Consider the 4×4 case

$$B = \begin{pmatrix} 0 & a & 0 & 0 \\ -a & 0 & b & 0 \\ 0 & -b & 0 & c \\ 0 & 0 & -c & 0 \end{pmatrix}.$$

Its determinant $\det B = a^2 c^2$ is the square of a single product. The same is true in general: the determinant of a tridiagonal antisymmetric matrix B of even order is the square of a single product,

$$\det B = \left(\prod b_{2k,2k-1} \right)^2. \tag{3}$$

Using the multiplicative property of determinants, and that $\det C^T = \det C$, we deduce from (1) that

$$\det B = (\det C)^2 \det A;$$

combining this with (2) and (3) we deduce that $\det A$ is the square of a *rational function* in the entries of A. To conclude we need therefore Lemma 2.

Lemma 2. If a polynomial P in n variables is the square of a rational function R, R is a polynomial.

Proof. For functions of one variable this follows by elementary algebra; so we can conclude that for each fixed variable x, R is a polynomial in x, with coefficients from the field of rational functions in the remaining variables. It follows that there exists a k such that the kth partial derivative of R with respect to any of the variables is zero. From this it is easy to deduce, by induction on the number of variables, that R is a polynomial in all variables. $\square$

APPENDIX 3

Symplectic Matrices

In Chapter 7 we studied *orthogonal* matrices O; they preserve a scalar product:

$$(Ox, Oy) = (x, y).$$

Scalar products are *symmetric* bilinear functions; in this appendix we investigate linear maps that preserve a nonsingular bilinear *alternating* function of the form (x, Ay), A a real anti-self-adjoint matrix, $\det A \neq 0$. It follows that A must be of even order $2n$. It suffices to specialize to $A = J$, where, in block notation,

$$J = \begin{pmatrix} O & I \\ -I & O \end{pmatrix}. \tag{1}$$

I is the $n \times n$ unit matrix.

EXERCISE 1. Prove that any real $2n \times 2n$ anti-self-adjoint matrix A, $\det A \neq 0$, can be written in the form

$$A = FJF^T,$$

J defined by (1), F some real matrix, $\det F \neq 0$.

The matrix J has the following properties, which will be used repeatedly:

$$J^2 = -I, \qquad J^{-1} = -J = J^T. \tag{2}$$

Theorem 1. A matrix S that preserves (x, Jy):

$$(Sx, JSy) = (x, Jy) \tag{3}$$

Linear Algebra and Its Applications, Second Edition, by Peter D. Lax
Copyright © 2007 John Wiley & Sons, Inc.

for all x and y, satisfies

$$S^T J S = J \tag{4}$$

and conversely.

Proof. $(Sx, JSy) = (x, S^T JSy)$. If this is equal to (x, Jy) for all x, y, $S^T SJy = J_y$ for all y. $\quad\square$

A real matrix S that satisfies (4) is called a *symplectic matrix*. The set of all symplectic matrices is denoted as $Sp(n)$. $\quad\square$

Theorem 2. (i) Symplectic matrices form a group under matrix multiplication.
(ii) If S is symplectic, so is its transpose S^T.
(iii) A symplectic matrix S similar to its inverse S^{-1}.

Proof. (i) It follows from (4) that every symplectic matrix is invertible. That they form a group follows from (3). To verify (ii), take the inverse of (4); using (2) we get

$$S^{-1} J (S^T)^{-1} = J.$$

Multiplying by S on the left, S^T on the right shows that S^T satisfies (4).

To deduce (iii) multiply (4) by S^{-1} on the right and J^{-1} on the left. We get that $J^{-1} S^T J = S^{-1}$, that is, that S^{-1} is similar to S^T. Since S^T is similar to S, (iii) follows. $\quad\square$

Theorem 3. Let $S(t)$ be a differentiable function of the real variable t, whose values are symplectic matrices. Define $G(t)$ by

$$\frac{d}{dt} S = GS. \tag{5}$$

Then G is of the form

$$G = JL(t), \qquad L \text{ self-adjoint.} \tag{6}$$

Conversely, if $S(t)$ satisfies (5) and (6) and $S(0)$ is symplectic, then $S(t)$ is a family of symplectic matrices.

Proof. For each t (4) is satisfied; differentiate it with respect to t:

$$\left(\frac{d}{dt} S^T \right) JS + S^T J \frac{d}{dt} S = 0.$$

Multiply by S^{-1} on the right, $(S^T)^{-1}$ on the left:

$$(S^T)^{-1}\frac{d}{dt}S^T J + J\left(\frac{d}{dt}S\right)S^{-1} = 0. \tag{7}$$

We use (5) to define G:

$$G = \left(\frac{d}{dt}S\right)S^{-1}.$$

Taking the transpose we get

$$G^T = (S^T)^{-1}\frac{d}{dt}S^T.$$

Setting these into (7) gives

$$G^T J + JG = 0,$$

from which (6) follows. $\qquad\qquad\square$

EXERCISE 2.   Prove the converse.

We turn now to the spectrum of a symplectic matrix S. Since S is real, its complex eigenvalues come in conjugate pairs, that is, if λ is an eigenvalue, so is $\bar{\lambda}$. According to part (iii) of Theorem 2, S and S^{-1} are similar; since similar matrices have the same spectrum, it follows that if λ is an eigenvalue of S, so is λ^{-1} and it has the same multiplicity. Thus the eigenvalues of a symplectic matrix S come in groups of four: $\lambda, \bar{\lambda}, \lambda^{-1}\overline{\lambda^{-1}}$, with three exceptions:

(a) When λ lies on the unit circle, that is, $|\lambda| = 1$, then $\lambda^{-1} = \bar{\lambda}$, so we only have a group of two.
(b) When λ is real, $\bar{\lambda} = \lambda$, so we only have a group of two.
(c) $\lambda = 1$ or -1.

The possibility is still open that $\lambda = \pm 1$ are simple eigenvalues of S; but this cannot occur according to

Theorem 4.   For a symplectic matrix S, $\lambda = 1$ or -1 cannot be a simple eigenvalue.

Proof. We argue indirectly: suppose, say, that $\lambda = -1$ is a simple eigenvalue, with eigenvector h:

$$Sh = -h. \tag{8}$$

Multiplying both sides by $S^T J$ and using (4) we get

$$Jh = -S^T Jh, \tag{8}'$$

which shows that Jh is eigenvector of S^T with eigenvalue -1.

We choose any self-adjoint, positive matrix L, and set $G = JL$. We define the one-parameter family of matrices $S(t)$ as $e^{tG}S$; it satisfies

$$\frac{d}{dt}S(t) = GS(t), \qquad S(0) = S. \tag{9}$$

According to Theorem 3, $S(t)$ is symplectic for all t.

If $S(0)$ has -1 as eigenvalue of multiplicity one, then for t small, $S(t)$ has a single eigenvalue near -1. This eigenvalue λ equals -1, for otherwise λ^{-1} would be another eigenvalue near -1. According to Theorem 8 of Chapter 9, the eigenvector $h(t)$ is a differentiable function of t. Differentiating $Sh = -h$ yields

$$\left(\frac{d}{dt}S\right)h + Sh_t = -h_t, \qquad h_t = \frac{d}{dt}h.$$

Using (9) and (8) we get

$$Gh = h_t + Sh_t.$$

Form the scalar product with Jh; using (8)$'$ we get

$$\begin{aligned}
(Gh, Jh) &= (h_t, Jh) + (Sh_t, Jh) = (h_t, Jh) + (h_t, S^T Jh) \\
&= (h_t, Jh) - (h_t, Jh) = 0.
\end{aligned} \tag{10}$$

According to (6), $G = JL$; set this into (10); since by (2), $J^T J = I$, we have

$$(JLh, Jh) = (Lh, J^T Jh) = (Lh, h) = 0. \tag{10}'$$

Since L was chosen to be self-adjoint and positive, $h = 0$, a contradiction. $\square$

EXERCISE 3.   Prove that plus or minus 1 cannot be an eigenvalue of odd multiplicity of a symplectic matrix.

Taking the determinant of (4), using the multiplicative property, and that $\det S^T = \det S$ we deduce that $(\det S)^2 = 1$ so that $\det S = 1$ or -1. More is true.

Theorem 5.   The determinant of a symplectic matrix S is 1.

Proof. Since we already know that $(\det S)^2 = 1$, we only have to exclude the possibility that $\det S$ is negative. The determinant of a matrix is the product of its eigenvalues. The complex eigenvalues come in conjugate pairs; their product is positive. The real eigenvalues $\neq 1, -1$ come in pairs λ, λ^{-1}, and their product is positive. According to Exercise 3, -1 is an eigenvalue of even multiplicity; so the product of the eigenvalues is positive.   □

We remark that it can be shown that the space $Sp(n)$ of symplectic matrices is connected. Since $(\det S)^2 = 1$ and since $S = I$ has determinant 1, it follows that $\det S = 1$ for all S in $Sp(n)$.

Symplectic matrices first appeared in *Hamiltonian mechanics*, governed by equations of the form

$$\frac{d}{dt}u = JH_u, \tag{11}$$

where $u(t)$ lies in $\mathbb{R}^{2n}$, H is some smooth function in $\mathbb{R}^{2n}$, and H_u is its gradient.

Definition.   A nonlinear mapping $u \to v$ is called a canonical transformation if its Jacobian matrix $\partial v/\partial u$ is symplectic.

Theorem 6.   A canonical transformation changes every Hamiltonian equation (11) into another equation of Hamiltonian form:

$$\frac{d}{dt}v = JK_v,$$

where $K(v(u)) = H(u)$.

EXERCISE 4.   Verify Theorem 6.

APPENDIX 4

Tensor Product

For an analyst, a good way to think of the tensor product of two linear spaces is to take one space as the space of polynomials in x of degree less than n, the other as the polynomials in y of degree less than m. Their tensor product is the space of polynomials in x and y, of degree less than n in x, less than m in y. A natural basis for polynomials are the powers $1, x, \ldots, x^{n-1}$ and $1, y, \ldots, y^{m-1}$, respectively; a natural basis for polynomials in x and y is $x^i y^j, i < n, j < m$.

This sets the stage for defining the tensor product of two linear spaces U and V as follows: Let $\{e_i\}$ be a basis of the linear space U, $\{f_j\}$ a basis for the linear space V. Then $\{e_i \otimes f_j\}$ is a basis for their tensor product $U \otimes V$.

It follows from this definition that

$$\dim U \otimes V = (\dim U)(\dim V). \tag{1}$$

The definition, however, is ugly, since it uses basis vectors.

EXERCISE I. Establish a natural isomorphism between tensor products defined with respect to two pairs of distinct bases.

Happily, we can define $U \otimes V$ in an invariant manner.
Take the collection of all formal sums

$$\sum u_i \otimes v_i, \tag{2}$$

where u_i and v_i are arbitrary vectors in U and V, respectively. Clearly, these sums form a linear space.

Sums of the form

$$(u_1 + u_2) \otimes v - u_1 \otimes v - u_2 \otimes v \tag{3}$$

Linear Algebra and Its Applications, Second Edition, by Peter D. Lax
Copyright © 2007 John Wiley & Sons, Inc.

and

$$u \otimes (v_1 + v_2) - u \otimes v_1 - u \otimes v_2 \qquad (3)'$$

are special cases of (2). These, and all their linear combinations, are called *null sums*. Using these concepts we can give a basis-free definition of tensor product.

Definition.   The tensor product $U \otimes V$ of two finite-dimensional linear spaces U and V is the *quotient space* of the space of all formal sums (2) modulo all null sums (3), (3)'.

This definition is basis-free, but a little awkward. Happily, there is an elegant way of presenting it.

Theorem 1.   There is a natural isomorphism between $U \otimes V$ as defined above and $\mathscr{L}(U', V)$, the space of all linear mappings of U' into V, where U' is the dual of U.

Proof. Let $\Sigma u_i \otimes v_i$ be a representative of an equivalence class in the quotient space. For any l in U', assign to l the image $\Sigma l(u_i)v_i$ in V. Since every null sum is mapped into zero, this mapping depends only on the equivalence class.
    The mapping L,

$$l \to \sum l(u_i)v_i,$$

is clearly linear and the assignment

$$\left\{ \sum u_i \otimes v_i \right\} \to L \qquad (4)$$

also is linear. It is not hard to show that every L in $\mathscr{L}(U', V)$ is the image of some vector in $U \otimes V$.   □

    EXERCISE 2.   Verify that (4) maps $U \otimes V$ onto $\mathscr{L}(U', V)$.

Theorem 1 treats the spaces U and V asymmetrically. The roles of U and V can be interchanged, leading to an isomorphism of $U \otimes V$ and $\mathscr{L}(V', U)$. The dual of a map $L: U' \to V$ is of course a map $L': V' \to U$.
    When U and V are equipped with real Euclidean structure, there is a natural way to equip $U \otimes V$ with Euclidean structure. As before, there are two ways of going about it. One is to choose *orthonormal* bases $\{e_i\}, \{f_j\}$ in U and V respectively, and declare $\{e_i \otimes f_j\}$ to be an orthonormal basis for $U \otimes V$. It remains to be shown that this Euclidean structure is independent of the choice of the orthonormal bases; this is easily done, based on the following lemma.

Lemma 2.   Let u, z be a pair of vectors in U, v, w a pair of vectors in V. Then

$$(u \otimes v, z \otimes w) = (u, z)(v, w). \tag{5}$$

Proof. Expand u and z in terms of the e_i, v and w in terms of f_j.

$$u = \sum a_i e_i, \qquad z = \sum b_k e_k,$$
$$v = \sum c_j f_j, \qquad w = \sum d_l f_l.$$

Then

$$u \otimes v = \sum a_i c_j e_i \otimes f_j, \qquad z \otimes w = \sum b_k d_l e_k \otimes f_l;$$

so

$$(u \otimes v, z \otimes w) = \sum a_i c_j b_i d_j$$
$$= \left(\sum a_i b_i \right) \left(\sum c_j d_j \right) = (u, z)(v, w). \qquad \square$$

Take the example presented at the beginning, where V is the space of polynomials in x of degree $< n$, and V is the space of polynomials in y of degree $< m$. Define in U the square integral over an x-interval A as the Euclidean norm, and in V the square integral over a y-interval B. Then the Euclidean structure in $U \otimes V$ defined by (5) is the square integral over the rectangle A $\times$ B.

We show now how to use the representation of $U \otimes V$ as $\mathscr{L}(U', V)$ to derive the Euclidean structure in $U \otimes V$ from the Euclidean structure of U and V. Here $U' = U$, so $U \otimes V$ is $\mathscr{L}(U, V)$.

Let M and L belong to $\mathscr{L}(U, V)$, and let L^* be the adjoint of L. We define

$$(M, L) = \operatorname{tr} L^* M. \tag{6}$$

Clearly this depends bilinearly on M and L. In terms of orthonormal bases, M and L can be expressed as matrices (m_{ij}) and (l_{ij}), and L^* as the transpose (l_{ji}). Then

$$\operatorname{tr} L^* M = \sum l_{ji} m_{ji}.$$

Setting L = M, we get

$$\| M \|^2 = (M, M) = \sum m_{ji}^2,$$

consistent with our previous definition.

Complex Euclidean structures can be handled the same way.

All of the foregoing is pretty dull stuff. To liven it up, here is a one-line proof of Schur's peculiar theorem from Chapter 10, Theorem 7: if $A = (A_{ij})$ and $B = (B_{ij})$ are positive symmetric $n \times n$ matrices then so is their entry by entry product $M = (A_{ij}B_{ij})$.

Proof. It was observed in Theorem 6 of Chapter 10 that every positive symmetric matrix can be written as a Gram matrix:

$$A_{ij} = (u_i, u_j), \qquad u_i \subset U, \text{ linearly independent,}$$
$$B_{ij} = (v_i, v_j), \qquad v_i \subset V, \text{ linearly independent.}$$

Now define g_i in $U \otimes V$ to be $u_i \otimes v_i$; by (5), $(g_i, g_j) = (u_i, u_j)(v_i, v_j) = A_{ij}B_{ij}$. This shows that M is a Gram matrix, therefore nonnegative.   □

EXERCISE 3.   Show that if $\{u_i\}$ and $\{v_j\}$ are linearly independent, so are $u_i \otimes v_i$. Show that M_{ij} is positive.

EXERCISE 4.   Let u be a twice differentiable function of $x_1, \ldots, x_n$ defined in a neighborhood of a point p, where u has a local minimum. Let (A_{ij}) be a symmetric, nonnegative matrix. Show that

$$\sum A_{ij} \frac{\partial^2 u}{\partial x_i \partial x_j}(p) \geq 0.$$

APPENDIX 5

Lattices

Definition. A lattice is a subset L of a linear space X over the reals with the following properties:

(i) L is closed under addition and subtraction; that is, if x and y belong to L, so do $x + y$ and $x - y$.

(ii) L is discrete, in the sense that any bounded (as measured in any norm) set of X contains only a finite number of points of L.

An example of a lattice in $\mathbb{R}^n$ is the collection of points $x = (x_1, \ldots, x_n)$ with integer components x_i. The basic theorem of the subject says that this example is typical.

Theorem 1. Every lattice has an integer basis, that is, a collection of vectors in L such that every vector in the lattice can be expressed uniquely as a linear combination of basis vectors with integer coefficients.

Proof. The dimension of a lattice L is the dimension of the linear space it spans. Let L be k-dimensional, and let $p_1, \ldots, p_k$ be a basis in L for the span of L; that is, every vector t in L can be expressed uniquely as

$$t = \sum a_j p_j, \qquad a_j \text{ real.} \tag{1}$$

Consider now the subset of those vectors t in L which are of form (1) with a_j between 0 and 1:

$$0 \leq a_j \leq 1, \qquad j = 1, \ldots, k. \tag{2}$$

This set is not empty, for its contains all vectors with $a_j = 0$ or 1. Since L is discrete, there are only a finite number of vectors t in L of this form; denote by q_1 that vector t of form (1), (2) for which a_1 is positive and as small as possible.   □

EXERCISE 1.   Show that a_1 is a rational number.

Now replace p_1 by q_1 in the basis; every vector t in L can be expressed uniquely as

$$t = b_1 q_1 + \sum_2^k b_j p_j, \qquad b_j \text{ real.} \tag{3}$$

We claim that b_1 occurring in (3) is an integer; for if not, we can subtract a suitable integer multiple of q_1 from t so that the coefficient b_1 of q_1 lies *strictly* between 0 and 1:

$$0 < b_1 < 1.$$

If then we substitute into (3) the representation (1) of q_1 in terms of $p_1, \ldots, p_k$ and add or subtract suitable integer multiples of $p_2, \ldots, p_k$, we find that the p_1 coefficient of t is positive and *less* than the p_1 coefficient of q_1. This contradicts our choice of q_1.

We complete our proof by an induction on k, the dimension of the lattice. Denote by L_0 the subset of L consisting of those vectors t in L whose representation of form (3), b_1 is zero. Clearly L_0 is a sublattice of L of dimension $k - 1$; by induction hypothesis L_0 has an integer basis $q_2, \ldots, q_k$. By (3), $q_1, \ldots, q_k$ is an integer basis of L.   □

An integer basis is far from unique as is shown in the following theorem.

Theorem 2.   Let L be an n-dimensional lattice in $\mathbb{R}^n$. Let $q_1, \ldots, q_n$ and $r_1, \ldots, r_n$ be two integer bases of L; denote by Q and R the matrices whose columns are q_1 and r_i, respectively. Then

$$Q = MR,$$

where M is a *unimodular* matrix, that is, a matrix with integer entries whose determinant is plus or minus 1.

EXERCISE 2.   (i) Prove Theorem 2.
(ii) Show that unimodular matrices form a group under multiplication.

Definition.   Let L be a lattice in a linear space X. The dual of L, denoted as L', is the subset of the dual X' of X consisting of those vectors ξ for which (t, ξ) is an integer for all t in L.

Theorem 3. (i) The dual of an n-dimensional lattice in an n-dimensional linear space is an n-dimensional lattice.

(ii) $L'' = L$.

EXERCISE 3.   Prove Theorem 3.

EXERCISE 4.   Show that L is discrete if and only if there is a positive number d such that the ball of radius d centered at the origin contains no other point of L.

APPENDIX 6

Fast Matrix Multiplication

How many scalar multiplications are needed to form the product C of two $n \times n$ matrices A and B? Since each entry of C is the product of a row of A with a column of B, and since C has n^2 entries, we need n^3 scalar multiplications, as well as $n^3 - n^2$ additions. It was a great discovery of Volker Strassen that there is a way of multiplying matrices that uses many fewer scalar multiplications and additions. The crux of the idea lies in a clever way of multiplying 2×2 matrices:

$$A = \begin{pmatrix} a_{11} & a_{12} \\ a_{21} & a_{22} \end{pmatrix}, \qquad B = \begin{pmatrix} b_{11} & b_{12} \\ b_{21} & b_{22} \end{pmatrix},$$

$$AB = C = \begin{pmatrix} c_{11} & c_{12} \\ c_{21} & c_{22} \end{pmatrix},$$

$c_{11} = a_{11}b_{11} + a_{12}b_{21}, c_{12} = a_{11}b_{12} + a_{12}b_{22}$, and so on. Define

$$
\begin{aligned}
\mathrm{I} &= (a_{11} + a_{22})(b_{11} + b_{22}), \\
\mathrm{II} &= (a_{21} + a_{22})b_{11}, \\
\mathrm{III} &= a_{11}(b_{12} - b_{22}), \\
\mathrm{IV} &= a_{22}(b_{21} - b_{11}), \\
\mathrm{V} &= (a_{11} + a_{12})b_{22}, \\
\mathrm{VI} &= (a_{21} - a_{11})(b_{11} + b_{12}), \\
\mathrm{VII} &= (a_{12} - a_{22})(b_{21} + b_{22}).
\end{aligned}
\tag{1}
$$

Linear Algebra and Its Applications, Second Edition, by Peter D. Lax
Copyright © 2007 John Wiley & Sons, Inc.

A straightforward but tedious calculation shows that the entries of the product matrix C can be expressed as follows:

$$c_{11} = \text{I} + \text{IV} - \text{V} + \text{VII}, \qquad c_{12} = \text{III} + \text{V},$$
$$c_{21} = \text{II} + \text{IV}, \qquad c_{22} = \text{I} + \text{III} - \text{II} + \text{VI}. \tag{2}$$

The point is that whereas the standard evaluation of the entries in the product matrix uses two multiplications per entry, therefore a total of eight, the seven quantities in (1) need only seven multiplications. The total number of additions and subtractions needed in (1) and (2) is 18.

The formulas (1) and (2) in no way use the commutativity of the quantities a and b. Therefore, (1) and (2) can be used to multiply 4×4 matrices by interpreting the entries a_{ij} and b_{ij} as 2×2 block entries. Proceeding recursively in this fashion, we can use (1) and (2) to multiply any two matrices A and B of order 2^k.

How many scalar multiplications $M(k)$ have to be carried out in this scheme? In multiplying two square matrices of order 2^k we have to perform seven multiplications of blocks of size $2^{k-1} \times 2^{k-1}$. This takes $7M(k-1)$ scalar multiplications. So

$$M(k) = 7M(k-1).$$

Since $M(0) = 1$, we deduce that

$$M(k) = 7^k = 2^{k \log_2 7} = n^{\log_2 7}, \tag{3}$$

where $n = 2^k$ is the order of the matrices to be multiplied.

Denote by $A(k)$ the number of scalar additions–subtractions needed to multiply two matrices of order 2^k using Strassen's algorithm. We have to perform 18 additions and 7 multiplications of blocks of size $2^{k-1} \times 2^{k-1}$; the latter takes $7A(k-1)$ additions, the former $18(2^{k-1})^2 = 9 \cdot 2^{2k-1}$. So altogether

$$A(k) = 9 \cdot 2^{2k-1} + 7A(k-1).$$

Introduce $B(k) = 7^{-k}A(k)$; then the above recursion can be rewritten as

$$B(k) = \frac{9}{2}\left(\frac{4}{7}\right)^k + B(k-1).$$

Summing with respect to k we get, since $B(0) = 0$,

$$B(k) = \frac{9}{2}\sum_{1}^{k}\left(\frac{4}{7}\right)^j < \left(\frac{9}{2}\right)\left(\frac{4}{3}\right) = 6;$$

therefore

$$A(k) \leq 6 \times 7^k = 6 \times 2^{k \log_2 7} = 6n^{\log_2 7} \tag{4}$$

Since $\log_2 7 = 2.807 \cdots$ is less than 3, the number of scalar multiplications required in Strassen's algorithm is for n large, very much less than n^3 the number of scalar multiplications required in the standard way of multiplying matrices.

Matrices whose order is not a power of 2 can be turned into one by adjoining a suitable number of 1 s on the diagonal.

Refinements of Strassen's idea have led to further reduction of the number of scalar multiplications needed to multiply two matrices. It has been conjectured that for any positive ϵ there is an algorithm that computes the product of two $n \times n$ matrices using cost $n^{2+\epsilon}$ scalar multiplication, where the contant depends on ϵ.

APPENDIX 7

Gershgorin's Theorem

This result can be used to give very simple estimates on the location of the eigenvalues of a matrix, crude or accurate depending on the circumstances.

Gershgorin Circle Theorem.   Let A be an $n \times n$ matrix with complex entries. Decompose it as

$$A = D + F, \tag{1}$$

where D is the diagonal matrix equal to the diagonal of A; F has zero diagonal entries. Denote by d_i the ith diagonal entry of D, and by f_i the ith row of F. Define the circular disc C_i to consist of all complex numbers z satisfying

$$|z - d_i| \leq |f_i|_1, \qquad i = 1, \ldots, n. \tag{2}$$

The 1-norm of a vector f is the sum of the absolute values of its components; see Chapter 14. Claim: every eigenvalue of A is contained in one of the discs C_i.

Proof. Let u be an eigenvector of A,

$$Au = \lambda u, \tag{3}$$

normalized as $|u|_\infty = 1$, where the ∞-norm is the maximum of the absolute value of the components u_j of u. Clearly, $|u_j| \leq 1$ for j and $u_i = 1$ for some i. Writing A = D + F in (3), the ith component can be written as $d_i + f_i u = \lambda$, which can be rewritten as

$$\lambda - d_i = f_i u.$$

Linear Algebra and Its Applications, Second Edition, by Peter D. Lax
Copyright © 2007 John Wiley & Sons, Inc.

The absolute value of the product fu is $\leq |f|_1 |u|_\infty$, so

$$|\lambda - d_i| \leq |f_i|_1 |u|_\infty = |f_i|_1. \qquad \Box$$

EXERCISE.   Show that if C_i is disjoint from all the other Gershgorin discs, then C_i contains exactly one eigenvalue of A.

In many iterative methods for finding the eigenvalues of a matrix A, A is transformed by a sequence of similarity transformations into A_k so that A_k tends to a diagonal matrix. Being similar to A, each A_k has the same eigenvalues as A. Gershgorin's theorem can be used to estimate how closely the diagonal elements of A_k approximate the eigenvalues of A.

APPENDIX 8

The Multiplicity of Eigenvalues

The set of $n \times n$ real, self-adjoint matrices forms a linear space of dimension $N = n(n+1)/2$. We have seen at the end of Chapter 9 that the set of degenerate matrices, that is, ones with multiple eigenvalues, form a surface of codimension 2, that is, of dimension $N - 2$. This explains the phenomenon of "avoided crossing," that is, in general, self-adjoint matrices in a one-parameter family have all distinct eigenvalues. By the same token a two-parameter family of self-adjoint matrices ought to have a good chance of containing a matrix with a multiple eigenvalue. In this appendix we state and prove such a theorem about two parameter families of the following form:

$$aA = bB + cC, \quad a^2 + b^2 + c^2 = 1. \tag{1}$$

Here A, B, C are real, self-adjoint $n \times n$ matrices, and a, b, c are real numbers.

Theorem (Lax). If $n \equiv 2 \pmod 4$, then there exist a, b, c such that (1) is degenerate, that is, has a multiple eigenvalue.

Proof. Denote by $\mathcal{N}$ the set of all nondegenerate matrices. For any N in $\mathcal{N}$ denote by $k_1 < k_2 < \cdots < k_n$ the eigenvalues of N arranged in increasing order and by u_j the corresponding normalized eigenvectors:

$$Nu_j = k_j u_j, \quad \| u_j \| = 1, j = 1, \ldots, n. \tag{2}$$

Note that each u_j is determined only up to a factor ± 1.

Let $0 \le t \le 2\pi$, be a closed curve in $\mathcal{N}$. If we fix $u_j(0)$, then the normalized eigenvector $u_j(t)$ can be determined uniquely as continuous functions of t. Since for a closed curve $N(2\pi) = N(0)$,

$$u_j(2\pi) = \tau_j u(0), \quad \tau_j = \pm 1. \tag{3}$$

Linear Algebra and Its Applications, Second Edition, by Peter D. Lax
Copyright © 2007 John Wiley & Sons, Inc.

The quantities $\tau_j, j = 1, \ldots, m$ are functionals of the curve $N(t)$. Clearly:

(i) Each τ_j is invariant under homotopy, that is, continuous deformation in $\mathcal{N}$.

(ii) For a constant curve, that is, $N(t)$ independent of t, each $\tau_j = 1$.

$$N(t) = \cos tA + \sin tB, \qquad 0 \le t \le 2\pi \tag{4}$$

is a closed curve in $\mathcal{N}$. Note that N is periodic, and

$$N(t + \pi) = -N(t).$$

It follows that

$$\lambda_j(t + \pi) = -\lambda_{n-j+1}(t)$$

and that

$$u_j(t + \pi) = \rho_j u_{n-j+1}(t), \tag{5}$$

where $\rho_j = \pm 1$. Since u_j is a continuous function of t, so is ρ_j; but since ρ_j can only take on discrete values, it is independent of t.

For each value of t, the eigenvectors $u_1(t), \ldots, u_n(t)$ form an ordered basis. Since they change continuously they retain their orientation. Thus the two ordered bases

$$u_1(0), \ldots, u_n(0) \quad \text{and} \quad u_1(\pi), \ldots, u_n(\pi) \tag{6}$$

have the same orientation. By (5),

$$u_1(\pi), \ldots, u_n(\pi) = \rho_1 u_n(0), \ldots, \rho_n u_1(0). \tag{6'}$$

Reversing the order of a basis for n even is the same as $n/2$ transpositions. Since each transposition reverses orientation, for $n \equiv 2 \pmod 4$ we have an odd number of transpositions. So in order for (6) and (6)' to have the same orientation,

$$\prod_1^n \rho_j = -1.$$

Writing this product as

$$\prod_1^{n/2} \rho_j \rho_{n-j+1} = -1,$$

we conclude there is an index k for which

$$\rho_k \rho_{n-k+1} = -1. \tag{7}$$

Using (5) twice, we conclude that

$$u_k(2\pi) = \rho_k u_{n-k+1}(\pi) = \rho_k \rho_{n-k+1} u_k(0).$$

This shows, by (3), that $\tau_k = \rho_k \rho_{n-k+1}$, and so by (7) that $\tau_k = -1$. This proves that the curve (4) cannot be deformed continuously in $\mathcal{N}$ to a point. But the curve (4) is the equator on the unit sphere $a^2 + b^2 + c^2 = 1$; if all matrices of form (1) belonged to $\mathcal{N}$, the equator could be contracted on the unit sphere to a point, contradicting $\tau_k = -1$ ☐

EXERCISE.   Show that if $n \equiv 2 (\mathrm{mod}\, 4)$, there are no $n \times n$ real matrices A, B, C not necessarily self-adjoint, such that all their linear combinations (1) have real and distinct eigenvalues.

Friedland, Robbin, and Sylvester have extended the theorem to all $n \equiv \pm 3, 4 (\mathrm{mod}\, 8)$, and have shown that it does not hold when $n \equiv 0, \pm 1 (\mathrm{mod}\, 8)$.

These results are of interest in the theory of hyperbolic partial differential equations.

BIBLIOGRAPHY

Friedland, S., Robbin, J., and Sylvester, J. On the crossing rule, *CPAM* **37** (1984), 19–37.

Lax, P. D. The multiplicity of eigenvalues, *Bull. AMS* **6** (1982), 213–214.

APPENDIX 9

The Fast Fourier Transform

In this appendix we study functions defined on a finite number of points arranged periodically. A function on these points is a sequence of numbers, generally complex, $u_1, \ldots, u_n$. In many applications, u_k stands for the value at $x = k/n$ of a periodic function $u(x)$ defined on the unit interval $[0, 1]$. The vector $(u_1, \ldots, u_n) = u$ is a discrete approximation of the function $u(x)$.

In this appendix we shall analyze the cyclic shift mapping S defined by

$$S(u_1, \ldots, u_n) = (u_2, u_3, \ldots, u_n, u_1).$$

The mapping S preserves the Euclidean norm

$$\|u_1\|^2 = \sum |u_k|^2;$$

such mappings are called unitary (see Chapter 8). According to Theorem 9 in Chapter 8, unitary mappings have a complete set of pairwise orthogonal eigenvectors e_j, with eigenvalues λ_j that are complex numbers of absolute value 1. We shall now calculate the eigenvector and eigenvalues of 5.

From the eigenvalue equation $Se = \lambda e$, we deduce that the components $(u_1, \ldots, u_n)$ of e satisfy

$$u_2 = \lambda u_1, \qquad u_3 = \lambda u_2, \ldots, u_n = \lambda u_{n-1}, \ u_1 = \lambda u_n.$$

We set $u_1 = \lambda$ and then deduce from the first $n - 1$ equations above that

$$u_2 = \lambda^2, \qquad u_3 = \lambda^3, \ldots, u_n = \lambda^n,$$

Linear Algebra and Its Applications, Second Edition, by Peter D. Lax
Copyright © 2007 John Wiley & Sons, Inc.

and from the last equation that $1 = \lambda^n$. This shows that the eigenvalues λ are the n roots of unity $\lambda_j = \exp\left(\frac{2\pi i}{n} j\right)$, and the corresponding eigenvectors

$$e_j = (\lambda_j, \lambda_j^2, \lambda_j^3, \ldots, \lambda_j^n). \tag{1}$$

Each eigenvector e_k has norm $\|e_j\| = \sqrt{n}$.

Every vector $u = (u_1, \ldots, u_n)$ can be expressed as a linear combination of eigenvectors:

$$u = \sum_1^n a_j e_j. \tag{2}$$

Using the explicit expression (1) of the eigenvector, we can rewrite (2) as

$$u_k = \sum_1^n a_j \exp\left(\frac{2\pi i}{n} jk\right). \tag{2'}$$

Using the orthogonality of the eigenvectors, along with their norm $\|e_j\| = \sqrt{n}$, we can express the coefficients a_j as

$$a_j = (u, e_j)/n = \frac{1}{n}\sum_1^n u_k \exp\left(-\frac{2\pi i}{n} jk\right). \tag{3}$$

It is instructive to compare formulas (2)$'$ and (3) to the expansion of periodic functions in terms of their Fourier series. But first we have to rewrite equation (2)$'$ as follows. Suppose n is odd; we rewrite the last $(n-1)/2$ terms in the sum (2)$'$ by introducing a new index of summation l related to j by $j = n - l$. Then

$$\exp\left(\frac{2\pi i}{n} jk\right) = \exp\left(\frac{2\pi i}{n}(n-l)k\right) = \exp\left(-\frac{2\pi i}{n} lk\right).$$

Setting this into (2)$'$, we can rewrite it as

$$u_k = \sum_{-(n-1)/2}^{(n-1)/2} a_j \exp\left(\frac{2\pi i}{n} jk\right), \tag{4}$$

where we have reverted to denote the index of summation by j and where a_j is defined by (3). A similar formula holds when n is even.

Let $u(x)$ be a periodic function of period 1. Its Fourier series representation is

$$u(x) = \sum b_j \exp(2\pi ijx), \tag{5}$$

where the Fourier coefficients are given by the formula

$$b_j = \int_0^1 u(x) \exp(-2\pi ijx)dx. \tag{6}$$

Setting $x = k/n$ into (5) gives

$$u\left(\frac{k}{n}\right) = \sum b_j \exp\left(\frac{2\pi i}{n}jk\right). \tag{5'}$$

The integral (6) can be approximated by a finite sum at the equidistant points $x_k = k/n$:

$$b_j \simeq \frac{1}{n}\sum_1^n u\left(\frac{k}{n}\right) \exp\left(-\frac{2\pi i}{n}jk\right). \tag{6'}$$

Suppose u is a smooth periodic function, say d times differentiable. Then the $(n-1)/2$ section of the Fourier series (5) is an excellent approximation to $u(x)$:

$$u(x) = \sum_{-(n-1)/2}^{(n-1)/2} b_j \exp(2\pi ijx) + O(n^{-d}) \tag{7}$$

Similarly, the approximating sum on the right-hand side of (6)' differs by $O(n^{-d})$ from b_n. It follows that if in (3) we take $u_k = u\left(\frac{k}{n}\right)$, a_j differs from b_j by $O(n^{-d})$.

When u is a smooth periodic function, its derivatives may be calculated by differentiating its Fourier series term by term:

$$\partial_x^m u(x) = \sum b_j(2\pi ij)^m \exp(2\pi ijx).$$

The truncated series is an excellent approximation to $\partial_x^m u$. It follows therefore that

$$\sum_{(n-1)/2}^{(n-1)/2} a_j(2\pi ij)^m \exp\left(\frac{2\pi i}{n}jk\right)$$

is an excellent approximation to $\partial_x^m u\left(\frac{k}{n}\right)$, provided that u_k in (3) is taken as $u\left(\frac{k}{n}\right)$. Therein lies the utility of the finite Fourier transform: It can be used to obtain highly accurate approximations to derivatives of smooth periodic functions, which can then be used to construct very accurate approximate solutions of differential equations.

On the other hand, operations such as multiplication of u by a given smooth function can be carried out fast and accurately when u is represented by its values at the points k/n. Since in the course of a calculation the operation of differentiation and multiplication may have to be carried out alternately many times, it is of greatest importance that the finite Fourier transform, the passage from the point values $(u_1, \ldots, u_k)$ to the set of Fourier coefficients $(a_1, \ldots, a_n)$ and its inverse, be carried out as fast as possible. The rest of this appendix is devoted to this task.

Using formula (3), n multiplications are needed for the computation of each Fourier coefficient a_j, provided that the roots of unity $\exp\left(-\frac{2\pi i}{n}l\right), \ldots, l = 1, \ldots, n$, are precomputed. Since there are n coefficients to compute, we need a total of n^2 multiplications to carry out the finite Fourier transform.

It is therefore a considerable surprise that the finite Fourier transform can be carried out performing only $n \log n$ multiplications, where the log is to base 2. Here is how it works:

Assume that n is even. Denote the array of coefficients $(a_1, \ldots, a_n)'$ by A and the array $(u_1, \ldots, u_n)'$ by U. Then relation (3) can be written in matrix form

$$A = \frac{1}{n}F_nU, \tag{8}$$

where the $n \times n$ matrix F_n is $F_{ij} = \omega^{jk}$, where $\omega = \exp\left(-\frac{2\pi i}{n}\right)$. The colums of F are $\bar{e}_1, \ldots, \bar{e}_n$, so we can write F as

$$F_n = (\bar{e}_1, \ldots, \bar{e}_n).$$

Reorder the columns of F by first listing all of even order $\bar{e}_2, \ldots, \bar{e}_n$ and then the odd ones, $\bar{e}_1, \ldots, \bar{e}_{n-1}$. Denote the rearranged matrix by F_n^r:

$$F_n^r = (\bar{e}_2, \ldots, \bar{e}_n, \bar{e}_1, \ldots, \bar{e}_{n-1}).$$

Fully written out, F_n^r looks like this; here we have used the facts that $\omega^n = 1$, $\omega^{n/2} = -1$:

$$
\left(
\begin{array}{cccc|cccc}
\omega^2 & \omega^4 & \cdots & 1 & \omega & \omega^3 & \cdots & \omega^{-1} \\
\omega^4 & \omega^8 & \cdots & 1 & \omega^2 & \omega^6 & \cdots & \omega^{-2} \\
\vdots & \vdots & & \vdots & \vdots & \vdots & & \vdots \\
1 & 1 & \cdots & 1 & -1 & -1 & & -1 \\
\hline
\omega^2 & \omega^4 & \cdots & 1 & -\omega & & \cdots & -\omega^{-1} \\
\omega^4 & \omega^8 & \cdots & 1 & -\omega^2 & & \cdots & -\omega^{-2} \\
\vdots & \vdots & & \vdots & \vdots & & & \vdots \\
1 & 1 & \cdots & 1 & 1 & & \cdots & 1
\end{array}
\right) = F_n^r.
$$

This suggests that we represent F_n^r in 2×2 block form

$$F_n^r = \left(\begin{array}{c|c} B_{11} & B_{12} \\ \hline B_{21} & B_{22} \end{array}\right). \tag{9}$$

We can immediately identify B_{11} and B_{21} as $F_{n/2}$. We claim that

$$B_{12} = DF_{n/2}, \qquad B_{22} = -DF_{n/2},$$

where D is an $n/2 \times n/2$ diagonal matrix with diagonal entries $\omega^{-1}, \omega^{-2}, \ldots, \omega^{-n/2}$. To see this, note that the first row of B_{12} can be written as $\omega^{-1}(\omega^2, \omega^4, \ldots, 1)$, the second row as $\omega^{-2}(\omega^4, \omega^8, \ldots, 1)$, and so on. This shows that $B_{12} = DF_{n/2}$; since $B_{22} = -B_{12}$, the rest follows.

Rearrange the components of the vector A and U, putting first all the even ones, followed by the odd ones:

$$U^r = \begin{pmatrix} U_{even} \\ U_{odd} \end{pmatrix}, \qquad A^r = \begin{pmatrix} A_{even} \\ A_{odd} \end{pmatrix}.$$

Then, (8) can be rewritten as

$$A^r = \frac{1}{n} F_n^r U^r.$$

Break A^r and U^r into blocks; by (9):

$$\begin{pmatrix} A_{even} \\ A_{odd} \end{pmatrix} = \frac{1}{n} \begin{pmatrix} F_{n/2} & DF_{n/2} \\ F_{n/2} & -DF_{n/2} \end{pmatrix} \begin{pmatrix} U_{even} \\ U_{odd} \end{pmatrix}.$$

So

$$\begin{aligned} A_{even} &= \frac{1}{n}(F_{n/2}U_{even} + DF_{n/2}U_{odd}), \\ A_{odd} &= \frac{1}{n}(F_{n/2}U_{even} - DF_{n/2}U_{odd}). \end{aligned} \tag{10}$$

Denote by $M(n)$ the number of multiplications needed to calculate the vector A, given U. Using formula (10), see that we have to apply $F_{n/2}$ to two different vectors which takes $2 M(n/2)$ multiplications; we have to apply D twice, another n multiplications, and we have to multiply by $1/n$, another n multiplications. So altogether

$$M(n) = 2M(n/2) + 2n. \tag{11}$$

If n is a power of 2, we have similar relations between $M(n/2)$ and $M(n/4)$, and so on; putting these relations together, we get

$$M(n) = n \log_2 n, \tag{12}$$

where $\log_2$ denotes logarithm to base 2.

There are some additional computational expenses; additions and rearrangements of vectors. The total amount of work is $5n \log_2 n$ flops (floating point operations).

The inverse operation, expressing u_k in terms of the a_j [see (2)'], is the same, except that ω is replaced by $\bar{\omega}$, and there is no division by n.

There is an interesting discussion of the history of the Fast Fourier Transform in the 1968 Arden House Workshop.

When Cooley and Tukey's paper on the Fast Fourier Transform appeared, *Mathematical Reviews* reviewed it by title only; the editors did not grasp its importance.

BIBLIOGRAPHY

Cooley, J. W., and Tukey, J. W. An algorithm for the machine calculation of complex Fourier Series, *Math. Comp.* **19** (1965), 297–301.

The 1968 Arden House Workshop on Fast Fourier Transform, *IEEE Trans.* Audio Electroacoust. **AU-17**, (1969).

Van Loan, C. F. *Introduction to Scientific Computing*, Prentice Hall, Englewood Cliffs, NJ.

The Spectral Radius

Let X be a finite-dimensional Euclidean space, and let A be a linear mapping of X into X. Denote by $r(A)$ the spectral radius of A:

$$r(A) = \max_i |a_i|, \tag{1}$$

where a_i rangers over all eigenvalues of A. We claim that

$$\lim_{j \to \infty} ||A^j||^{1/j} = r(A), \tag{2}$$

where $||A^j||$ denotes the norm of the jth power of A.

Proof. A straightforward estimate [see inequality $(48)_j$ of Chapter 7] shows that

$$||A^j||^{1/j} \geq r(A). \tag{3}$$

We shall show that

$$\lim \sup ||A^j||^{1/j} \leq r(A). \tag{4}$$

Combining (3) and (4) gives (2).

We can introduce an orthonormal basis in X, thereby turning X into $\mathbb{C}^n$, with the standard Euclidean norm $(|x_1|^2 + \cdots + |x_n|^2)^{1/2}$, and A into an $n \times n$ matrix with complex entries. We start with the *Schur factorization* of A: $\qquad \square$

Theorem 1. Every square matrix A with complex entries can be factored as

$$A = QTQ^*, \tag{5}$$

Linear Algebra and Its Applications, Second Edition, by Peter D. Lax
Copyright © 2007 John Wiley & Sons, Inc.

where Q is unitary and T is upper triangular:

$$t_{ij} = 0 \qquad \text{for } i > j.$$

(5) is called a Schur factorization of A.

Proof. If A is a normal matrix, then according to Theorem 8 of Chapter 8 it has a complete set of orthogonal eigenvectors $q_1, \ldots, q_n$, with eigenvalues $a_1, \ldots, a_n$:

$$A q_k = a_k q_k. \tag{6}$$

Choose the q_k to have norm 1, and define the matrix Q as

$$Q = (q_1, \ldots, q_n).$$

Since the columns of Q are pairwise orthogonal unit vectors, Q is unitary. Equations (6) can be expressed in terms of Q as

$$AQ = QD, \tag{6}'$$

where D is the diagonal matrix with $D_{kk} = a_k$. Multiplying $(6)'$ by Q^* on the right gives a Schur factorization of A, with $T = D$.

For arbitrary A we argue inductively on the order n of A.

We have shown at the beginning of Chapter 6 that every $n \times n$ matrix A has at least one eigenvalue a, possibly complex:

$$A_q = a_q. \tag{7}$$

Choose q to have norm 1, and complete it to an orthonormal basis $q_1, \ldots, q_n, q_1 = q$, and define the matrix U to have columns $q_1, \ldots, q_n$:

$$U = (q_1, \ldots, q_n).$$

Clearly, U is a unitary matrix, and the first column of AU is aq:

$$AU = (aq, c_2, \ldots, c_n). \tag{7}'$$

The adjoint U^* of U has the form

$$U^* = \begin{pmatrix} \bar{q}'_1 \\ \vdots \\ \bar{q}'_n \end{pmatrix},$$

where the row vector q_j' is the transpose of the column vector q_j, and $\bar{q}_j'$ is its complex conjugate. Take the product of (7)$'$ with U*; since the rows $\bar{q}_2', \ldots, \bar{q}_n'$ of U* are orthogonal to q_1, we get, in block notation,

$$U^*AU = \begin{pmatrix} a & B \\ 0 & C \end{pmatrix}, \tag{8}$$

where B is a $1 \times (n-1)$ matrix and C an $(n-1) \times (n-1)$ matrix.

By the induction hypothesis, C has a Schur factorization

$$C = RT_1R^*, \tag{9}$$

where R is a unitary $(n-1) \times (n-1)$ matrix, and T_1 an upper triangular $(n-1) \times (n-1)$ matrix. Set

$$Q = U\begin{pmatrix} 1 & 0 \\ 0 & R \end{pmatrix}; \tag{10}$$

where Q, the product of two unitary matrices, is unitary, and

$$Q^* = \begin{pmatrix} 1 & 0 \\ 0 & R^* \end{pmatrix}U^*. \tag{10*}$$

Using (10) and (10)*, we get

$$Q^*AQ = \begin{pmatrix} 1 & 0 \\ 0 & R^* \end{pmatrix}U^*AU\begin{pmatrix} 1 & 0 \\ 0 & R \end{pmatrix}. \tag{11}$$

Using (8), we can write the right-hand side of (11) as

$$\begin{pmatrix} 1 & 0 \\ 0 & R^* \end{pmatrix}\begin{pmatrix} a & B \\ 0 & C \end{pmatrix}\begin{pmatrix} 1 & 0 \\ 0 & R \end{pmatrix}.$$

Carrying out the block multiplication of the three matrices gives

$$\begin{pmatrix} a & BR \\ 0 & R^*CR \end{pmatrix}.$$

It follows from (9) that $R^*CR = T_1$; so we conclude from (11) that

$$Q^*AQ = \begin{pmatrix} a & RR \\ 0 & T_1 \end{pmatrix} = T,$$

an upper triangular $n \times n$ matrix. Multiplying this equation by Q on the left and Q* on the right gives the Schur factorization (5) of A. $\square$

The Schur factorization (5) shows that A and T are similar. We have seen in Chapter 6 that similar matrices have the same eigenvalues. Since the eigenvalues of an upper triangular matrix are its diagonal entries, T in (5) is of the form

$$T = D_A + S, \tag{12}$$

where D_A is a diagonal matrix whose diagonal entries are the eigenvalues $a_1, \dots, a_n$ of A, and S is an upper triangular matrix whose diagonal entries are zero.

EXERCISE 1. Prove that the eigenvalues of a upper triangular matrix are its diagonal entries.

We show now how to use Schur factorization of A to obtain the estimate (3) of the norm of A^j. Let D be a diagonal matrix

$$D = \begin{pmatrix} d_1 & \cdot & 0 \\ & \ddots & \\ 0 & & d_n \end{pmatrix},$$

whose diagonal entries, all nonzero, will be chosen later. According to the rules of matrix multiplication,

$$(D^{-1}TD)_{ij} = t_{ij} \frac{d_j}{d_i}. \tag{13}$$

We choose $d_j = e^j$, where ϵ is some positive number, and denote by D_ϵ the diagonal matrix with diagonal entries $d_j = e^j$. We denote by T_ϵ the matrix

$$T_\epsilon = D_\epsilon^{-1}TD_\epsilon. \tag{14}$$

According to (13), the entries of T_ϵ are $t_{ij}\,e^{j-i}$. This shows that the diagonal entries of T_ϵ are the same as those of T and that the off-diagonal entries are those of T, multiplied by a *positive* integer power of ϵ. We split T_ϵ into the sum of its diagonal and off-diagonal part; we get, analogous to (12),

$$T_\epsilon = D_A + S_\epsilon. \tag{12}\epsilon$$

The Euclidean norm of the diagonal matrix D_A is the maximum of the absolute value of its diagonal entries:

$$\|D_A\| = \max |a_j| = r(A).$$

Since each entry of S_ϵ contains ϵ raised to a positive integer power, for $\epsilon < 1$,

$$\|S_\epsilon\| \le c\epsilon,$$

where c is some positive number. Using the triangle inequality, we conclude that

$$||T_\epsilon|| = ||D_A + S_\epsilon|| \le r(A) + c\epsilon. \tag{15}$$

It follows from (14) that

$$T = D_\epsilon T_\epsilon D_\epsilon^{-1}.$$

Set this in the Schur factorization (5) of A:

$$A = QD_\epsilon T_\epsilon D_\epsilon^{-1} Q^*. \tag{16}$$

Denote QD_ϵ by M. Since Q is unitary, $Q^* = Q^{-1}$, and (16) can be rewritten as

$$A = MT_\epsilon M^{-1}. \tag{16}'$$

It follows that

$$A^j = MT_\epsilon^j M^{-1}.$$

Using the multiplicative inequality for the matrix norm, we obtain the inequality

$$||A^j|| \le ||M|| \, ||M^{-1}|| \, ||T_\epsilon||^j.$$

Taking the jth root, we get

$$||A^j||^{1/j} \le m^{1/j} ||T_\epsilon||,$$

where $m = ||M|| \, ||M^{-1}||$. Using the estimate (15) on the right gives

$$||A^j||^{1/j} \le m^{1/j}(r(A) + c\epsilon).$$

Now let j tend to ∞; we get that

$$\limsup ||A^j||^{1/j} \le r(A) + c\epsilon.$$

Since this holds for all positive $\epsilon < 1$, we have

$$\limsup ||A^j||^{1/j} \le r(A),$$

as asserted in (4). As noted there, (4) and (3) simply (2).   □

EXERCISE 2. Show that the Euclidean norm of a diagonal matrix is the maximum of the absolute value of its eigenvalues.

Note 1. The crucial step in the argument is to show that every matrix is similar to an upper triangular matrix. We could have appealed to the result in Chapter 6 that every matrix is similar to a matrix in Jordan form, since matrices in Jordan form are upper triangular. Since the proof of the Jordan form is delicate, we preferred to base the argument on Schur factorization, whose proof is more robust.

EXERCISE 3.   Prove the analogue of relation (2),

$$\lim_{j \to \infty} |A^j|^{1/j} = r(A), \tag{17}$$

when A is a linear mapping of any finite-dimensional *normed*, linear space X (see Chapters 14 and 15).

Note 2. Relation (17) holds for mappings in infinite-dimensional spaces as well. The proof given above relies heavily on the spectral theory of linear mappings in finite-dimensional spaces, which has no infinite-dimensional analogue. We shall therefore sketch another approach to relation (17) that has a straightforward extension to infinite-dimensional normed linear spaces. This approach is based on the notion of matrix-valued analytic functions.

Definition 1.   Let $z = x + iy$ be a complex variable, A(z), an $n \times n$ matrix-valued function of z. A(z) is an *analytic function* of z in a domain G of the z plane if all entries $a_{ij}(z)$ of A(z) are analytic functions of z in G.

Definition 2.   X is a finite-dimensional normed linear space, and A(z) is a family of linear mappings of X into X, depending on the complex parameter z. A(z) depends analytically on z in a domain G if the limit

$$\lim_{h \to 0} \frac{A(z + h) - A(z)}{h} = A'(z)$$

exists in the sense of convergence defined in equation (16) of Chapter 15.

EXERCISE 4.   Show that the two definitions are equivalent.

EXERCISE 5.   Let A(z) be an analytic matrix function in a domain G, invertible at every point of G. Show that then $A^{-1}(z)$, too, is an analytic matrix function in G.

EXERCISE 6.   Show that the Cauchy integral theorem holds for matrix-valued functions.

The analytic functions we shall be dealing with are *resolvents*. The resolvent of A is defined as

$$R(z) = (zI - A)^{-1} \tag{18}$$

for all z not an eigenvalue of A. It follows from Exercise 5 that $R(z)$ is an analytic function.

Theorem 2.   For $|z| > |A|$ $R(z)$ has the expansion

$$R(z) = \sum_0^\infty \frac{A^n}{z^{n+1}}. \tag{19}$$

Proof. By the multiplicative estimate $|A^n| \le |A|^n$, it follows that the series on the right-hand of (19) converges for $|z| > |A|$.

Multiply (19) by $(zI - A)$; term-by-term multiplication gives I on the right-hand side. This proves that (19) is the inverse of $(zI - A)$. $\square$

Multiply (19) by z^j and integrate it over any circle $|z| = s > |A|$. On the right-hand side we integrate term by term; only the jth integral is $\ne 0$, so we get

$$\int_{|z|=s} R(z)z^j \, dz = 2\pi i A^j. \tag{20}$$

Since $R(z)$ is an analytic function outside the spectrum of A, we can, according to Exercise 6, deform the circle of integration to any circle of radius $s = r(A) + \epsilon$, $\epsilon > 0$:

$$\int_{|z|=r(A)+\epsilon} R(z)z^j \, dz = 2\pi i A^j. \tag{20}'$$

To estimate the norm of A^j from its integral representation (20)', we rewrite the dz integration in terms of $d\theta$ integration, where θ is the polar angle, $z = se^{i\theta}$ and $dz = sie^{i\theta} \, d\theta$:

$$A^j = \frac{1}{2\pi} \int_0^{2\pi} R(se^{i\theta})s^{j+1} e^{i\theta(j+1)} d\theta. \tag{21}$$

The norm of an integral of linear maps is bounded by the maximum of the integrand times the length of the interval of integration. Since $R(z)$ is an analytic function, it is continuous on the circle $|z| = r(A) + \epsilon$, $\epsilon > 0$; denote the maximum of $|R(z)|$ on this circle by $c(\epsilon)$. We can then estimate the norm of A^j from its integral representation (21), with $s = r(A) + \epsilon$, as follows:

$$|A^j| \le (r(A) + \epsilon)^{j+1} c(\epsilon).$$

Take the jth root:

$$|A^j|^{1/j} \leq m(\epsilon)^{1/j}(r(A) + \epsilon), \tag{22}$$

where $m(\epsilon) = (r(A) + \epsilon)\, c(\epsilon)$. Let j tend to ∞ in (22); we get

$$\limsup_{j \to \infty} |A^j|^{1/j} \leq r(A) + \epsilon.$$

Since this holds for all positive ϵ, no matter how small,

$$\limsup |A^j|^{1/j} \leq r(A).$$

On the other hand, analogously to (3),

$$|A^j| \geq r(A)^j$$

for any norm. Taking the jth gives

$$|A^j|^{1/j} \geq r(A).$$

Combining this with (21), we deduce (17). $\square$

This proof nowhere uses the finite dimensionality of the normed linear space on which A acts.

APPENDIX 11

The Lorentz Group

1. In classical mechanics, particles and bodies are located in absolute, motionless space equipped with Euclidean structure. Motion of particles is described by giving their position in absolute space as a function of an absolute time.

In the relativistic description, there is no absolute space and time, because space and time are inseparable. The speed of light is the same in two coordinate systems moving with constant velocity with respect to each other. This can be expressed by saying that the *Minkowski metric* $t^2 - x^2 - y^2 - z^2$ is the same in both coordinate systems—here we have taken the speed of light to have the numerical value 1.

A linear transformation of four-dimensional space–time that preserves the quadratic form $t^2 - x^2 - y^2 - z^2$ is called a *Lorentz transformation*. In this chapter we shall investigate their properties.

We start with the slightly simpler $(2 + 1)$-dimensional space–time. Denote by u the space–time vector $(t, x, y)'$, and denote by M the matrix

$$M = \begin{pmatrix} 1 & 0 & 0 \\ 0 & -1 & 0 \\ 0 & 0 & -1 \end{pmatrix}. \tag{1}$$

Clearly,

$$t^2 - x^2 - y^2 = (u, Mu), \tag{2}$$

where $(,)$ denotes the standard scalar product in $\mathbb{R}^3$.

The condition that the Lorentz transformation L preserve the quadratic form (2) is that for all u,

$$(Lu, MLu) = (u, Mu) \tag{3}$$

Linear Algebra and Its Applications, Second Edition, by Peter D. Lax
Copyright © 2007 John Wiley & Sons, Inc.

Since this holds for all u, v, and $u + v$, it follows that

$$(Lu, MLv) = (u, Mv) \qquad (3)'$$

for all pairs u, v. The left-hand side can be rewritten as $(u, L'MLv)$; since $(3)'$ holds for all u and v, we conclude that

$$L'ML = M. \qquad (4)$$

Take the determinant of (4); using the multiplicative property of determinants, along with the fact that $\det M = 1$, we get

$$(\det L')(\det L) = (\det L)^2 = 1. \qquad (4)'$$

This shows that every Lorentz transformation is invertible.

EXERCISE 1.   Show that if L is a Lorentz transformation, so is L'.

The Lorentz transformations form a group. For clearly, the composite of two transformations that preserve the quadratic form (2) also preserves (2). Similarly, if L preserves (2), so does its inverse.

A 3×3 matrix has nine elements. Both sides of $(3)'$ are symmetric 3×3 matrices; their equality imposes six conditions on the entries of L. Therefore, roughly speaking, the Lorentz group forms a three-dimensional manifold.

Definition.   The *forward light cone* (flc) is the set of points in space–time for which $t^2 - x^2 - y^2$ is positive, and t is positive. The *backward light cone* (blc) consists of points for which $t^2 - x^2 - y^2$ is positive and t is negative.

Theorem 1.   A Lorentz transformation L either maps each light cone onto itself or onto each other.

Proof. Take any point, say $(1, 0, 0)'$ in the flc; since L preserves the Minkowski metric (2), the image of this point belongs to either the forward or backward light cone, say the flc. We claim that then L maps every point u in the flc into the flc. For suppose on the contrary that some point u in the flc is mapped by L into the blc. Consider the interval connecting $(1, 0, 0)'$ and u; there would be some point v on this interval that is mapped by L into a point w whose t-component is zero. For such a point w, $(w, Mw) \leq 0$. Since all point v on this interval lies in the flc $(v, Mv) > v$. But $w = Lv$, a contradiction to (3).

We show next that every point z in the flc is the image of some point u in the flc. For, since L is invertible, $z = Lu$ for some u; since L preserves the quadratic form (2), u belongs to one of the light cones. If it were the backward light cone, then by the argument given above L would map all points of the blc into the flc. But L maps $(-1, 0, 0)'$ into $-L(1, 0, 0)'$, which lies in the blc, a contradiction.

If L maps $(1, 0, 0)'$ into the blc, we argue analogously; this completes the proof of Theorem 1.   □

Definition.   It follows from $(4)'$ that $\det L = \pm 1$. The Lorentz transformations that map the flc onto itself, and for which $\det L = 1$, form the *proper Lorentz group*.

Theorem 2.   Suppose L belongs to the proper Lorentz group, and maps the point $e = (1, 0, 0)$ onto itself. Then L is rotation around the t axis.

Proof. $Le = e$ implies that the first column of L is $(1, 0, 0)'$. According to (4), $L'ML = M$; since $Me = e$, $L'e = e$; therefore the first column of L' is $(1, 0, 0)'$. So the first row of L is $(1, 0, 0)$. Thus L has the form

$$L = \begin{pmatrix} 1 & 0 & 0 \\ 0 & & \\ 0 & & R \end{pmatrix}.$$

Since L preserves the Minkowski metric, M is an isometry. Since $\det L = 1$, $\det M = 1$; so R is a rotation.   □

EXERCISE 2.   Show that Lorentz transformations preserve solutions of the wave equation. That is, if $f(t, x, y)$ satisfies

$$f_{tt} - f_{xx} - f_{yy} = 0,$$

then $f(L(t, x, y))$ satisfies the same equation.

Next we shall present an explicit description of proper Lorentz transformations. Given any point $u = (t, x, y)'$, we represent it by the 2×2 symmetric matrix U:

$$U = \begin{pmatrix} t - x & y \\ y & t + x \end{pmatrix} \tag{5}$$

Clearly, U is real and symmetric and

$$\det U = t^2 - x^2 - y^2, \qquad \operatorname{tr} U = 2t. \tag{6}$$

Let W be any 2×2 real matrix whose determinant equals 1. Define the 2×2 matrix V by

$$WUW' = V. \tag{7}$$

Clearly, V is real and symmetric and

$$\det V = (\det W)(\det U)(\det W') = \det U, \tag{8}$$

since we have assumed that det W = 1. Denote the entries of V as

$$V = \begin{pmatrix} t' - x' & y' \\ y' & t' + x' \end{pmatrix}. \tag{9}$$

Given W, (5), (7), and (9) defines a linear transformation $(t, x, y) \rightarrow (t', x', y')$. It follows from (6) and (8) that $t^2 - x^2 - y^2 = t'^2 - x'^2 - y'^2$. That is, each W generates a Lorentz transformation. We denote it by L_W. Clearly, W and $-W$ generate the same Lorentz transformation. Conversely,

EXERCISE 3.   Show that if W and Z generate the same Lorentz transformation, then $Z = W$ or $Z = -W$.

The 2×2 matrices W with real entries and determinant 1 form a group under matrix multiplication, called the special linear group of order 2 over the reals. This group is denoted as SL(2, $\mathbb{R}$).

EXERCISE 4.   Show that SL(2, $\mathbb{R}$) is connected—that is, that every W in SL(2, $\mathbb{R}$) can be deformed continuously within SL(2, $\mathbb{R}$) into I.

Formulas (5), (7), and (9) define a two-to-one mapping of SL(2, $\mathbb{R}$) into the $(2 + 1)$-dimensional Lorentz group. This mapping is a *homomorphism*, that is,

$$L_{WZ} = L_W L_Z. \tag{10}$$

EXERCISE 5.   Verify (10).

Theorem 3.   (a) For W in SL(2, $\mathbb{R}$), L_W belongs to the proper Lorentz group.
(b) Given any two points u and v in the flc, satisfying $(u, Mu) = (v, Mv)$, there is a Y in SL(2, $\mathbb{R}$) such that $L_Y u = v$.
(c) If Z is a rotation, L_Z is a rotation around the t axis.

Proof. (a) A symmetric matrix U representing a point $u = (t, x, y)'$ in the flc is positive, and the converse is also true. For according to (6), det U $= t^2 - x^2 - y^2$, tr U $= 2t$, and the positivity of both is equivalent to the positivity of the symmetric matrix U.

By definitions (7) and (9), the matrix V representing $v = L_W u$ is V $= WUW'$; clearly, if U is a positive symmetric matrix, so is V. This shows that L_W maps the flc into the flc.

According to (4)', the determinant of the Lorentz transformation L_W is 1 or -1. When W is the identity I, L_W is the identity, and so det $L_I = 1$. During a continuous deformation of W, det L_W changes continuously, so it doesn't change at all. Therefore, det $L_W = 1$ for all W that can be deformed continuously into I. According to Exercise 4, all W can be deformed into I; this shows that for all W, L_W is a proper Lorentz transformation.

(b) We shall show that given any v in the flc, there is a W in SL(2, $\mathbb{R}$) such that L_W maps $te = t(1,0,0)'$ into v, where t is the positive square root of (v, Mv). The matrix representing e is I; the matrix representing $L_W te$ is $WtIW' = tWW'$. So we have to choose W so that $tWW' = V$, where V represents v. Since $t^2 = (v, Mv) = \det V$ and since by (a) V is positive, we can satisfy the equation for W by setting W as the positive square root of $t^{-1}V$.

Similarly, for any other point u in the flc for which $(u, Mu) = (v, Mv)$, there is a Z in SL(2, $\mathbb{R}$), for which $L_Z te = u$. Then $L_W L_Z^{-1}$ maps u into v. Since $W \to L_W$ is a homomorphism, $L_W L_Z^{-1} = L_{WZ^{-1}}$.

(c) Suppose that Z is a rotation in $\mathbb{R}^2$; then $Z'Z = I$. Using the commutativity of trace, we get from $V = ZUZ'$ that

$$\operatorname{tr} V = \operatorname{tr} ZUZ' = \operatorname{tr} UZ'Z = \operatorname{tr} U$$

for all U. For U of form (5) and V of form (9), $\operatorname{tr} U = 2t$, $\operatorname{tr} V = 2t'$, so $t = t'$ for all U. Since $t^2 - x^2 - y^2 = t'^2 - x'^2 - y'^2$, it follows that L_Z maps $(t, 0, 0)$ into itself. We appeal to Theorem 2 to conclude that L_Z is rotation around the t axis. $\square$

EXERCISE 6.   Show that if Z is rotation by angle θ, L_Z is rotation by angle 2θ.

Theorem 4.   Every proper Lorentz transformation L is of the form L_Y, Y in SL(2, $\mathbb{R}$).

Proof. Denote by u the image of $e = (1,0,0)$ under L:

$$Le = u.$$

Since e lies in the flc, so does u. According to part (b) of Theorem 3, $L_W e = u$ for some W in SL(2, $\mathbb{R}$). Therefore $L_W^{-1} L e = e$; according to Theorem 2, $L_W^{-1} L$ is rotation around the t axis. By part (c) of Theorem 3, along with Exercise 6, there is a rotation Z in SL(2, $\mathbb{R}$) such that $L_W^{-1} L = L_Z$; it follows that $L = L_W L_Z = L_{WZ}$. $\square$

EXERCISE 7.   Show that a symmetric 2×2 matrix is positive iff its trace and determinant are both positive.

EXERCISE 8.   **(a)** Let L(s) be a one-parameter family of Lorentz transformations that depends differentiably on s. Show that L(s) satisfies a differential equation of the form

$$\partial_s L = AML, \tag{11}$$

where A(s) is anti-self-adjoint.

(b) Let A(s) be a continuous function whose values are anti-self-adjoint matrices. Show that every solution L(s) of (11), whose initial value is I, belongs to the proper Lorentz group.

2. In this section we shall outline a construction of a model of non-Euclidean geometry using the Lorentz group. By a geometry of the plane, we mean a set of elements, called points, with the following structures:

(i) a collection of subsets called lines.

(ii) a group of motions that map the plane onto itself, and map lines into lines.

(iii) Angles of intersecting lines.

(iv) Distance of pairs of points.

All axioms of Euclidean geometry, except the parallel postulate, have to be satisfied.

Definition. The model, denoted as $\mathbb{H}$, consists of the points u located in the positive half of the hyperboloid

$$(u, Mu) = t^2 - x^2 - y^2 = 1, \qquad t > 0. \tag{12}$$

The *lines* are intersection of $\mathbb{H}$ with planes through the origin $(u, p) = 0$, where p satisfies

$$(p, Mp) < 0. \tag{13}$$

The *group of motions* is the proper Lorentz group.

Theorem 5. **(a)** Every plane $(u, p) = 0$ that satisfies (13) has a nonempty intersection with $\mathbb{H}$.

(b) Any two distinct points of $\mathbb{H}$ lie on one and only one line.

(c) Every proper Lorentz transformation maps lines onto lines.

Proof. **(a)** Set $p = (s, a, b)$; condition (13) means that $s^2 < a^2 + b^2$. The point $u = (a^2 + b^2, -as, -bs)$ satisfies $(u, p) = 0$. We claim that u belongs to the flc; for

$$(u, Mu) = (a^2 + b^2)^2 - (a^2 + b^2)s^2$$

is positive, and so is $a^2 + b^2$. It follow that u/k, where k is the positive square root of (u, Mu), belongs to $\mathbb{H}$.

(b) Conversely, suppose that the plane $(u, p) = 0$, $p = (s, a, b)$, contains a point $u = (t, x, y)$ that belongs to the flc; then p satisfies (13). For suppose not—that is, $s^2 \geq a^2 + b^2$. Since $u = (t, x, y)$ belongs to the flc, we have $t^2 > x^2 + y^2$. Multiplying these inequalities gives

$$s^2 t^2 > (a^2 + b^2)(x^2 + y^2).$$

Since $(u,p) = 0$,

$$st = -(ax + by)$$

Applying the Schwarz inequality on the right gives the opposite of the previous inequality, a contradictions.

Given two distinct points u and v in $\mathbb{H}$, there is a unique p, except for a constant multiple, that satisfies $(u,p) = 0$, $(v,p) = 0$. According to what we have shown above, p satisfies (13) when u or v belongs to $\mathbb{H}$.

(c) Take the line consisting of all points u in $\mathbb{H}$ that satisfy $(u,p) = 0$, where p is a given vector, $(p, Mp) < 0$. Let L be a proper Lorentz transformation; the inverse image of the line under L consists of all points v such that $u = Lv$. These points v satisfy $(Lv, p) = (v, Lp) = 0$. We claim that the points v lie on $\mathbb{H}$, and that $q = Lp$ satisfies $(q, Mq) < 0$. Both of these assertions follow from the properties of proper Lorentz transformations.   □

Next we verify that our geometry in non-Euclidean. Take all lines $(u,p) = 0$ through the point $u = (1,0,0)$. Clearly, such a p is of the form $p = (0, a, b)$, and the points $u = (t, x, y)$ on the line $(u,p) = 0$ satisfy

$$ax + by = 0. \tag{14}$$

Take $q = (1, 1, 1)$; points $u = (t, x, y)$ on the line $(u, q) = 0$ satisfy $t + x + y = 0$. For such a point u,

$$(u, Mu) = t^2 - x^2 - y^2 = (x + y)^2 - x^2 - y^2 = 2xy. \tag{15}$$

The points u on the intersection of the two lines satisfy both $(u,p) = 0$ and $(u, q) = 0$. If a and b are of the same sign, it follows from (14) that x and y are of the opposite sign; it follows from (15) that such a u does not lie in the flc.

Thus there are *infinitely many* lines through the point $(1, 0, 0)$ that *do not* intersect the line $t + x + y = 0$ in $\mathbb{H}$; this violates the parallel postulate of Euclidean geometry.

In our geometry, the proper Lorentz transformation are the analogues of Euclidean motion translations combined with rotations. Both objects form a three-parameter family.

We turn now to the definition of distance in our geometry. Take two nearby points in $\mathbb{H}$, denoted as (t, x, y) and $(t + dt, x + dx, y + dy)$. Their image under a proper Lorentz transformation L is (t', x', y') and $(t' + dt', x' + dx', y' + dy')$. Since L is linear, (dt', dx', dy') is the image of (dt, dx, dy) under L, and therefore

$$dt'^2 - dx'^2 - dy'^2 = dt^2 - dx^2 - dy^2,$$

an invariant quadratic form. Since for points of $\mathbb{H}$, $t^2 - x^2 - y^2 = 1$, we have $dt = \frac{x}{t}dx + \frac{y}{t}dy$. So we choose as invariant metric

$$dx^2 + dy^2 - dt^2 = \frac{1+y^2}{t^2}dx^2 + \frac{1+x^2}{t^2}dy^2 - \frac{2xy}{t^2}dxdy. \tag{16}$$

Once we have a metric, we can define the angle between lines at their point of intersection using the metric (16).

3. In this section we shall briefly outline the theory of Lorentz transformations in $(3+1)$-dimensional space–time. The details of the results and their proofs are analogous to those described in Section 1.

The Minkowski metric in $3+1$ dimensions is $t^2 - x^2 - y^2 - z^2$, and a Lorentz transformation is a linear map that preserves this metric. The Minkowski metric can be expressed, analogously with (2), as

$$(u, Mu), \tag{17}$$

where M is the 4×4 diagonal matrix whose diagonal entries are $1, -1, -1$, and -1, and u denotes a point $(t, x, y, z)'$ in $(3+1)$-dimensional space–time. The forward light cone is defined, as before, as the set of points u for which $(u, Mu) > 0$ and $t > 0$.

A Lorentz transformation is represented by a matrix L that satisfies the four-dimensional analogue of equation (4):

$$L'ML = M. \tag{18}$$

A *proper* Lorentz transformation is one that maps the flc onto itself and whose determinant $\det L$ equals 1. The proper Lorentz transformations form a group.

Just as in the $(2+1)$-dimensional case, proper Lorentz transformations in $3+1$ space can be described explicitly. We start by representing vectors $u = (t, x, y, z)'$ in $3+1$ space by complex-valued self-adjoint 2×2 matrices

$$U = \begin{pmatrix} t-x & y+iz \\ y-iz & t+x \end{pmatrix} \tag{19}$$

The Minkowski metric of u can be expressed as

$$t^2 - x^2 - y^2 - z^2 = \det U. \tag{20}$$

Let W be any complex-valued 2×2 matrix of determinant 1. Define the 2×2 matrix V by

$$V = WUW^*, \tag{21}$$

where W^* is the adjoint of W, and U is defined by (19). Clearly, V is self-adjoint, so it can be written as

$$V = \begin{pmatrix} t' - x' & y' + iz' \\ y' - iz' & t' + x' \end{pmatrix}. \tag{22}$$

given W, (19), (21), and (22) define a linear map $(t, x, y, z) \rightarrow (t', x', y', z')$. Take the determinant of (21):

$$\det V = (\det W)(\det U)(\det W^*).$$

Using (20), (20)', and $\det W = 1$, it follows that

$$t^2 - x^2 - y^2 - z^2 = t'^2 - x'^2 - y'^2 - z'^2.$$

This shows that each W generates a Lorentz transformation. We denote it as L_W.

The complex-valued 2×2 matrices of determinant 1 form a *group* denoted as SL(2, $\mathbb{C}$).

Theorem 6.   **(a)** For every W in SL(2, $\mathbb{C}$), L_W defined above is a proper Lorentz transformation.

(b) The mapping $W \rightarrow L_W$ is a homomorphic map of SL(2, $\mathbb{C}$) onto the proper Lorentz group. This mapping is 2 to 1.

We leave it to the reader to prove this theorem using the techniques developed in Section 1.

4. In this section we shall establish a relation between the group SU(2, $\mathbb{C}$) of 2×2 unitary matrices and SO(3, $\mathbb{R}$), the group of rotations in $\mathbb{R}^3$.

We represent a point $(x, y, z)'$ of $\mathbb{R}^3$ by the 2×2 matrix

$$\begin{pmatrix} x & y + iz \\ y - iz & -x \end{pmatrix} = U. \tag{23}$$

Clearly,

$$-\det U = x^2 + y^2 + z^2. \tag{24}$$

The matrices U are 2×2 self-adjoint matrices, trace of $U = 0$.

Theorem 7.   Z is a 2×2 unitary matrix of determinant 1, and U is as above. Then

$$V = ZUZ^* \tag{25}$$

is

(i) self-adjoint,
(ii) $\det V = \det U$,
(iii) $\operatorname{tr} V = 0$.

Proof. (i) is clearly true. To see (ii) take the determinant of (25). To deduce (iii), we use the commutativity of trace—that is, that $\operatorname{tr} AB = \operatorname{tr} BA$. Thus

$$\operatorname{tr} V = \operatorname{tr} ZUZ^* = \operatorname{tr} UZ^*Z = \operatorname{tr} U = 0;$$

here we have used the fact that $Z^*Z = I$ for Z unitary. $\qquad\square$

Combining (23), (24), and (25) shows that the mapping $U \to V$ defined by (25) engenders an orthogonal transformation of $\mathbb{R}^3$. We denote it as O_Z.

The mapping defined by (25) is analogous to the mapping of $SL(2, \mathbb{C})$ onto the $(3 + 1)$-dimensional proper Lorentz group described in Section 3.

Theorem 8. The mapping $Z \to O_Z$ defined above is a homomorphic map of $SU(2, \mathbb{C})$ onto $SO(3, \mathbb{R})$. This mapping is 2 to 1.

We leave it to the reader to prove this theorem.

We shall now show how to use this representation of rotation in $\mathbb{R}^3$ to rederive Euler's theorem (see Section 1 of Chapter 11) that every rotation in three space is around a uniquely defined axis. Representing the rotation as in (25), we have to show that given any 2×2 unitary matrix, $\det Z = 1$, there is a self-adjoint 2×2 matrix U, $\operatorname{tr} U = 0$, that satisfies

$$ZUZ^* = U.$$

Multiplying this relation by Z on the right, we get $ZU = UZ$. To find a U that commutes with Z, we use the eigenvector e and f of Z. They are orthogonal and satisfy, since $\det Z = 1$,

$$Ze = \lambda e, \qquad Zf = \bar{\lambda}f, \qquad |\lambda| = 1.$$

Define U by setting

$$Ue = e, \qquad Uf = -f.$$

Clearly, U is self-adjoint; since its eigenvalues are 1 and -1, $\operatorname{tr} U = 0$. The axis of rotation consists of all real multiples of U.

APPENDIX 12

Compactness of the Unit Ball

In this Appendix we shall present examples of Euclidean spaces X whose unit ball is compact—that is, where every sequence $\{x_k\}$ of vectors in X, $\| x_k \| \le 1$, has a convergent subsequence. According to Theorem 17 of Chapter 7, such spaces are finite dimensional. Thus compactness of the unit ball is an important criterion for finite dimensionality.

Let G be a bounded domain in the x, y plane whose boundary is smooth. Let $u(x, y)$ be a twice differentiable function that satisfies in G the partial differential equation

$$au + \Delta u = 0, \tag{1}$$

where a is a positive constant, and Δ is the Laplace operator:

$$\Delta u = u_{xx} + u_{yy}; \tag{2}$$

here subscripts x, y denote partial derivatives with respect to these variables.

Denote by S the set of solutions of equation (1) which in addition are zero on the boundary of G:

$$u(x, y) = 0 \qquad \text{for } (x, y) \text{ in } \partial G. \tag{3}$$

Clearly, the set S of such solutions form a linear space.

We define for u in S the norm

$$||u||^2 = \int_G u^2(x, y) dx dy. \tag{4}$$

Linear Algebra and Its Applications, Second Edition, by Peter D. Lax
Copyright © 2007 John Wiley & Sons, Inc.

Theorem 1.   The unit ball in S is compact.

Proof. Multiply equation (1) by u and integrate over G:

$$0 = \int_G (au^2 + u\Delta u)\, dxdy. \tag{5}$$

Using the definition (2) of Δ, we can integrate the second term by parts:

$$\int_G u(u_{xx} + u_{yy})\, dxdy = -\int (u_x^2 + u_y^2)\, dxdy \tag{6}$$

There are no boundary terms resulting from the integration by parts because u is zero on the boundary of G. Setting (6) into (5), we get

$$\int_G (u_x^2 + u_y^2)\, dxdy = a \int_G u^2\, dxdy. \tag{7}$$

We appeal now to a *compactness criterion* due to Franz Rellich: □

Theorem 2.   (Rellich) Let R be a set of smooth functions in a bounded domain G with smooth boundary whose square integrals, along with the square integrals of their first derivatives over G, is uniformly bounded:

$$\int_G u^2 \le m, \qquad \int u_x^2 \le m, \qquad \int u_y^2 \le m. \tag{8}$$

Every sequence of functions in R contains a subsequence that converges in the square integral norm.

The unit ball $\| u \| \le 1$ in S has the properties (8), since according to inequality (7) the square integral of their first derivatives is bounded by a. So Theorem 1 is a consequence of Theorem 2.   □

It follows therefore from Theorem 17 of Chapter 7 that the solutions of equation (1) that satisfy the boundary conditions (3) form a *finite-dimensional* space.

Theorem 2, and therefore Theorem 1, is not restricted to functions of two variables or to the specific differential equation (1).

The proof of Rellich's compactness criterion would take us too far afield. But the analogous result where the square integral norm is replaced by the maximum norm is simpler:

Theorem 3. Let G be a bounded domain with a smooth boundary, and let D be a set of functions in G whose values and the values of their first derivatives are uniformly bounded in G by a common bound m. Every sequence of functions in D contains a subsequence that converges in the maximum norm.

EXERCISE 1. **(i)** Show that a set of functions whose first derivatives are uniformly bounded in G are equicontinuous in G.

(ii) Use (i) and the Arzela–Ascoli theorem to prove Theorem 3.

A Characterization of Commutators

We shall prove the following result.

Theorem 1. An $n \times n$ matrix X is the commutator of two $n \times n$ matrices A and B,

$$X = AB - BA, \tag{1}$$

iff the trace of X is zero.

Proof. We have shown in Theorem 7 of Chapter 5 that trace is commutative—that is, that

$$\operatorname{tr} AB = \operatorname{tr} BA.$$

It follows that for X of form (1), $\operatorname{tr} X = 0$. We show now the converse. □

Lemma 2. Every matrix X all of whose diagonal entries are zero can be represented as a commutator.

Proof. We shall construct explicitly a pair of matrices A and B so that (1) holds. We choose arbitrarily n distinct numbers $a_1, \ldots, a_n$ and define A to be the diagonal matrix with diagonal entries a_i:

$$A_{ij} = \begin{cases} 0 & \text{for } i \neq j, \\ a_i & \text{for } i = j. \end{cases}$$

We define B as

$$B_{ij} = \begin{cases} \dfrac{X_{ij}}{a_i - a_j} & \text{for } i \neq j, \\ \text{anything} & \text{for } i = j. \end{cases}$$

Linear Algebra and Its Applications, Second Edition, by Peter D. Lax
Copyright © 2007 John Wiley & Sons, Inc.

Then for $i \neq j$

$$
\begin{aligned}
(AB - BA)_{ij} &= a_i B_{ij} - B_{ij} a_j \\
&= (a_i - a_j) B_{ij} = X_{ij},
\end{aligned}
$$

while

$$
(AB - BA)_{ii} = a_i B_{ii} - B_{ii} a_i = 0.
$$

This verifies (1).   □

To complete the proof of Theorem 9 we make use of the observation that if X can be represented as a commutator, so can any matrix similar to X. This can be seen formally by multiplying equation (1) by S on the left and S^{-1} on the right:

$$
\begin{aligned}
SXS^{-1} &= SABS^{-1} - SBAS^{-1} \\
&= (SAS^{-1})(SBS^{-1}) - (SBS^{-1})(SAS^{-1}).
\end{aligned}
$$

Conceptually, we are using the observation that similar matrices represent the same mapping but in different coordinate systems.

Lemma 3.   Every matrix X whose trace is zero is similar to a matrix all whose diagonal entries are zero.

Proof. Suppose not all diagonal entries of X are zero, say $x_{11} \neq 0$. Then, since $\text{tr}\, X = 0$, there must be another diagonal entry, say x_{22}, that is neither zero nor equal to x_{11}. Therefore the 2×2 minor in the upper left corner of X,

$$
\begin{pmatrix} x_{11} & x_{12} \\ x_{21} & x_{22} \end{pmatrix} = Y,
$$

is *not* a multiple of the identity. Therefore there is a vector h with two components such that Yh is not a multiple of h. We introduce now h and Yh as new basis in $\mathbb{R}^2$; with respect to this basis Y is represented by a matrix whose first diagonal element is zero.

Continuing in this fashion we make changes of variables in two-dimensional subspaces that introduce a new zero on the diagonal of the matrix representing X, without distroying any of the zeros that are already there, until there are $n - 1$ zeros on the diagonal. But since $\text{tr}\, X = 0$, the remaining diagonal element is zero too.   □

Combining Lemma 2 and Lemma 3 gives Theorem 1.

APPENDIX 14

Liapunov's Theorem

In this Appendix we give a far-reaching extension of Theorem 20 in Chapter 10. We start by replacing Z in that result by its negative, $W = -Z$,

Theorem 1. Let W be a mapping of a finite-dimensional Euclidean space into itself whose self-adjoint part is negative:

$$W + W^* < 0. \tag{1}$$

Then the eigenvalues of W have negative real part.

This can be proved the same way as Theorem 20 was in Chapter 10. We state now a generalization of this result.

Theorem 2. Let W be a mapping of a finite-dimensional Euclidean space X into itself. Let G be a positive self-adjoint map of X into itself that satisfies the inequality

$$GW + W^*G < 0. \tag{2}$$

Then the eigenvalues of W have negative real part.

Proof. Let h be an eigenvector of W, where w is the corresponding eigenvalue:

$$Wh = wh. \tag{3}$$

Let the left-hand side of (2) act on h, and take the scalar product with h; according to (2),

$$((GW + W^*G)h, h) < 0. \tag{4}$$

Linear Algebra and Its Applications, Second Edition, by Peter D. Lax
Copyright © 2007 John Wiley & Sons, Inc.

We can rewrite this as

$$(GWh, h) + (Gh, Wh) < 0.$$

Using the eigenvalue equation (3), this can be restated as

$$\lambda(Gh, h) + \bar{\lambda}(Gh, h) < 0.$$

We have assumed that G is positive; this means that the quadratic form (Gh, h) is positive. So we conclude that

$$\lambda + \bar{\lambda} < 0. \qquad \square$$

A theorem of Liapunov says that the converse of Theorem 2 holds:

Theorem 3 (Liapunov). Let W be a mapping of a finite-dimensional Euclidean space X into itself, whose eigenvalues have negative real part. Then there exists a positive self-adjoint mapping G such that inequality (2),

$$GW + W^*G < 0, \qquad (2)$$

holds.

Proof. We recall from Chapter 9 the definition of the exponential of a matrix:

$$e^W = \sum_0^\infty \frac{W^k}{k!}. \qquad (5)$$

According to Exercise 7 there, the eigenvalues of e^W are the exponentials of the eigenvalue of W.

Let t be any real number. The eigenvalues of e^{Wt} are e^{wt}, where w is an eigenvalue of W.

Lemma 4. If the eigenvalues of W have negative real part, $||e^{Wt}||$ tends to zero at an exponential rate as $t \to +\infty$.

Proof. According to Theorem 18 of Chapter 7, whose proof appears in Appendix 10, for any mapping A of a finite-dimensional Euclidean space X into itself,

$$\lim ||A^j||^{1/j} = r(A), \qquad (6)$$

where $r(A)$ is the spectral radius of A. We apply this to $A = e^W$. By assumption, the eigenvalues w of W have negative real part. It follows that the eigenvalues of e^W, which are of the form e^w, are less than 1 in absolute value. But then the spectral radius of e^W, the maximum of all e^w, is also less than 1:

$$r(e^W) < 1. \tag{7}$$

We conclude from (6) applied to $A = e^W$ that

$$\| e^{Wj} \| < (r(e^W) + \epsilon)^j, \tag{8}$$

where ϵ tends to zero as $j \to \infty$. It follows from (7) and (8) that $\|e^{Wt}\|$ decays exponentially as $t \to \infty$ through integer values.

For t not an integer, we decompose t as $t = j + f$, where j is an integer and f is between 0 and 1, and we factor e^{Wt} as

$$e^{Wt} = e^{Wj}e^{Wf}.$$

So

$$\| e^{Wt} \| \le \| e^{Wj} \| \, \| e^{Wf} \|. \tag{9}$$

To estimate $\| e^{Wf} \|$ we replace in (5) W by Wf,

$$e^{Wf} = \sum_0^\infty \frac{W^k f^k}{k!},$$

and apply the additive and multiplicative estimates for the norm of a matrix:

$$\| e^{Wf} \| \le \sum_0^\infty \| W^k \| f^k/k!$$

$$\le \sum \frac{\| W \|^k f^k}{k!} = e^{\| W \| f}$$

Since f lies between 0 and 1,

$$e^{\| W \| f} \le e^{\| W \|}. \tag{10}$$

We can use (8) and (10) to estimate the right-hand side of (9); using $j = t - f$, we get

$$\| e^{Wt} \| \le (r + \epsilon)^{t-1} e^{\| W \|}, \tag{11}$$

where ϵ tends to zero as $t \to \infty$. According to (7), $r = r(e^W) < 1$; thus it follows from (11) that $\| e^{Wt} \|$ decays to zero at an exponential rate as t tends to ∞. $\square$

The adjoint of e^{Wt} is e^{W^*t}. Since adjoint mappings have the same norm, it follows from Lemma 4 that $||e^{W^*t}||$ tends to zero at an exponential rate as $t \to \infty$.

We are now ready to prove Theorem 3 by writing down an explicit formula for G:

$$G = \int_0^\infty e^{W^*t} e^{Wt} dt, \tag{12}$$

and verifying that it has the properties required in Theorem 3. But first a word about integrals of functions whose values are mappings:

$$\int_a^b A(t) \, dt. \tag{13}$$

There are two ways of defining integrals of form (13), where $A(t)$ is a continuous function of t.

Definition 1. Express $A(t)$ as a matrix $(a_{ij}(t))$ with respect to some basis. Each of the entries $a_{ij}(t)$ is a continuous scalar-valued function, real or complex, whose integral is well-defined in calculus.

Definition 2. Form approximating sums

$$\sum A(t_j)|\Delta_j|; \tag{14}$$

their limit as the subdivision $[a, b] = u\Delta j$ is refined is the integral (13).

EXERCISE 1. Show that the sums (14) tend to a limit as the size of the subintervals Δ_j tends to zero. (*Hint*: Imitate the proof for the scalar case.)

EXERCISE 2. Show that the two definitions are equivalent.

Note. The advantage of Definition 2 is that it can be carried over to infinite-dimensional normed linear spaces.

The integral (12) is over an infinite interval. We define it, as in the scalar case, as the limit of the integral over [0, T] as T tends to ∞.

EXERCISE 3. Show, using Lemma 4, that for the integral (12)

$$\lim_{T \to \infty} \int_0^T e^{W^*t} e^{Wt} dt$$

exists.

We show now that G, as defined by (12), has the three properties required in Theorem 3:

(i) G is self-adjoint.

(ii) G is positive.

(iii) $GW + W^*G$ is negative.

To show (i), we note that the adjoint of a mapping defined by an integral of the form (13) is

$$\int_a^b A^*(t)\, dt,$$

It follows that if the integrand $A(t)$ is self-adjoint for each value of t, then so is the integral (13). Since the integrand in (12) is self-adjoint, so is the integral G, as asserted in (i).

To show (ii), we make use of the observation that for an integral of form (13) and for any vector h,

$$\left(h, \int_a^b A(t)dt\, h\right) = \int (h,\, A(t)h)\, dt\,.$$

It follows that if the integrand $A(t)$ is self-adjoint and positive, so is the integral (13). Since the integrand in (12) is self-adjoint and positive,

$$(h,\, e^{W^*t}e^{Wt}h) = (e^{Wt}h,\, e^{Wt}h) = \|e^{Wt}h\|^2 > 0,$$

so is the integral G, as asserted in (ii). To prove (iii), we apply the factors W and W^* under the integral sign:

$$GW + W^*G = \int_0^\infty (e^{W^*t}e^{Wt}W + W^*e^{W^*t}e^{Wt})\, dt. \tag{15}$$

Next we observe that the integrand on the right is a derivative:

$$\frac{d}{dt}e^{W^*t}e^{Wt}. \tag{16}$$

To see this, we use the rule for differentiating the product of e^{W^*t} and e^{Wt}:

$$e^{W^*t}\frac{d}{dt}e^{Wt} + \left(\frac{d}{dt}e^{W^*t}\right)e^{Wt}. \tag{17}$$

We combine this with the rule for differentiating exponential functions (see Theorem 5 of Chapter 9):

$$\frac{d}{dt}e^{Wt} = e^{Wt}\,W, \quad \frac{d}{dt}e^{W^*t} = e^{W^*t}\,W^* = W^*e^{W^*t}.$$

Setting these into (17) shows that (16) is indeed the integrand in (15):

$$GW + W^*G = \int_0^{\infty} \frac{d}{dt}e^{W^*t}e^{Wt}dt. \tag{15$'$}$$

We apply now the fundamental theorem of calculus to evaluate the integral on the right of (15)$'$ as

$$e^{W^*t}e^{Wt}\Big|_0^{\infty} = -I.$$

Thus we have

$$GW + W^*G = -I,$$

a negative self-adjoint mapping, as claimed in (iii). This completes the proof of Theorem 3.   □

    The proof, and therefore the theorem, holds in infinite-dimensional Euclidean spaces.

APPENDIX 15

The Jordan Canonical Form

In this appendix we present a proof of the converse part of Theorem 12 in Chapter 6:

Theorem 1. Let A and B be a pair of $n \times n$ matrices with the following properties.

(i) A and B have the same eigenvalues $c_1, \ldots, c_k$.

(ii) The dimension of the nullspaces

$$N_m(c_j) = \text{nullspace of } (A - c_j I)^m$$

and

$$M_m(c_j) = \text{nullspace of } (B - c_j I)^m$$

are equal for all c_j and all m:

$$\dim N_m(c_j) = \dim M_m(c_j). \tag{1}$$

Then A and B are similar.

Proof. In Theorem 12 of Chapter 6, we have shown that these conditions are necessary for A and B to be similar. We show now that they are sufficient by introducing a special basis in which the action of A is particularly simple and depends only on the eigenvalues c_j and the dimensions (1). We shall deal with each eigenvalue c_j separately; for simplicity we take $c = c_j$ to be zero. This can be accomplished by subtracting cI from A; at the end we shall add cI back.

Linear Algebra and Its Applications, Second Edition, by Peter D. Lax
Copyright © 2007 John Wiley & Sons, Inc.

The nullspaces of A^m are nested:

$$N_1 \subset N_2 \subset \cdots \subset N_d,$$

where d is the index of the eigenvalue $c = 0$.

Lemma 2. A maps the quotient space N_{i+1}/N_i into N_i/N_{i-1}, and this mapping is one-to-one.

Proof. A maps N_{i+1} into N_{ij}; therefore A maps N_{i+1}/N_i into N_i/N_{i-1}. Let $\{x\}$ be a nonzero equivalence class in N_{i+1}/N_{ij}; this means that no x belongs to N_i. It follows that Ax does not belong to N_{i-1}; this shows that $A\{x\} = \{Ax\}$ is not the zero class in N_i/N_{i-1}. This proves Lemma 2. $\square$

It follows from Lemma 2 that

$$\dim (N_{i+1}/N_i) \leq \dim (N_i/N_{i-1}). \tag{2}$$

The special basis for A in N_d will be introduced in batches. The first batch,

$$x_1, \ldots, x_{l_0}, \qquad l_0 = \dim (N_d/N_{d-1}), \tag{3}$$

are any l_0 vectors in N_d that are linearly independent mod N_{d-1}. The next batch is

$$Ax_1, \ldots, Ax_{l_0}; \tag{4}$$

these belong to N_{d-1}, and are linearly independent mod N_{d-2}. According to (2), with $i = d - 1$, $\dim (N_{d-1}/N_{d-2}) \geq \dim (N_d/N_{d-1}) = l_0$. We choose the next batch of basis vectors in N_{d-1},

$$x_{l_0+1}, \ldots, x_{l_1}, \tag{5}$$

where $l_1 = \dim (N_{d-1}/N_{d-2})$, to complete the vectors (4) to a basis of N_{d-1}/N_{d-2}. The next batch is

$$A^2 x_1, \ldots, A^2 x_{l_0}, Ax_{l_0+1}, \ldots, Ax_{l_1}. \tag{6}$$

The next batch,

$$x_{l_1+1}, \ldots, x_{l_2}, \tag{7}$$

is chosen in N_{d-2} to complete the vectors (6) to a basis of N_{d-2}/N_{d-3}, where $l_2 = \dim(N_{d-2}/N_{d-3})$. We continue this process until we reach N_1.

It is illuminating to arrange the basis elements in a table:

$$x_1, Ax_1, \ldots, A^{d-1}x_1$$

$$\vdots \qquad \qquad \vdots$$

$$x_{l_0}, Ax_{l_0}, \ldots, A^{d-1}x_{l_0}$$
$$x_{l_0+1}, Ax_{l_0+1}, \ldots, A^{d-2}x_{l_0+1}$$

$$\vdots$$

$$x_{l_1}, Ax_{l_1}, \ldots, A^{d-2}x_{l_1}$$

$$\vdots$$

$$x_{l_{d-2}+1}$$

$$\vdots$$

$$x_{l_{d-1}}$$

Here l_j denotes $\dim(N_{d-j}/N_{d-j-1})$. There are d basis elements in each of the first l_0 rows, for a total of dl_0. There are $d-1$ basis elements in the next $l_1 - l_0$ rows, for a total of $(d-1)(l_1 - l_0)$, and so on. The total number of basis elements is therefore

$$dl_0 + (d-1)(l_1 - l_0) + \cdots + (l_{d-1} - l_{d-2}).$$

This sum equals

$$l_0 + \cdots + l_{d-1} = \sum_0^{d-1} \dim(N_{d-j}/N_{d-j-1}) = \dim N_d.$$

When we write A in matrix form referring to the special basis constructed before, the action of A on the components in each of the first l_0 rows is of the form

$$\begin{pmatrix} 0 1 & \cdots & 0 \\ \vdots & & i \\ 0 & \cdots & 0 \end{pmatrix}$$

that is, a $d \times d$ matrix with $1-s$ along the superdiagonal directly above the main diagonal, but zeros everywhere else. The action of A on the components in each of the next $l_1 - l_0$ rows is similar, except that the matrices have dimension $(d-1) \times (d-1)$, and so on.

Recall that in order to simplify the presentation of the special basis we have replaced A by $A - cI$. Putting back what we have subtracted leads to the following matrix form of the action of A:

$$\begin{pmatrix} c1 & \cdots & 0 \\ \vdots & & i \\ 0 & \cdots & c \end{pmatrix}$$

that is, each entry along the main diagonal is the eigenvalue c, $1 - s$ along the superdiagonal directly above it, and zeros everywhere else. A matrix of this form is called a *Jordan block*; when all Jordan blocks are put together, the resulting matrix is called a Jordan representation of the mapping A.

The Jordan representation of A depends only on the eigenvalues of A and the dimension of the generalized eigenspaces $N_j(a_k)$, $j = 1, \ldots, d_k$, $k = 1, \ldots$. Therefore two matrices that have the same eigenvalues and the same-dimensional eigenspaces and generalized eigenspaces have the same Jordan representation. This shows that they are similar. □

APPENDIX 16

Numerical Range

Let X be a Euclidean space over the complex numbers, and let A be a mapping of X into X.

Definition. The *numerical range* of A is the set of complex numbers

$$(Ax, x), \qquad ||x|| = 1.$$

Note that the eigenvalues of A belong to its numerical range.

Definition. The *numerical radius* $w(A)$ of A is the supremum of the absolute values in the numerical range of A:

$$w(A) = \sup_{||x||=1} |(Ax, x)|. \tag{1}$$

Since the eigenvalues of A belong to its numerical range, the numerical radius of A is $\geq$ its spectral radius:

$$r(A) \leq w(A). \tag{2}$$

EXERCISE 1. Show that for A normal, equality holds in (2).

EXERCISE 2. Show that for A normal,

$$w(A) = ||A||. \tag{3}$$

Linear Algebra and Its Applications, Second Edition, by Peter D. Lax
Copyright © 2007 John Wiley & Sons, Inc.

Lemma 1. **(i)** $w(A) \leq \|A\|.$

          (ii) $\|A\| \leq 2w(A).$ (4)

Proof. By the Schwarz inequality we have

$$|(Ax, x)| \leq \|Ax\| \|x\| \leq \|A\| \|x\|^2;$$

since $\|x\| = 1$, part **(i)** follows.

(ii) Decompose A into its self-adjoint and anti-self-adjoint parts:

$$A = S + iT.$$

Then

$$(Ax, x) = (Sx, x) + i(Tx, x)$$

splits (Ax, x) into its real and imaginary parts; therefore

$$|(Ax, x)| \geq (Sx, x), \quad |(Ax, x)| \geq (Tx, x).$$

Taking the supremum over all unit vectors x gives

$$w(A) \geq w(S), \, w(A) \geq w(T).$$ (5)

Since S and T are self-adjoint,

$$w(S) = \|S\|, \, w(T) = \|T\|.$$

Adding the two inequalities (5), we get

$$2w(A) \geq \|S\| + \|T\|.$$

Since $\|A\| \leq \|S\| + \|T\|$, (4) follows.   □

Paul Halmos conjectured the following:

Theorem 2. For A as above,

$$w(A^n) \leq w(A)^n$$ (6)

for every positive integer n.

The first proof of this result was given by Charles Berger. The remarkable simple proof presented here is Carl Pearcy's.

Lemma 3. Denote by $r_k, k = 1, \ldots, n$, the nth roots of unity: $r_k = e^{2\pi i k/n}$. For all complex numbers z,

$$1 - z^n = \prod_k (1 - r_k z), \tag{7}$$

and

$$1 = \frac{1}{n} \sum_j \prod_{k \neq j} (1 - r_k z). \tag{8}$$

EXERCISE 3. Verify (7) and (8).
Set A in place of z, we get

$$I - A^n = \prod_k (I - r_k A) \tag{9}$$

and

$$I = \frac{1}{n} \sum_j \prod_{k \neq j} (I - r_k A). \tag{10}$$

Let x be any unit vector, $\|x\| = 1$; denote

$$\prod_{k \neq j} (I - r_k A)x = x_j. \tag{11}$$

Letting (9) act on x and using (11), we get

$$x - A^n x = (I - r_j A)x_j, \qquad j = 1, \ldots, n. \tag{12}$$

From (10) acting on x, we get

$$x = \frac{1}{n} \sum x_j. \tag{13}$$

Take the scalar product of (12) with x; since $\|x\| = 1$, we get on the left

$$1 - (A^n x, x) = (x - A^n x, x) = \frac{1}{n}\left(x - A^n x, \sum x_j\right); \tag{14}_1$$

in the last step we have used (13). Next use (12) on the right:

$$\frac{1}{n}\sum_j (x_j - r_j A x_j, \, x_j) = \frac{1}{n}\left(\sum_j \|x_j\|^2 - r_j(Ax_j, x_j)\right). \qquad (14)_2$$

By the definition of $w(A)$,

$$|(Ax_j, x_j)| \le w(A)\|x_j\|^2. \qquad (15)$$

Suppose that $w(A) \le 1$. Then, since $|r_j| = 1$, it follows from (15) that the real part of $(14)_2$ is nonnegative. Since $(14)_2$ equals $(14)_1$, it follows that the real part of $1 - (A^n x, x)$ is nonnegative.

Let ω be any complex number, $|\omega| = 1$. From the definition of numerical radius, it follows that $wu(\omega A) = w(A)$; therefore, by the above argument, if $w(A) \le 1$, we obtain

$$1 - \mathrm{Re}(\omega^n A^n x, \, x) = 1 - \mathrm{Re}\,\omega^n(A^n x, \, x) \ge 0.$$

Since this holds for all $\omega, |\omega| = 1$, it follows that

$$|(A^n x, \, x)| \le 1$$

for all unit vectors x. It follows that $w(A^n) \le 1$ if $w(A) \le 1$.

Since $w(A)$ is a homogeneous function of degree 1,

$$w(zA) = |z|w(A),$$

and conclusion (6) of Theorem 2 follows. $\qquad\square$

Combining Theorem 2 with Lemma 1, we obtain the following:

Corollary 2′. Let A denote an operator as above for which $w(A) \le 1$. Then for all n, we have

$$\|A^n\| \le 2. \qquad (16)$$

This corollary is useful for studying the stability of difference approximations of hyperbolic equations.

Note 1. The proof of Theorem 2 nowhere makes use of the finite dimensionality of the Euclidean space X.

Note 2. Toeplitz and Hausdorff have proved that the numerical range of every mapping A is a convex subset of the complex plane.

EXERCISE 4. Determine the numerical range of $A = \begin{pmatrix} 1 & 1 \\ 0 & 1 \end{pmatrix}$ and of $A^2 = \begin{pmatrix} 1 & 2 \\ 0 & 1 \end{pmatrix}$.

BIBLIOGRAPHY

Halmos, P. R. *A Hilbert Space Problem Book*, Van Nostrand, New York, 1967.

Lax, P. D., and Wendroff, B. Difference schemes for hyperbolic equations with high order of accuracy, *CPAM* **17** (1964), 381–398.

Lax, P. D., and Nirenberg, L. On stability for difference schemes; a sharp form of Garding's inequality, *CPAM* **19** (1966), 473–492.

Pearcy, C. An elementary proof of the power inequality for the numerical radius, *Michigan Math. J.* **13** (1966), 289–291.

Index

Linear Algebra and Its Applications, Second Edition, by Peter D. Lax
Copyright © 2007 John Wiley & Sons, Inc.

9 780471 751564